Texts in Applied Mathematics

Volume 67

More information about this series at http://www.springer.com/series/1214

Hansjörg Kielhöfer

Calculus of Variations

An Introduction to the One-Dimensional
Theory with Examples and Exercises

 Springer

Hansjörg Kielhöfer
Rimsting, Bayern
Germany

ISSN 0939-2475 ISSN 2196-9949 (electronic)
Texts in Applied Mathematics
ISBN 978-3-319-89038-8 ISBN 978-3-319-71123-2 (eBook)
https://doi.org/10.1007/978-3-319-71123-2

Mathematics Subject Classification (2010): 49-01, 49J05

Printed on acid-free paper

This Springer imprint is published by Springer Nature
The registered company is Springer International Publishing AG
The registered company address is: Gewerbestrasse 11, 6330 Cham, Switzerland

Preface

This book is the translation of my book "Variationsrechnung," published by Vieweg + Teubner and Springer Fachmedien, Wiesbaden, Germany, in 2010. The German version is based on lectures that I gave many times at the University of Augsburg, Germany. The audience consisted of students of mathematics, and also of physics, having a solid background in calculus and linear algebra.

My goal is to offer students a fascinating field of mathematics, which emerged historically from concrete questions in geometry and physics. In order to keep the prerequisites as low as possible I confine myself to one independent variable, which is commonly called "one-dimensional calculus of variations." Some advanced mathematical tools, possibly not familiar to the reader, are given along with proofs in the Appendix or can be found in introductory textbooks on analysis given in the bibliography. Accordingly, this book is a textbook for an introductory course on the calculus of variations appropriate for students with no preliminary knowledge of functional analysis. The exercises with solutions make the text also appropriate for self-study.

I present several famous historical problems, for example, Dido's problem, the brachistochrone problem of Johann Bernoulli, the problem of the hanging chain, the problem of geodesics, etc. In order to find solutions, i.e., to determine minimizers, I start by establishing the Euler-Lagrange equation, first without and later with constraints, for which Lagrange's multiplier rule becomes crucial. Minimizers, whose existence is not questioned at this point, necessarily satisfy the Euler-Lagrange equations.

Apart from these, I also discuss questions arising in phase transitions and microstructures. These problems are typically nonconvex, and the Weierstraß-Erdmann corner conditions on broken extremals become relevant.

In the history of the calculus of variations the existence of a minimizer was questioned only in the second half of the 19th century by Weierstraß. We present his famous counterexample to Dirichlet's principle, which awakens the requirement for an existence theory. This leads to the "direct methods in the calculus of variations." Here one independent variable has the advantage that the Sobolev spaces and the functional analytic tools can be given without great difficulties in the text or in the Appendix. Some emphasis is put on quadratic functionals, since their Euler-Lagrange equations are linear. The above-mentioned Dirichlet's principle offers an elegant way to prove the existence of solutions of (linear) boundary value problems: simply obtain minimizers.

The text includes numerous figures to aid the reader. Exercises intended to deepen the coverage of topics are given at the end of each section. Solutions to the exercises are provided at the end of the book.

I thank Rita Moeller and Bernhard Gawron for preparing the manuscript, and I thank Ingo Blechschmidt for producing the figures. Finally, I thank my friend Tim Healey for many corrections and suggestions, which improved the book.

March 2017   Hansjörg Kielhöfer

Contents

Introduction

The calculus of variations was created as a tool to calculate minimizers (or maximizers) of certain mathematical quantities taking values in the set of real numbers. Historically these quantities have been derived from concrete problems in geometry and physics, some of which we study in this book.

In the 18th century mathematicians and physicists, such as Maupertuis, d'Alembert, Euler, and Lagrange, postulated variational principles stating that nature acts with minimal expenditure ("principle of least action"). The mathematical formulation of these principles led to the calculus of variations presented in this book in a modern language.

The notion "calculus of variations" goes back to the year 1744 and was introduced by Euler: For extremals of a function, the derivative vanishes, and in the 18th century that derivative was called the "first variation."

The "mathematical quantities" investigated in this book are not ranges of mappings from $\mathbb{R}^n$ into $\mathbb{R}$, in general, i.e., they are not ranges of functions of finitely many variables. This is a main difference between the calculus of variations and optimization theory, which has the same goal, namely finding extremals. Our examples show that the domain of definition of the mappings is infinite-dimensional. Traditionally, such mappings are called "functionals," a notion that reappears in the nomenclature "functional analysis." As a matter of fact, the calculus of variations and functional analysis became "siblings" who cross-fertilized each other up to present time. An essential step is the extension of the n-dimensional linear space $\mathbb{R}^n$ to an infinite-dimensional function space. Since the methods of linear algebra are not so effective in infinite-dimensional spaces one needs an additional topological structure.

Let us study some historical examples.

1. What is **the shortest distance between two points**? This question can be answered only if an admitted connection is defined. It must be a continuous curve that has a well-defined length. Moreover it must be determined where that curve runs: in a plane, in a space, on a sphere, on a manifold? The variables of this problem are admitted curves that connect two fixed points, and the real

quantities to be minimized are their lengths. Admitted curves cannot be determined by finitely many real variables, but they form a subset of an infinite-dimensional function space.

We assume that the admitted curves connect two points in a plane or in $\mathbb{R}^n$. For a definition of their length, we use the Euclidean distance between two points what defines also the length of the connecting line segment:

$$x = (x_1, x_2 \ldots, x_n), \quad y = (y_1, \ldots, y_n),$$

$$\|x - y\| = \left(\sum_{i=1}^{n} (x_i - y_i)^2\right)^{1/2}, \, n \in \mathbb{N}. \tag{1}$$

An admitted curve k connecting two points A and B has the length $L(k)$ which is defined as the supremum of the lengths of all inscribed polygonal chains, where, in turn, the length of a polygonal chain is the sum of the lengths of its line segments (Figure 1).

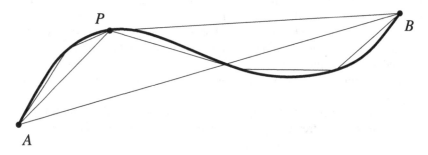

Fig. 1   On the Length of a Rectifiable Curve

A continuous curve having finite length is called rectifiable. Let k be a rectifiable curve connecting A and B and let P be a point on k different from A and B. Identifying points in the Euclidean space with their position vectors, respectively, the triangle inequality for the Euclidean distance yields

$$\|B - A\| \leq \|P - A\| + \|B - P\|, \tag{2}$$

where equality holds if and only if P is on the straight line connecting A and B. According to the definition of $L(k)$

$$\|B - A\| \leq \|P - A\| + \|B - P\| \leq L(k) \tag{3}$$

and

$$\|B - A\| < \|P - A\| + \|B - P\| \leq L(k), \tag{4}$$

if P is not on the straight line connecting A and B. This proves that all rectifiable curves connecting A and B are longer than $\|B-A\|$ if they contain a point not on the line segment between A and B. Moreover, the shortest connection is given by the line segment having length $\|B-A\|$.

2. **Heron's shortest distance problem** and **the law of reflection**. Consider two points A and B in the plane on one side of a straight line g. For which point P on g is $\|P-A\|+\|B-P\|$ minimal? (Figure 2)

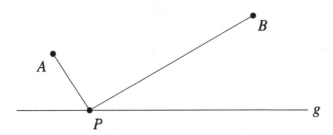

Fig. 2 Heron's Problem

We reflect A with respect to the line g, and we obtain A'. Obviously $\|P-A\| = \|P-A'\|$ and according to our example 1, $\|P-A'\|+\|B-P\|$ is minimal if the three points A', P, and B are on a line. The three angles in Figure 3 are equal

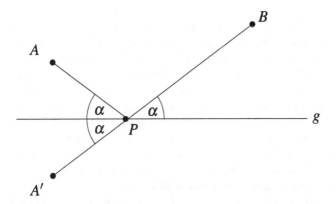

Fig. 3 The Law of Reflection

which proves the law of reflection: The reflected beam APB is shortest if the incidence angle equals the angle of reflection. This means also that on the path APB the light needs the shortest running time.

Fermat's principle in geometric optics (Fermat 1601–1665) states that a light beam chooses the path on which it has minimal running time. This principle implies the law of refraction.

3. **Snell's law of refraction:** (W. Snellius 1580–1626) Given two points A and B in a plane on opposite sides of a line g that divides the plane into two half planes in which a light beam has two different velocities v_1 and v_2. On which path APB does the light beam have the minimal running time?

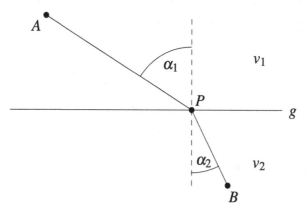

Fig. 4 The Law of Refraction

Snell's law of refraction reads:

$$\frac{\sin \alpha_1}{\sin \alpha_2} = \frac{v_1}{v_2}, \tag{5}$$

where the angles α_1 and α_2 are shown in Figure 4.

A different approach to the same problem is the following: A shoreline of a lake is represented by the line g, and a person at A observes a swimmer who threatens to drown at B. Therefore he needs help as quick as possible. Since the person at A runs faster than he can swim (v_1 is bigger than v_2) he chooses a path as sketched in Figure 4. Determine the point P where he jumps into the lake (Exercise 0.2). Hint: The respective distances between A and P and between P and B are expressed as a function of P (in suitable coordinates). The velocities v_1 and v_2 determine the times T_1 and T_2 to cover the respective distances. Minimize $T_1 + T_2$ as a function of P by setting its derivative to zero.

4. **The isoperimetric problem of queen Dido** (9th century b.c.) When the phoenician princess Dido had fled to North Africa, she asked the Berber king for a bit of land. She was offered as much land as could be encompassed by an oxhide. Dido cut the oxhide into fine stripes so that she got a long enough rope to encircle an entire hill named Byrsa, the center of Carthage. What shape was the region that she encircled so that area was maximal?

Let k be a closed curve having prescribed length $L(k)$ and let $F(k)$ denote the area in its interior. For which k is $F(k)$ maximal? We follow the arguments of J. Steiner (1796–1863): It is seen by reflection that the enclosed region is convex, cf. Figure 5. Choose points A and B on k that divide k into two arcs k_1 and k_2 of equal length. Then the regions F_1 and F_2 encircled by k_1, k_2, and the line segment g between A and B, respectively, should have the same size.

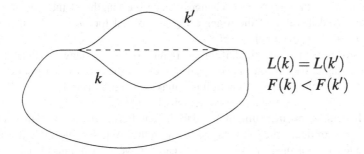

$$L(k) = L(k')$$
$$F(k) < F(k')$$

Fig. 5 Convexity of the Maximal Area

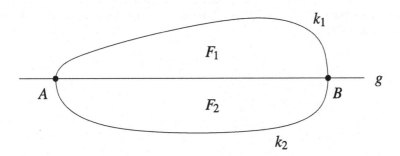

Fig. 6 Reduction to One Half

If F_1 is bigger than F_2, say, then replace F_2 by the reflection of F_1 with respect to the line g, cf. Figure 6. Thus the problem of maximal area is reduced to the following: Find the arc k_1 with prescribed length, whose endpoints A and B are on a straight line g, such that the area bounded by k_1 and g is maximal.

As seen before the region is convex. Choose a point P on k_1 and let α be the angle at P in the triangle APB, cf. Figure 7:

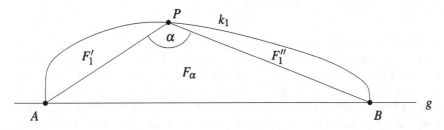

Fig. 7 On Steiner's Solution of Dido's Problem

Imagine a hinge in P such that the sides PA and PB can rotate increasing or decreasing the angle α. The areas F_1' and F_1'' are fixed and rotate as well. The length of the arc k_1 does not change under rotation either, only the area of the triangle F_α with corners APB changes. Since the lengths of the sides PA and PB are fixed the area of the triangle F_α is maximal for $\alpha_0 = 90°$. Then $F_1 = F_1' + F_1'' + F_{\alpha_0}$ is maximal as well.

This proves that F_1 is maximal if for any point P on k_1 the triangle APB is a right-angle triangle. By the converse of Thales' theorem the arc k_1 is a semicircle and therefore the closed curve k encircling a maximal area is a circle.

We return to Dido's problem in Paragraphs 1.7 and 2.2.

5. **The brachistochrone problem** posed by Johann Bernoulli (1667–1748). In June 1696, he introduced the following problem in the "Acta Eruditorum" (a Journal for Science published in Leipzig, Germany): Given two points A and B in a vertical plane, what is the curve traced out by a point mass M acted on only by gravity, which starts at A and reaches B in the shortest time?

   In January, 1697, Bernoulli published his solution that we discuss in Paragraph 1.8. It is commonly accepted that 1696 was the year of birth of modern calculus of variations (Figure 8).

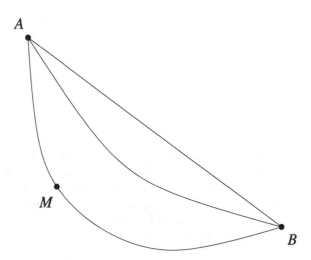

Fig. 8 Admitted Curves for Bernoulli's Problem

Not knowing so far Bernoulli's arguments, we realize in view of the historical examples that the methods to solve a variational problem have been diverse. In the middle of the 18th century the time was ripe for a systematic approach. As a matter of fact, about 1755 Euler (1707–1783) and Lagrange (1736–1813) postulated independently a differential equation that extremals must fulfill. That differential equation is now called Euler-Lagrange equation. In those days, the existence of extremals was presumed evident, and one was interested in necessary conditions for extremals: After Euler and Lagrange had given a necessary condition in terms of the first variation, Legendre (1752–1833) then gave one in terms of the second variation.

In 1860 Weierstraß (1815–1897) shook the faith in the hitherto existing methods by a counterexample that he published in 1869. It falsifies Dirichlet's (1805–1859) arguments to solve the boundary value problem for Laplace's (1749–1827) equation by a variational problem. Weierstraß' counterexample shows that the existence of a minimizer is not evident, and it initiated the need for an existence theory which is now known as "direct methods in the calculus of variations." These techniques developed hand in hand with the new field of "functional analysis," and consequently understanding the former requires some knowledge of the latter. Accordingly the "direct methods" are less appropriate for beginners than the Euler-Lagrange calculus. Nonetheless, we delve into it in Chapter 3.

Before the Euler-Lagrange calculus can be applied a variational problem has to be given a mathematical formulation. In case of the brachistochrone problem, the running time of the point mass needs to be expressed in terms of the curve. This "mathematical modeling" requires both physical knowledge and intuition. Namely the domain of definition of the time functional has to be given mathematically. A too narrow restriction of the class of admissible or competing curves might falsify the result. On the other hand, a too large class might burst open the mathematical possibilities.

In this book the mathematical modeling yields functionals of the form

$$J(y) = \int_a^b F(x,y,y')dx \tag{6}$$

or in parametric form

$$J(x,y) = \int_{t_a}^{t_b} \Phi(x,y,\dot{x},\dot{y})dt . \tag{7}$$

For real functions $y = y(x)$ where $x \in [a,b] \subset \mathbb{R}$ this reads as

$$J(y) = \int_a^b F(x,y(x),y'(x))dx \tag{8}$$

or for plane curves $(x,y) = (x(t),y(t))$ where $t \in [t_a,t_b] \subset \mathbb{R}$

$$J(x,y) = \int_{t_a}^{t_b} \Phi(x(t),y(t),\dot{x}(t),\dot{y}(t))dt . \tag{9}$$

Following a tradition which goes back to Newton (1643–1727) the dot indicates the derivative with respect to t (which is not necessarily the real time). Obviously admissible functions or curves need to be differentiable. The functions F and Φ depending on three or four variables, respectively, are called Lagrange functions or Lagrangians. Generalizations to curves in space or in $\mathbb{R}^n$ are apparent. However, for the theory presented in this book the admissible functions or curves must depend only on one independent variable, and thus all integrals are taken over an interval. The corresponding theory is commonly called the "one-dimensional theory."

As mentioned before, this book focuses on the Euler-Lagrange calculus with an introduction to direct methods.

Exercises

0.1. Let $A, B \in \mathbb{R}^n$ two points and $x(t) = (x_1(t), \ldots, x_n(t))$, $t \in [t_a, t_b]$, a continuously differentiable curve connecting A and B, i.e., $x(t_a) = A$ and $x(t_b) = B$. Prove

$$\|B - A\| \leq \int_{t_a}^{t_b} \|\dot{x}(t)\| dt = \text{length of the curve.} \tag{10}$$

Here $\|\dot{x}(t)\| = (\sum_{k=1}^n (\dot{x}_k(t))^2)^{1/2}$ and (10) says that the length of all continuously differentiable curves connecting A and B is at least as big as $\|B - A\|$ which is the length of the straight line segment $x(t) = A + t(B - A)$, $t \in [0, 1]$, connecting A and B. This proves that the line segment is the shortest among all continuously differentiable connections.

Hint: Prove (10) with an inequality for integrals that follows from the triangle inequality for approximating Riemann sums.

0.2. Prove Snell's refraction law (5).

Chapter 1
The Euler-Lagrange Equation

1.1 Function Spaces

In order to give the functionals

$$J(y) = \int_a^b F(x,y,y')dx \qquad (1.1.1)$$

a domain of definition, we need to introduce suitable function spaces. First of all we require that the Lagrange function or Lagrangian,

$$F : [a,b] \times \mathbb{R} \times \mathbb{R} \to \mathbb{R}, \qquad \text{is continuous.} \qquad (1.1.2)$$

Here $[a,b] = \{x | a \leq x \leq b\}$ is a compact interval in the real line.

Definition 1.1.1. $C[a,b] = \{y | y : [a,b] \to \mathbb{R} \text{ is continuous}\}$,
 $C^1[a,b] = \{y | y \in C[a,b], y \text{ is differentiable on } [a,b], y' \in C[a,b]\}$,
where in the boundary points the one-sided derivatives are taken. A function $y \in C^1[a,b]$ is called continuously differentiable on $[a,b]$.
 $C^{1,pw}[a,b] = \{y | y \in C[a,b], y \in C^1[x_{i-1},x_i], i = 1,\ldots,m\}$,
where $a = x_0 < x_1 < \cdots < x_m = b$ is a partition of $[a,b]$ depending on y. A function $y \in C^{1,pw}[a,b]$ is called piecewise continuously differentiable.
 Obviously $C^1[a,b] \subset C^{1,pw}[a,b] \subset C[a,b]$ and all three spaces are infinite-dimensional vector spaces over $\mathbb{R}$, provided that addition of functions and scalar multiplication are defined in the natural way: $(y_1 + y_2)(x) = y_1(x) + y_2(x)$, $(\alpha y)(x) = \alpha y(x)$ for $\alpha \in \mathbb{R}$.

The graph of a typical function $y \in C^{1,pw}[a,b]$ looks like the graph sketched in Figure 1.1.

© Springer International Publishing AG 2018
H. Kielhöfer, *Calculus of Variations*, Texts in Applied Mathematics 67,
https://doi.org/10.1007/978-3-319-71123-2_1

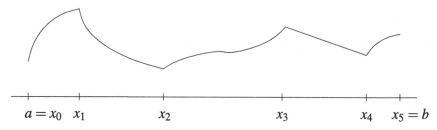

Fig. 1.1   The Graph of a Piecewise Continuously Differentiable Function

For $y \in C^{1,pw}[a,b]$, define

$$J(y) = \int_a^b F(x,y,y')dx = \sum_{i=1}^m \int_{x_{i-1}}^{x_i} F(x,y(x),y'(x))dx. \qquad (1.1.3)$$

Then $J : C^{1,pw}[a,b] \to \mathbb{R}$ is a well defined **functional** or a function of a function. The integrals in the sum (1.1.3) exist since in view of (1.1.2) the integrand is continuous on each interval $[x_{i-1},x_i]$, $i = 1,\dots,m$.

The real vector spaces of Definition 1.1.1 are normed as follows:

Definition 1.1.2. *For $y \in C[a,b]$ let $\|y\|_0 = \|y\|_{0,[a,b]} := \max_{x \in [a,b]} |y(x)|$,*

for $y \in C^1[a,b]$ let $\|y\|_1 = \|y\|_{1,[a,b]} := \|y\|_{0,[a,b]} + \|y'\|_{0,[a,b]}$, and

for $\quad y \in C^{1,pw}[a,b] \quad$ let $\quad \|y\|_{1,pw} = \|y\|_{1,pw,[a,b]} := \|y\|_{0,[a,b]} + \max_{i \in \{1,\dots,m\}} \{\|y'\|_{0,[x_{i-1},x_i]}\}$.

For two functions $y_1, y_2 \in X = C[a,b]$, $C^1[a,b]$ or $C^{1,pw}[a,b]$, the expression $\|y_1 - y_2\|$ is the distance between y_1 and y_2 with respect to the norm in X.

Any partition of a function $y \in C^{1,pw}[a,b]$ can clearly be increased by finitely many arbitrary points without changing the properties of Definition 1.1.1 and of the norm $\|y\|_{1,pw}$ in Definition 1.1.2. For two functions $y_1, y_2 \in C^{1,pw}[a,b]$, their distance can be defined via the union of their respective partitions.

The norm $\| \quad \|$ on a real vector space X has the following properties:

Definition 1.1.3. *A mapping $\| \quad \| : X \to \mathbb{R}$ is a norm provided*

1. $\|y\| \geq 0 \quad$ for all $y \in X$, $\|y\| = 0 \Leftrightarrow y = 0$,
2. $\|\alpha y\| = |\alpha|\|y\| \quad$ for all $\alpha \in \mathbb{R}$, $y \in X$,
3. $\|y_1 + y_2\| \leq \|y_1\| + \|y_2\| \quad$ for all $y_1, y_2 \in X$.

Inequality 3. is called triangle inequality.

Verify the properties of a norm for all norms given in Definition 1.1.2.

In a normed vector space X, convergence and continuity are defined as follows:

Definition 1.1.4. *We define for a sequence* $(y_n)_{n\in\mathbb{N}} \subset X$:
$$\lim_{n\to\infty} y_n = y_0 \Leftrightarrow \quad \text{for any } \varepsilon > 0 \text{ there exists an } n_0(\varepsilon) \text{ such that}$$
$$\|y_n - y_0\| < \varepsilon \text{ for all } n \geq n_0(\varepsilon).$$
Let $J : D \to \mathbb{R}$ *be a functional defined on a subset* $D \subset X$.
J *is continuous in* $y_0 \in D \Leftrightarrow \quad$ *for any* $\varepsilon > 0$ *there is a* $\delta(\varepsilon) > 0$,
such that $|J(y) - J(y_0)| < \varepsilon$ *for all*
$y \in D$ *with* $\|y - y_0\| < \delta(\varepsilon)$
$\Leftrightarrow \quad$ *for each sequence* $(y_n)_{n\in\mathbb{N}} \subset D \subset X$
satisfying $\lim_{n\to\infty} y_n = y_0$,
there holds $\lim_{n\to\infty} J(y_n) = J(y_0)$.
J *is continuous on* D *if* J *is continuous in all* $y_0 \in D$.

The last equivalence defining continuity via sequences is proved in the same way as for functions defined on subsets of $\mathbb{R}$ or $\mathbb{R}^n$.

The convergence in $X = C[a,b]$ or in $X = C^1[a,b]$ means uniform convergence of y_n or of y_n and y_n', respectively, on $[a,b]$. That uniform convergence guarantees that any limit function belongs to the same space as the sequence. This is not necessarily the case for convergence in the space $C^{1,pw}[a,b]$, cf. Exercise 1.1.1. The normed spaces $C[a,b]$ and $C^1[a,b]$ are each "complete" due to the above-mentioned property defining a Banach space. The space $C^{1,pw}[a,b]$ is not a Banach space, but this fact does not play a role in chapters 1 and 2.

Remark. *The introduction of the space* $C^{1,pw}[a,b]$ *is necessary to describe "broken extremals." Variational problems with a Lagrangian that is nonconvex as a function of* y' *typically have broken extremals. We discuss such nonconvex variational problems in Paragraphs 1.5, Example 6, 1.11, and 2.4. They are not only of mathematical interest, since extremals can have only specific "corners," but they also play an important role in the modeling of phase transitions, microstructures, and the decomposition of binary alloys.*

Exercises

1.1.1. A sequence $(y_n)_{n\in\mathbb{N}} \subset X$ in a normed vector space X is called a Cauchy sequence if for any $\varepsilon > 0$ there is an $n_0(\varepsilon)$ such that $\|y_m - y_n\| < \varepsilon$ for all $m, n \geq n_0(\varepsilon)$. Let

$$y(x) = \begin{cases} \dfrac{1}{k^2}(x - kx + 1) & \text{for} \quad x \in \left[\dfrac{1}{k}, \dfrac{1}{k-1}\right], \\ \dfrac{1}{k^2}(x + kx - 1) & \text{for} \quad x \in \left[\dfrac{1}{k+1}, \dfrac{1}{k}\right], \end{cases}$$

for $k = 2n$, $n \in \mathbb{N}$, and $y(0) = 0$. Sketch y and verify that $y \in C[0,1]$, but $y \notin C^{1,pw}[0,1]$. Let

$$y_n(x) = \begin{cases} y(x) & \text{for} \quad x \in \left[\dfrac{1}{2n+1}, 1\right], \\ 0 & \text{for} \quad x \in \left[0, \dfrac{1}{2n+1}\right], \end{cases}$$

for $n \in \mathbb{N}$. Show that $(y_n)_{n \in \mathbb{N}} \subset C^{1,pw}[0,1]$ is a Cauchy sequence with respect to the norm $\| \quad \|_{1,pw,[0,1]}$ but that there is no $y_0 \in C^{1,pw}[0,1]$ such that $\lim_{n \to \infty} y_n = y_0$ in $C^{1,pw}[0,1]$.

Hint: $\lim_{n \to \infty} y_n = y$ in $C[0,1]$.

1.1.2. Prove that $J : C^{1,pw}[a,b] \to \mathbb{R}$ defined by

$$J(y) = \int_a^b F(x,y,y')dx \quad \text{as in (1.1.3)}$$

is continuous if (1.1.2) holds.

1.2 The First Variation

Let $J : D \subset X \to \mathbb{R}$ be a functional defined on a subset D of a normed vector space X. We require that, for $y \in D$ and for some fixed $h \in X$, the vector $y + th$ stays within D for all $t \in (-\varepsilon, \varepsilon) \subset \mathbb{R}$. Then $g(t) = J(y + th)$ defines a function $g : (-\varepsilon, \varepsilon) \subset \mathbb{R} \to \mathbb{R}$.

Definition 1.2.1. *If*

$$g'(0) = \lim_{t \to 0} \frac{J(y + th) - J(y)}{t} \quad in \ \mathbb{R} \tag{1.2.1}$$

exists, then the functional J is Gâteaux differentiable in y in direction h and the derivative $g'(0)$ is denoted $dJ(y,h)$.

The Gâteaux differential $dJ(y,h)$ (Gâteaux, 1889–1914) satisfies $dJ(y, \alpha h) = \alpha dJ(y,h)$, but it is neither linear nor continuous in h, in general, as is shown by the following example: Let $X = \mathbb{R}^2$, $y = (y_1, y_2) \in D = \mathbb{R}^2$, and

$$J(y) = \begin{cases} y_1^2 \left(1 + \dfrac{1}{y_2}\right) & \text{for } y_2 \neq 0, \\ 0 & \text{for } y_2 = 0. \end{cases} \tag{1.2.2}$$

Then we get for $y = (0,0)$ and $h = (h_1, h_2)$ with $h_2 \neq 0$

$$\lim_{t \to 0} \frac{J(y + th) - J(y)}{t} = \lim_{t \to 0} \left(th_1^2 + \frac{h_1^2}{h_2} \right) = \frac{h_1^2}{h_2} \tag{1.2.3}$$

and $dJ(0, h) = \dfrac{h_1^2}{h_2}$ for $h_2 \neq 0$, $dJ(0, h) = 0$ for $h_2 = 0$.

If a functional $J : \mathbb{R}^n \to \mathbb{R}$ is totally or **Fréchet differentiable** (Fréchet, 1878–1973), then the Gâteaux differential $dJ(y, h)$ is linear and continuous in $h = (h_1, ..., h_n) \in \mathbb{R}^n$:

$$\frac{d}{dt} J(y + th)|_{t=0} = \sum_{k=1}^{n} \frac{\partial J}{\partial y_k}(y) h_k = (\nabla J(y), h), \tag{1.2.4}$$

by the chain rule. In $\mathbb{R}^n$, the Fréchet differential is represented by the gradient $\nabla J(y) = \left(\frac{\partial J}{\partial y_1}(y), ..., \frac{\partial J}{\partial y_n}(y) \right)$ and the Euclidean scalar product $(\ ,\ )$.

Definition 1.2.2. *If $dJ(y, h)$ exists in $y \in D \subset X$ for $h \in X$ and if $dJ(y, h)$ is linear in h, then $dJ(y, h)$ is called the first variation of J in y in direction h, and it is denoted*

$$dJ(y, h) = \delta J(y) h. \tag{1.2.5}$$

If (1.2.5) holds for all $h \in X_0 \subset X$, where X_0 is a linear subspace, then

$$\delta J(y) : X_0 \to \mathbb{R} \tag{1.2.6}$$

is a linear functional.

In case (1.2.4) we have $X_0 = X = \mathbb{R}^n$ and $\delta J(y) h = (\nabla J(y), h)$. Moreover $\delta J(y) : X_0 \to \mathbb{R}$ is continuous.

Remark. *If X is finite-dimensional, then any linear functional on X is continuous, as can be seen by its matrix representation (or its representation with a scalar product). However, if X is infinite-dimensional, this is not necessarily the case: Let $X = C^1[0, 1] \subset C[0, 1]$ be normed as $C[0, 1]$ in Definition 1.1.2 and define $Ty = y'(1)$ for $y \in X$. Then for $y_n(x) = \frac{1}{n} x^n$ we obtain $\|y_n\|_{0,[0,1]} = \frac{1}{n}$, $Ty_n = 1$, whence $\lim_{n \to \infty} y_n = 0$, $\lim_{n \to \infty} Ty_n = 1 \neq T0 = 0$, which shows that T is not continuous in $y = 0$.*

We now compute the Gâteaux differential of the functional (1.1.3):

Proposition 1.2.1. *Let the functional*

$$J(y) = \int_a^b F(x, y, y') dx \tag{1.2.7}$$

be defined on $D \subset C^{1,pw}[a,b]$. We assume that to each $y \in D$ and $h \in C_0^{1,pw}[a,b] :=$
$C^{1,pw}[a,b] \cap \{y(a) = 0, y(b) = 0\}$, there is some ε such that $y + th \in D$ for all $t \in$
$(-\varepsilon, \varepsilon)$. Moreover we assume that the Lagrange function $F : [a,b] \times \mathbb{R} \times \mathbb{R} \to \mathbb{R}$
is continuous and continuously partially differentiable with respect to the last two
variables. Then the Gâteaux differential exists in all $y \in D$ in all directions $h \in$
$C_0^{1,pw}[a,b]$ and is represented by

$$\delta J(y)h = \int_a^b F_y(x,y,y')h + F_{y'}(x,y,y')h' dx. \tag{1.2.8}$$

Here F_y and $F_{y'}$ denote the partial derivatives of F with respect to the second and
third variables, respectively.

Proof. We fix y and h and w.l.o.g. we assume the same partition $a = x_0 < x_1 < \cdots < x_m = b$ for both functions. We obtain for some $x \in [x_{i-1}, x_i] \subset [a,b]$ and for all $t \in (-\varepsilon, \varepsilon) \setminus \{0\}$

$$\frac{1}{t}(F(x,y(x)+th(x),\ y'(x)+th'(x)) - F(x,y(x),y'(x)))$$
$$= \frac{1}{t}\int_0^t \frac{d}{ds}F(x,y(x)+sh(x),\ y'(x)+sh'(x))ds$$
$$= F_y(x,y(x),y'(x))h(x) + F_{y'}(x,y(x),y'(x))h'(x) \tag{1.2.9}$$
$$+ \frac{1}{t}\int_0^t F_y(x,y(x)+sh(x),\ y'(x)+sh'(x)) - F_y(x,y(x),y'(x))dsh(x)$$
$$+ \frac{1}{t}\int_0^t F_{y'}(x,y(x)+sh(x),\ y'(x)+sh'(x)) - F_{y'}(x,y(x),y'(x))dsh'(x).$$

Since y and h are in $C^1[x_{i-1}, x_i]$,

$$\{(x,y(x)+sh(x),\ y'(x)+sh'(x))|x \in [x_{i-1},x_i], |s| \le \frac{\varepsilon}{2}\}$$
$$\subset [x_{i-1},x_i] \times [-c,c] \times [-c',c'] \subset [a,b] \times \mathbb{R} \times \mathbb{R} \tag{1.2.10}$$

for some positive constants c and c'. Uniform continuity of F_y on the compact set (1.2.10) implies that for all $x \in [x_{i-1}, x_i]$ and for any $\tilde{\varepsilon} > 0$, we have

$$|F_y(x,y(x)+sh(x),\ y'(x)+sh'(x)) - F_y(x,y(x),y'(x))| < \tilde{\varepsilon},$$
$$\text{provided} \quad |s|(|h(x)|+|h'(x)|) < \delta(\tilde{\varepsilon}) \quad \text{and} \quad |s| \le \frac{\varepsilon}{2}. \tag{1.2.11}$$

In view of $|h(x)| + |h'(x)| \le \|h\|_{0,[x_{i-1},x_i]} + \|h'\|_{0,[x_{i-1},x_i]} \le \|h\|_{1,pw,[a,b]}$, estimate (1.2.11) is fulfilled for

$$|s| < \min\left\{\frac{\varepsilon}{2}, \frac{\delta(\tilde{\varepsilon})}{\|h\|_{1,pw}}\right\}. \tag{1.2.12}$$

An analogous estimate holds for $F_{y'}$. Then for all $x \in [x_{i-1}, x_i]$ and for any $\hat{\varepsilon} > 0$ (1.2.9) implies

$$
\begin{aligned}
&\left| \frac{1}{t}(F(x,y(x)+th(x), y'(x)+th'(x)) - F(x,y(x), y'(x))) \right. \\
&\left. -(F_y(x,y(x), y'(x))h(x) + F_{y'}(x,y(x), y'(x))h'(x)) \right| \\
&\leq \frac{1}{|t|} |t| \tilde{\varepsilon}(|h(x)| + |h'(x)|) \leq \tilde{\varepsilon}\|h\|_{1,pw} < \hat{\varepsilon},
\end{aligned} \tag{1.2.13}
$$

$$
\text{provided} \quad 0 < |t| < \min\left\{ \frac{\varepsilon}{2}, \frac{\delta(\tilde{\varepsilon})}{\|h\|_{1,pw}} \right\} \quad \text{and} \quad 0 < \tilde{\varepsilon} < \frac{\hat{\varepsilon}}{\|h\|_{1,pw}}.
$$

Estimate (1.2.13) means that

$$
\begin{aligned}
\lim_{t \to 0} &\left(\frac{1}{t}(F(x,y(x)+th(x), y'(x)+th'(x)) - F(x,y(x), y'(x))) \right) \\
&= F_y(x,y(x), y'(x))h(x) + F_{y'}(x,y(x), y'(x))h'(x) \\
&\text{uniformly for} \quad x \in [x_{i-1}, x_i] , \ i = 1,\ldots,m.
\end{aligned} \tag{1.2.14}
$$

Uniform convergence allows the interchange of the limit and the integral, i.e., it implies the convergence of the integral to the integral of the limit. Hence, we obtain finally

$$
\begin{aligned}
\lim_{t \to 0} \frac{J(y+th) - J(y)}{t} &= \\
\lim_{t \to 0} \sum_{i=1}^{m} \int_{x_{i-1}}^{x_i} \frac{1}{t}(F(x,y(x)+th(x), & y'(x)+th'(x)) - F(x,y(x), y'(x)))dx \\
= \sum_{i=1}^{m} \int_{x_{i-1}}^{x_i} F_y(x,y(x), y'(x))h(x) &+ F_{y'}(x,y(x), y'(x))h'(x)dx \\
= \int_a^b F_y(x,y,y')h + F_{y'}(x,y,y')h'dx &= dJ(y,h) = \delta J(y)h,
\end{aligned} \tag{1.2.15}
$$

where the last equality follows from the linearity $h \mapsto dJ(y,h)$, cf. Definition 1.2.2. □

Proposition 1.2.2. *Under the same hypotheses as those of Proposition 1.2.1 the first variation*

$$
\delta J(y) : C_0^{1,pw}[a,b] \to \mathbb{R} \tag{1.2.16}
$$

is linear and continuous for each $y \in D$. In particular,

$$
|\delta J(y)h| \leq C(y)\|h\|_{1,pw} \quad \text{for all} \ h \in C_0^{1,pw}[a,b] , \tag{1.2.17}
$$

where the positive constant $C(y)$ depends on $y \in D$.

If the functional (1.2.7) is defined on all of $C^{1,pw}[a,b]$ we can choose any h in $C^{1,pw}[a,b]$, and the proof of Proposition 1.2.1 yields

$$\delta J(y)h = \int_a^b F_y(x,y,y')h + F_{y'}(x,y,y')h'dx,$$

$\delta J(y) : C^{1,pw}[a,b] \to \mathbb{R}$ is linear and continuous and   (1.2.18)

$|\delta J(y)h| \leq C(y)\|h\|_{1,pw}$ for all $y,h \in C^{1,pw}[a,b].$

The homogeneous boundary conditions on h play no role in the proof.

Exercises

1.2.1. Prove Proposition 1.2.2.

1.2.2. Assume in addition to the hypotheses of Proposition 1.2.1 that the Lagrange function F is two times continuously partially differentiable with respect to the last two variables. Then

$$g''(0) = \delta^2 J(y)(h,h)$$

exists and is called **the second variation** of J in y in direction h. Prove the representation

$$\delta^2 J(y)(h,h) = \int_a^b F_{yy}h^2 + 2F_{yy'}hh' + F_{y'y'}(h')^2 dx,$$

where $F_{yy} = F_{yy}(x,y(x),y'(x))$ and analogously $F_{yy'},F_{y'y'}$ are the second partial derivatives of F with respect to the second and third variables, respectively.

1.2.3. Under the same hypotheses as those of Exercise 1.2.2, prove that

$$\delta^2 J(y) : C^{1,pw}[a,b] \times C^{1,pw}[a,b] \to \mathbb{R}   \text{ defined by}$$

$$\delta^2 J(y)(h_1,h_2) = \int_a^b F_{yy}h_1 h_2 + F_{yy'}(h_1 h_2' + h_1' h_2) + F_{y'y'}h_1' h_2' dx$$

is bilinear and continuous for each $y \in D$. In particular

$$|\delta^2 J(y)(h_1,h_2)| \leq C(y)\|h_1\|_{1,pw}\|h_2\|_{1,pw}$$
$$\text{for all } h_1,h_2 \in C^{1,pw}[a,b].$$

1.3 The Fundamental Lemma of Calculus of Variations

The derivative of a piecewise continuously differentiable function is piecewise continuous. Therefore we introduce

Definition 1.3.1. $C^{pw}[a,b] = \{y|y : [a,b] \to \mathbb{R}, \ y \in C[x_{i-1},x_i], i = 1,\ldots,m\}$ *for a partition* $a = x_0 < x_1 < \cdots < x_m = b$ *depending on y.*

At a corner of a piecewise continuously differentiable function the two one-sided derivatives have two different values. Accordingly we allow that a piecewise continuous function has two values of the two one-sided limits at a point of its partition. This contradicts the definition of a function, but we afford that inaccuracy since accuracy in Definition 1.3.1 would be cumbersome.

Lemma 1.3.1. *If* $f \in C^{pw}[a,b]$ *and*

$$\int_a^b fh\,dx = 0 \quad \text{for all } h \in C_0^\infty(a,b), \tag{1.3.1}$$

then $f(x) = 0$ *for all* $x \in [a,b]$.

$C_0^\infty(a,b)$ is called the space of "test functions," and it consists of all infinitely differentiable functions having a compact support in the open interval (a,b). The support of some h is the closure of $\{x|h(x) \neq 0\}$. Apparently $h(a) = h(b) = 0$ and $C_0^\infty(a,b) \subset C_0^{1,pw}[a,b]$.

Proof. Assume $f(x) \neq 0$ for some $x \in [x_{i-1},x_i] \subset [a,b]$. Since f is continuous on $[x_{i-1},x_i]$ there is some open interval I in $[x_{i-1},x_i]$ such that $f(x) \neq 0$ for all $x \in I$. If f is positive in I, say, choose a function $h \in C_0^\infty(a,b)$ having its support in I and being positive in the interior of its support. (Such a function exists by the remark below.) Then $fh \geq 0$ in $[a,b]$ with $fh > 0$ in the support of h, and thus the integral of fh over $[a,b]$ is positive, contradicting (1.3.1). $\qquad\square$

Remark. *A function* $h \in C_0^\infty(a,b)$ *having a prescribed support, and only one sign is constructed as follows: The function* $g(x) = \pm exp(-(1-x^2)^{-1})$ *for* $|x| < 1$ *and* $g(x) = 0$ *for* $|x| \geq 1$ *is in* $C_0^\infty(\mathbb{R})$ *and has support* $[-1,1]$. *Then* $h(x) = g((x-x_0)/r)$ *has for any* $x_0 \in \mathbb{R}$ *and* $r > 0$ *the support* $[x_0 - r, x_0 + r]$, *and* h *is positive or negative in* $(x_0 - r, x_0 + r)$, *depending on the sign of g.*

Lemma 1.3.2. *If* $f \in C^{pw}[a,b]$ *and*

$$\int_a^b fh'\,dx = 0 \quad \text{for all } h \in C_0^{1,pw}[a,b], \tag{1.3.2}$$

then $f(x) = c$ *for all* $x \in [a,b]$.

Proof. Choose $c = \frac{1}{b-a}\int_a^b f(x)dx = \frac{1}{b-a}\sum_{i=1}^m \int_{x_{i-1}}^{x_i} f(x)dx$ and $h(x) = \int_a^x (f(s) - c)ds$. Then $h \in C[a,b]$, $h(a) = 0$, $h(b) = 0$, and for $x \in [x_{i-1}, x_i]$ we obtain $h'(x) = f(x) - c$ when taking the respective one-sided derivatives in the partition points.

Thus $h \in C_0^{1,pw}[a,b]$, and in view of (1.3.2) and the choice of c, we find

$$\int_a^b (f - c)h'dx = \int_a^b fh'dx - c\int_a^b h'dx = 0. \tag{1.3.3}$$

On the other hand

$$\int_a^b (f - c)h'dx = \int_a^b (f - c)^2 dx = \sum_{i=1}^m \int_{x_{i-1}}^{x_i} (f(x) - c)^2 dx = 0, \tag{1.3.4}$$

and the continuity of $f - c$ on $[x_{i-1}, x_i]$ implies $f(x) = c$ for all $x \in [a,b]$. $\square$

Lemma 1.3.3. *For f and h in $C^{1,pw}[a,b]$, the formula for integration by parts is valid:*

$$\int_a^b fh'dx = -\int_a^b f'hdx + fh\big|_a^b. \tag{1.3.5}$$

For $h \equiv 1$ formula (1.3.5) gives the fundamental theorem of calculus for piecewise continuously differentiable functions.

Proof. Assuming identical partitions for f and h, formula (1.3.5) reads as follows:

$$\sum_{i=1}^m \int_{x_{i-1}}^{x_i} fh'dx = -\sum_{i=1}^m \int_{x_{i-1}}^{x_i} f'hdx + fh\big|_a^b. \tag{1.3.6}$$

Since $f, h \in C^1[x_{i-1}, x_i]$ the classical formula of integration by parts holds on each interval $[x_{i-1}, x_i]$. In the sum (1.3.6), the boundary values at interior partition points x_i, $i = 1, \ldots, m-1$ drop out and only the boundary values at $x_0 = a$ and $x_m = b$ are left. $\square$

Next we state and prove the **fundamental lemma of calculus of variations** due to DuBois-Reymond (1831–1889):

Lemma 1.3.4. *If $f, g \in C^{pw}[a,b]$ and*

$$\int_a^b fh + gh'dx = 0 \quad \text{for all } h \in C_0^{1,pw}[a,b], \tag{1.3.7}$$

then $g \in C^{1,pw}[a,b] \subset C[a,b]$ and $g' = f$ piecewise on $[a,b]$, i.e., $g'(x) = f(x)$ for $x \in [x_{i-1}, x_i]$ if $f \in C[x_{i-1}, x_i]$.

Proof. Defining $F(x) = \int_a^x f(s)ds$, the function $F \in C[a,b]$, and for $x \in [x_{i-1}, x_i]$ we obtain $F'(x) = f(x)$ when taking one-sided derivatives at partition points. Consequently $F \in C^{1,pw}[a,b]$, and due to Lemma 1.3.3, we find

$$\int_a^b fh\,dx = \int_a^b F'h\,dx = -\int_a^b Fh'\,dx \quad \text{for all } h \in C_0^{1,pw}[a,b]. \tag{1.3.8}$$

By (1.3.7) this implies

$$\int_a^b fh + gh'\,dx = \int_a^b (-F+g)h'\,dx = 0 \quad \text{for all } h \in C_0^{1,pw}[a,b]. \tag{1.3.9}$$

Since $g \in C^{pw}[a,b]$, Lemma 1.3.2 is applicable. Hence $-F(x) + g(x) = c$ or $g(x) = c + F(x)$ for all $x \in [a,b]$. Therefore $g \in C^{1,pw}[a,b]$ and $g' = F' = f$ piecewise in the sense explained before. $\square$

Lemma 1.3.4 is a regularity theorem in the sense that under validity of (1.3.7), g is more regular than assumed. In particular g is continuous on the entire interval $[a,b]$. A special case is the following:

$$\begin{array}{l} \text{If } f, g \in C[a,b] \text{ and (1.3.7) holds,} \\ \text{then } g \in C^1[a,b] \text{ and } g' = f \text{ on } [a,b]. \end{array} \tag{1.3.10}$$

1.4 The Euler-Lagrange Equation

For the functional

$$J(y) = \int_a^b F(x, y, y')dx, \tag{1.4.1}$$

defined on $D \subset C^{1,pw}[a,b]$, we give now the most important necessary condition that a minimizing or maximizing function $y \in D$ has to fulfill.

Definition 1.4.1. *A function $y \in D$ is a local minimizer of the functional (1.4.1) if*

$$J(y) \leq J(\tilde{y}) \quad \text{for all } \tilde{y} \in D \text{ satisfying} \quad \|y - \tilde{y}\|_{1,pw} < d \tag{1.4.2}$$

with a constant $d > 0$.

A local maximizer is defined in the analog way. It is sufficient to study minimizers since maximizers of J are minimizers of $-J$.

A local minimizer can be defined using a different norm: If in (1.4.2) one requires $\|y - \tilde{y}\|_0 < d$, then the inequality in (1.4.2) has to be fulfilled by more functions, which implies that the definition of a local minimizer is stronger. Accordingly, one calls y a **strong local minimizer** in this case. We shall not use this definition. In Exercise 1.4.7 we see that a local minimizer is not necessarily a strong local minimizer.

We assume that for $y \in D$ the first variation $\delta J(y)h$ exists for all $h \in C_0^{1,pw}[a,b]$ according to Definition 1.2.2. If the domain of definition is characterized by boundary conditions, i.e., if $D = C^{1,pw}[a,b] \cap \{y(a) = A, \ y(b) = B\}$, this is true.

Proposition 1.4.1. *Let $y \in D \subset C^{1,pw}[a,b]$ be a local minimizer of the functional (1.4.1), and assume that the Lagrange function $F : [a,b] \times \mathbb{R} \times \mathbb{R} \to \mathbb{R}$ is continuous and continuously partially differentiable with respect to the last two variables. Then*

$$F_{y'}(\cdot,y,y') \in C^{1,pw}[a,b] \subset C[a,b] \quad and$$
$$\frac{d}{dx}F_{y'}(\cdot,y,y') = F_y(\cdot,y,y') \quad piecewise \ on \ [a,b]. \tag{1.4.3}$$

For $y \in C^1[a,b]$ we have $F_{y'}(\cdot,y,y') \in C^1[a,b]$, and $(1.4.3)_2$ holds on the entire interval $[a,b]$.

Proof. For any fixed $h \in C_0^{1,pw}[a,b]$, $g(t) = J(y+th)$ is defined for $t \in (-\varepsilon,\varepsilon)$ and g is locally minimal at $t = 0$, by assumption. Furthermore the first variation (= the Gâteaux differential) $dJ(y,h) = g'(0)$ exists, and due to a well-known result of calculus, $g'(0) = 0$. Hence by (1.2.8),

$$\delta J(y)h = \int_a^b F_y(x,y,y')h + F_{y'}(x,y,y')h'dx = 0 \tag{1.4.4}$$
$$\text{for all } h \in C_0^{1,pw}[a,b].$$

By the assumptions on y and F, the functions $F_y(\cdot,y,y')$ and $F_{y'}(\cdot,y,y')$ are in $C^{pw}[a,b]$, and therefore Lemma 1.3.4 is applicable and implies (1.4.3). The last claim is covered by (1.3.10). $\square$

Equation $(1.4.3)_2$ is the **Euler-Lagrange equation**. It is an ordinary differential equation that has to be fulfilled by local minimizers piecewise.

The local minimizer is not necessarily two times (piecewise) differentiable. If this is the case and if the second partial derivatives $F_{y'y'}$, $F_{y'y}$, and $F_{y'x}$ exist, then by differentiation of the left-hand side of $(1.4.3)_2$, we find

$$F_{y'y'}(\cdot,y,y')y'' + F_{y'y}(\cdot,y,y')y' + F_{y'x}(\cdot,y,y') = F_y(\cdot,y,y') \tag{1.4.5}$$
$$\text{(piecewise) on } [a,b].$$

Equation (1.4.5) is a quasilinear ordinary differential equation of second order since the highest (second) derivative of y appears linearly in it. Equation (1.4.4) is called the weak version of the Euler-Lagrange equation (1.4.3). Lemma 1.3.3 implies:

Proposition 1.4.2. *The weak version (1.4.4) and the strong version (1.4.3) of the Euler-Lagrange equation are equivalent.*

Proposition 1.4.2 is exceptional for one independent variable x. It is not valid, in general, for more than one independent variable, i.e., for partial differential equations.

A solution $y \in D \subset C^{1,pw}[a,b]$ of the Euler-Lagrange equation is not necessarily a (local) minimizer (or maximizer). Here is an example:

$$J(y) = \int_0^1 (y')^3 dx \quad \text{on } D = C^{1,pw}[0,1] \cap \{y(0) = 0, y(1) = 0\} \qquad (1.4.6)$$

A function $y \in D$ solves the Euler-Lagrange equation if

$$3(y')^2 \in C^{1,pw}[0,1] \subset C[0,1] \quad \text{and} \quad \frac{d}{dx} 3(y')^2 = 0 \quad \text{piecewise on } [0,1]. \quad (1.4.7)$$

This means

$$(y')^2 = c_1 \geq 0 \quad \text{and } y' = \pm\sqrt{c_1} \text{ on } [0,1]. \qquad (1.4.8)$$

There are infinitely many solutions that fulfill the boundary conditions. Some of them are sketched in Figure 1.2. A special solution is $y \equiv 0$.

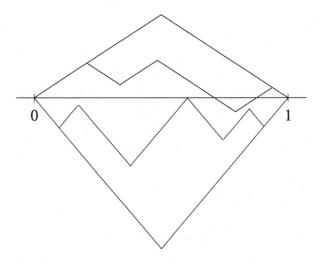

Fig. 1.2   Solutions of the Euler-Lagrange Equation

No solution is a local minimizer or maximizer, cf. Exercise 1.4.8.

However, if the Lagrange function is convex with respect to the last two variables, then any solution of the Euler-Lagrange equation is a local minimizer.

Definition 1.4.2. *Let the Lagrange function $F : [a,b] \times \mathbb{R} \times \mathbb{R} \to \mathbb{R}$ be continuous and continuously partially differentiable with respect to the last two variables. F is convex with respect to the last two variables if*

$$F(x,\tilde{y},\tilde{y}') \geq F(x,y,y') + F_y(x,y,y')(\tilde{y}-y) + F_{y'}(x,y,y')(\tilde{y}'-y')$$
$$\text{for all} \quad (x,y,y'), (x,\tilde{y},\tilde{y}') \in [a,b] \times \mathbb{R} \times \mathbb{R}. \tag{1.4.9}$$

Geometrically (1.4.9) means that the graph of $F(x,\cdot,\cdot)$ is above the tangent plane spanned by the two tangents to the graphs of $F(x,\cdot,y')$ and of $F(x,y,\cdot)$, respectively, cf. Figure 1.3.

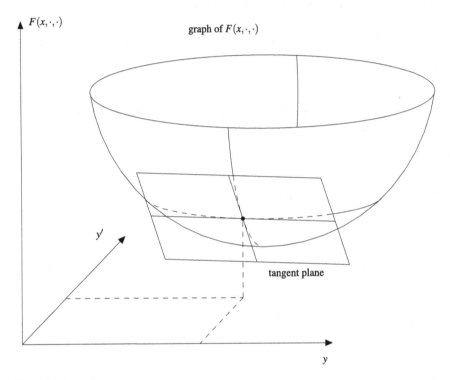

Fig. 1.3 Convexity

Proposition 1.4.3. *Assume that the Lagrange function F of the functional J given by (1.4.1) and defined on $D = C^{1,pw}[a,b] \cap \{y(a) = A, y(b) = B\}$ is continuous, continuously partially differentiable, and convex with respect to the last two variables. Then any solution $y \in D$ of its Euler-Lagrange equation is a global minimizer of J.*

Proof. Let $y \in D$ any solution of the Euler-Lagrange equation, and let $\tilde{y}$ be any element in D. Then $h = \tilde{y} - y \in C_0^{1,pw}[a,b]$ since both functions fulfill the same boundary conditions. Convexity (1.4.9) implies

$$J(\tilde{y}) = \int_a^b F(x,\tilde{y},\tilde{y}')dx = \int_a^b F(x,y+h,y'+h')dx$$

$$\geq \int_a^b F(x,y,y')dx + \int_a^b F_y(x,y,y')h + F_{y'}(x,y,y')h'dx \qquad (1.4.10)$$

$$= J(y) \quad \text{due to (1.4.4).}$$

Therefore y is not only a local but also a global minimizer. □

Let $y \in D \subset C^{1,pw}[a,b]$ be a local minimizer of J, and assume that the second variation of J exists in y as defined in Exercise 1.2.2. Then by a result of calculus,

$$\delta^2 J(y)(h,h) \geq 0 \quad \text{for all } h \in C_0^{1,pw}[a,b]. \qquad (1.4.11)$$

Under additional assumptions (1.4.11) implies a necessary condition on a local or global minimizer, cf. Exercise 1.4.3. Exercises 1.4.4 and 1.4.6 give sufficient conditions on a global and a local minimizer, respectively.

Exercises

1.4.1. The support supp(h) of a function $h : \mathbb{R} \to \mathbb{R}$ is the closure of the set $\{x|h(x) \neq 0\}$. Prove that to any compact interval $I \subset (a,b)$ there is a sequence $(h_n)_{n \in \mathbb{N}} \subset C_0^{1,pw}[a,b]$ having the properties

a) supp$(h_n) \subset I$ for all $n \geq n_0$,

b) $\lim\limits_{n \to \infty} \int_a^b h_n^2 dx = 0$,

c) $\lim\limits_{n \to \infty} \int_a^b (h_n')^2 dx = \infty$.

1.4.2. Under the hypotheses of Exercise 1.2.2 there exists the second variation of J in $y \in D \subset C^{1,pw}[a,b]$ in direction $h \in C_0^{1,pw}[a,b]$. We assume in addition that $F_{yy'}(\cdot,y,y') \in C^{1,pw}[a,b]$. Prove

$$\delta^2 J(y)(h,h) = \int_a^b Ph^2 + Q(h')^2 dx \quad \text{where}$$

$$P = F_{yy} - \frac{d}{dx}F_{yy'} \in C^{pw}[a,b] \quad \text{and} \quad Q = F_{y'y'} \in C^{pw}[a,b].$$

1.4.3. Prove that under the hypotheses of Exercise 1.4.2, a local minimizer $y \in D \subset C^{1,pw}[a,b]$ of J given by (1.4.1) has to satisfy

$$F_{y'y'}(x,y(x),y'(x)) \geq 0 \quad \text{for all } x \in [a,b].$$

Hint: Use (1.4.11) and Exercises 1.4.2 and 1.4.1.

This **necessary condition, due to Legendre,** can be proven without the additional assumption given in Exercise 1.4.2, cf. reference [3], page 57.

1.4.4. Assume the existence of the second variation of J in $y \in D = C^{1,pw}[a,b] \cap \{y(a) = A, \, y(b) = B\}$ in direction $h \in C_0^{1,pw}[a,b]$, cf. Exercise 1.2.2. Furthermore assume that

$$\delta J(y)h = 0 \quad \text{and}$$
$$\delta^2 J(\tilde{y})(h,h) \geq 0 \quad \text{for all } \tilde{y} \in D \quad \text{and for all } h \in C_0^{1,pw}[a,b].$$

Prove that y is a global minimizer of J on D.

Hint: Use for two times continuously differentiable function $g : \mathbb{R} \to \mathbb{R}$ the identity $g(1) - g(0) = g'(0) + \int_0^1 (1-t)g''(t)dt$.

1.4.5. Under the hypotheses of Exercise 1.2.2 the second variation of J exists in $y \in D = C^{1,pw}[a,b] \cap \{y(a) = A, \, y(b) = B\}$ in direction $h \in C_0^{1,pw}[a,b]$. Prove

$$J(y+h) = J(y) + \delta J(y)h + \frac{1}{2}\delta^2 J(y)(h,h) + R(y,h)$$
$$\text{where } R(y,h)/\|h\|_{1,pw}^2 \to 0 \quad \text{if } \|h\|_{1,pw} \to 0.$$

The continuous linear mapping $\delta J(y)$ is also called (first) **Fréchet derivative**, and the continuous bilinear mapping $\delta^2 J(y)$ is also called second Fréchet derivative of J in y.

1.4.6. Assume that the first and second Fréchet derivatives of J exist in $y \in D \subset C^{1,pw}[a,b]$ and that the following relations hold:

$$\delta J(y)h = 0$$
$$\delta^2 J(y)(h,h) \geq C\|h\|_{1,pw}^2$$

for all $h \in C_0^{1,pw}[a,b]$ for a constant $C > 0$. Prove that y is a local minimizer of J on D.

1.4.7. Consider the functional

$$J(y) = \int_0^1 (y')^2 + (y')^3 dx.$$

a) Prove that $y = 0$ is a local minimizer of J on $D = C^{1,pw}[0,1] \cap \{y(0) = 0, \, y(1) = 0\}$.
b) Prove that $y = 0$ is not a strong local minimizer according to the definition after Definition 1.4.1.

Hint: Define for $b \in (0,1)$ and $n \in \mathbb{N}$

$$
y_{n,b}(x) =
\begin{cases}
\dfrac{1}{nb}x & \text{for} \quad x \in [0,b], \\[2ex]
-\dfrac{1}{n(1-b)}x + \dfrac{1}{n(1-b)} & \text{for} \quad x \in [b,1],
\end{cases}
$$

then $y_{n,b} \in D$ and for any $d > 0$ there is an $n \in \mathbb{N}$ and some suitable $b_n \in (0,1)$ with $\|y_{n,b_n}\|_0 < d$ and $J(y_{n,b_n}) < 0$.

1.4.8. Show that no solution of the Euler-Lagrange equation for the functional (1.4.6) is a local minimizer or maximizer.

Hint: Use also (1.4.11).

1.5 Examples of Solutions of the Euler-Lagrange Equation

1. $J(y) = \int_{-1}^{1} y^2(2x - y')^2 dx$ is defined on $D = C^1[-1,1] \cap \{y(-1) = 0, y(1) = 1\}$.
The function

$$
y(x) =
\begin{cases}
0 & \text{for } x \in [-1,0], \\
x^2 & \text{for } x \in [0,1],
\end{cases}
$$

is in D and $J(y) = 0$. By $J(y) \geq 0$ for all $y \in D$ that function is a global minimizer. Observe that y is not in $C^2[-1,1]$, but it fulfills the Euler-Lagrange equation on the entire interval $[-1,1]$.

2. The Dirichlet integral $J(y) = \int_a^b (y')^2 dx$ is defined on $D = C^{1,pw}[a,b] \cap \{y(a) = A, y(b) = B\}$.
Omitting the factor 2, the Euler-Lagrange equation reads

$$
\frac{d}{dx}y' = y'' = 0 \quad \text{piecewise on } [a,b],
$$

which is solved by $y(x) = c_1^i x + c_2^i$ for $x \in [x_{i-1}, x_i]$, $i = 1, \ldots, m$. This is not the whole truth: The definition of D and $(1.4.3)_1$ imply that y as well as y' is in $C[a,b]$, and thus the solution is the line $y(x) = c_1 x + c_2$ with $c_1 = (B-A)/(b-a)$ and $c_2 = (bA - aB)/(b-a)$, fulfilling the boundary conditions.
In order to decide on a minimizing property of that line we can apply Proposition 1.4.3 or we can compute the second variation:

$$
\delta^2 J(\tilde{y})(h,h) = \int_a^b 2(h')^2 dx \geq 0 \quad \text{for all } \tilde{y} \in D \text{ and for all } h \in C_0^{1,pw}[a,b].
$$

Exercise 1.4.4 shows that the line is a global minimizer.
Finally we can argue directly as follows: Let $\tilde{y} \in D$ and $\tilde{y} = y + \tilde{y} - y = y + h$ where $h \in C_0^{1,pw}[a,b]$. Then

$$J(\tilde{y}) = J(y+h) = \int_a^b (y')^2 dx + 2 \int_a^b y'h' dx + \int_a^b (h')^2 dx$$

$$\geq J(y) + 2c_1 \int_a^b h' dx = J(y),$$

since the second integral vanishes by the homogeneous boundary conditions.

3. **The counterexample of Weierstraß** reads as follows: $J(y) = \int_{-1}^1 x^2 (y')^2 dx$
defined on $D = C^1[-1,1] \cap \{y(-1) = -1, y(1) = 1\}$. Apparently $J(y) \geq 0$ for all $y \in D$, and upon inserting

$$y_n(x) = \frac{\arctan nx}{\arctan n},$$

we find

$$J(y_n) = \int_{-1}^1 \frac{n^2 x^2}{(\arctan n)^2 (1+n^2 x^2)^2} dx$$

$$< \frac{1}{(\arctan n)^2} \int_{-1}^1 \frac{dx}{1+n^2 x^2} = \frac{2}{n \arctan n}.$$

Since $\lim_{n \to \infty} J(y_n) = 0$, we deduce that $\inf_{y \in D} J(y) = 0$. But there is no function $y \in D$ such that $J(y) = 0$. Such a function has to fulfill $xy'(x) = 0$ for all $x \in [-1,1]$, which means $y'(x) = 0$ for $x \in [-1,0) \cup (0,1]$, and in view of the boundary conditions $y(x) = -1$ for $x \in [-1,0)$ and $y(x) = 1$ for $x \in (0,1]$. However, such a function is not in the domain of definition D of J.
The Euler-Lagrange equation

$$\frac{d}{dx}(2x^2 y') = 0 \quad \text{or} \quad x^2 y' = c_1 \quad \text{has the solutions}$$
$$y(x) = -\frac{c_1}{x} + c_2.$$

None of these functions is in D either.
That example contradicts Dirichlet's argument that there exists an admissible function for which a functional, which is convex and bounded from below, attains its minimum.

4. The length of the curve $\{(x,y(x)) | x \in [a,b]\}$ between (a,A) and (b,B) is given by the functional $J(y) = \int_a^b \sqrt{1+(y')^2} dx$ on $D = C^{1,pw}[a,b] \cap \{y(a) = A, y(b) = B\}$.
The Euler-Lagrange equation reads

$$\frac{d}{dx} \frac{y'}{\sqrt{1+(y')^2}} = 0 \quad \text{piecewise on } [a,b] \text{ or}$$

$$y'(x) = \frac{c_i}{\sqrt{1-c_i^2}} = c_1^i \quad \text{for } x \in [x_{i-1}, x_i], \, i = 1, \ldots, m, \quad \text{having the solutions}$$

$$y(x) = c_1^i x + c_2^i \quad \text{for } x \in [x_{i-1}, x_i] \subset [a,b].$$

As in example 2 the continuity of $F_{y'}(\cdot, y, y')$ and of y on the entire interval $[a,b]$ imply $c_1^i = c_1$ and $c_2^i = c_2$ for $i = 1, \ldots, m$. With the constants of example 2 we obtain the line segment between (a,A) and (b,B), and by Exercise 0.1 of the Introduction, it minimizes J on D.

A different argument uses Exercise 1.4.4: By

$$F_{y'y'}(\tilde{y}') = \frac{1}{(1+(\tilde{y}')^2)^{3/2}} > 0, \ F_{yy}(\tilde{y}') = F_{yy'}(\tilde{y}') = 0$$

the representation in Exercise 1.2.2 implies $\delta^2 F(\tilde{y})(h,h) \geq 0$ for all $\tilde{y} \in D$ and for all $h \in C_0^{1,pw}[a,b]$. Therefore the only solution of the Euler-Lagrange equation is a global minimizer.

5. $J(y) = \int_a^b y^2 + (y')^2 dx$ is defined on $D = C^{1,pw}[a,b] \cap \{y(a) = A, y(b) = B\}$. The Euler-Lagrange equation reads (omitting the factors 2)

$$y'' = y \quad \text{piecewise on } [a,b],$$

having the solutions $y(x) = c_1^i \cosh x + c_2^i \sinh x$ for $x \in [x_{i-1}, x_i]$. Continuity of y as well as of y' by $(1.4.3)_1$ means

$$(c_1^i - c_1^{i+1}) \cosh x_i + (c_2^i - c_2^{i+1}) \sinh x_i = 0,$$
$$(c_1^i - c_1^{i+1}) \sinh x_i + (c_2^i - c_2^{i+1}) \cosh x_i = 0,$$

whence $c_1^i = c_1^{i+1} = c_1$ and $c_2^i = c_2^{i+1} = c_2$ for $i = 1, \ldots, m-1$. The boundary conditions determine the constants in a unique way (exercise). Proposition 1.4.3 answers the question whether the solution is an extremal. We can also compute the second variation:

$$\delta^2 J(\tilde{y})(h,h) = 2 \int_a^b h^2 + (h')^2 dx \geq 0$$
$$\text{for all } \tilde{y} \in D \text{ and for all } h \in C_0^{1,pw}[a,b].$$

By Exercise 1.4.4 the solution $y(x) = c_1 \cosh x + c_2 \sinh x$ is a global minimizer.

6. $J(y) = \int_0^1 ((y')^2 - 1)^2 dx$ is defined on $D = C^{1,pw}[0,1]$. Apparently $J(y) \geq 0$ for all $y \in D$, and $J(y) = 0$ for all piecewise continuously differentiable functions with $y' = \pm 1$. All sawtooth functions with slopes ± 1 are global minimizers of J on D, cf. Figure 1.4. Even with boundary conditions there might be infinitely many, for instance for $y(0) = y(1) = 0$. The Euler-Lagrange equation reads (omitting the factor 4)

$$\frac{d}{dx}((y')^2 - 1)y' = 0 \quad \text{piecewise on } [a,b], \text{ and}$$
$$((y')^2 - 1)y' \quad \text{is continuous on } [a,b].$$

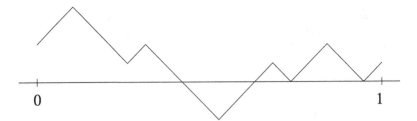

Fig. 1.4 A Global Minimizer

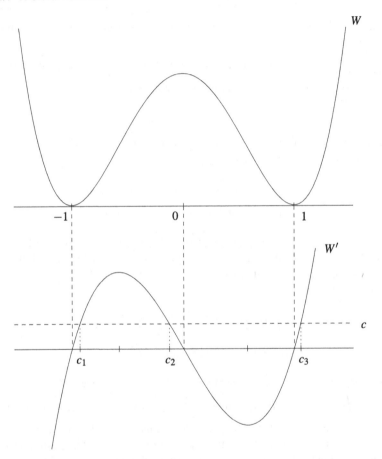

Fig. 1.5 A W-Potential

Solutions are given by piecewise lines but that do not necessarily have slopes ± 1. We consider the graphs of $W(z) = (z^2 - 1)^2$ and of $W'(z) = 4(z^2 - 1)z$, in Figure 1.5. The Euler-Lagrange equation requires $4((y')^2 - 1)y' = c$ having the solutions $y' = c_i$, $i = 1, 2, 3$, cf. Figure 1.5. For $c \neq 0$ we obtain $W(y') > 0$ and

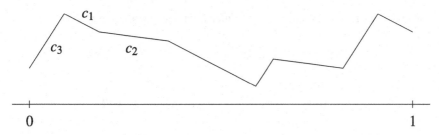

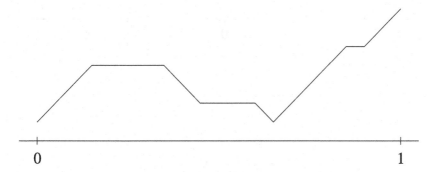

Fig. 1.6   A Solution of the Euler-Lagrange Equation

Fig. 1.7   A Solution of the Euler-Lagrange Equation

therefore a function with slopes $y' = c_i$ as sketched in Figure 1.6 is not a global minimizer. For $c = 0$ we obtain the slopes $y' = -1, 0, 1$, and because $W(0) = 1$, a function as sketched in Figure 1.7 is not a minimizer as well.

The second variation reads

$$\delta^2 J(y)(h,h) = \int_0^1 W''(y')(h')^2 dx,$$

where $W''(z) = 4(3z^2 - 1)$ is sketched in Figure 1.8.

Sawtooth functions with slopes $y' = c_i$ where $|c_i| \geq \frac{1}{\sqrt{3}}$ fulfill the necessary Legendre condition $W''(y') \geq 0$ for a local minimizer given in Exercise 1.4.3. The sufficient conditions given in Exercises 1.4.4 and 1.4.6 are not applicable. To summarize, global minimizers of a nonconvex variational problem are not determined by the first and second variations alone. One needs additional necessary conditions known as the "Weierstraß-Erdmann corner conditions," cf. Paragraph 1.11.

Next we discuss three **special cases**.

7. The Lagrange function does not depend explicitly on x: $J(y) = \int_a^b F(y,y')dx$ is defined on $D \subset C^{1,pw}[a,b]$, and let $y \in D \cap C^2(a,b)$ be a local minimizer. We need the additional assumption on the regularity of y for technical reasons. For global minimizers it is not required (cf. Proposition 1.11.2), and for an "elliptic" Euler-Lagrange equation, it is automatically fulfilled (cf. Exercise 1.5.1 and

Proposition 1.11.4). In any case the Euler-Lagrange equation reads

$$\frac{d}{dx}F_{y'}(y,y') = F_y(y,y') \quad \text{on } [a,b],$$

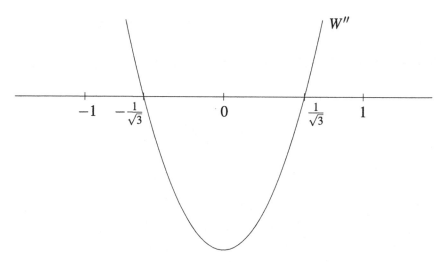

Fig. 1.8 The W-Potential is not Convex

and due to the additional regularity of y we may compute:

$$\frac{d}{dx}(F(y,y') - y'F_{y'}(y,y'))$$

$$= F_y(y,y')y' + F_{y'}(y,y')y'' - y''F_{y'}(y,y') - y'\frac{d}{dx}F_{y'}(y,y')$$

$$= (F_y(y,y') - \frac{d}{dx}F_{y'}(y,y'))y'.$$

This shows that any solution of the Euler-Lagrange equation and any constant function solves the differential equation of first order

$$F(y,y') - y'F_{y'}(y,y') = c_1 \quad \text{on } [a,b].$$

Any solution of this differential equation solves the Euler-Lagrange equation or it is constant. If it can be solved for y' one obtains $y' = f(y;c_1)$, which, in turn, is solved by "separating the variables":

Let $h(y;c_1)$ be a primitive of $\dfrac{1}{f(y;c_1)}$, i.e.,

$$\frac{d}{dy}h(y;c_1) = \frac{1}{f(y;c_1)} \neq 0.$$

There exists the inverse function h^{-1} and

$$y(x) = h^{-1}(x + c_2; c_1) \quad \text{solves } y' = f(y; c_1), \text{ since}$$

$$\frac{d}{dx} y(x) = \frac{1}{\frac{d}{dy} h(y(x); c_1)} = f(y(x); c_1).$$

8. The Lagrange function does no depend explicitly on y: $J(y) = \int_a^b F(x, y')dx$ is defined on $D \subset C^{1,pw}[a,b]$. The Euler-Lagrange equation reads

$$\frac{d}{dx} F_{y'}(\cdot, y') = 0 \quad \text{piecewise on } [a,b], \text{ whence}$$

$$F_{y'}(x, y'(x)) = c_1 \quad \text{on } [a,b].$$

Observe that due to $(1.4.3)_1$, $F_{y'}(\cdot, y') \in C[a,b]$. If the latter equation can be solved for y', i.e.,

$$y'(x) = f(x; c_1), \quad \text{then by integration}$$

$$y(x) = \int f(x; c_1)dx + c_2.$$

If $D = C^{1,pw}[a,b]$ then $c_1 = 0$. Indeed, for any $h \in C^{1,pw}[a,b]$ the weak version of the Euler-Lagrange equation (1.4.4) and Lemma 1.3.3 give

$$\int_a^b F_{y'}(x, y')h'dx = 0 = \int_a^b c_1 h'dx = c_1(h(b) - h(a)).$$

Hence $c_1 = 0$ since $h(a)$ and $h(b)$ are arbitrary.

9. The Lagrange function does not depend explicitly on y': $J(y) = \int_a^b F(x, y)dx$ is defined on $D \subset C^{pw}[a,b]$. The Euler-Lagrange equation reads

$$F_y(x, y(x)) = 0 \quad \text{on } [a,b]$$

which needs to be solved for y. This is not a differential equation.

Remark. *Cases 8 and 9 seem to be identical upon substituting $y' = u$ in case 8. However, one obtains according to case 9*

$$F_u(x, u(x)) = 0, \quad u = y',$$

whereas case 8 gives

$$F_{y'}(x, y'(x)) = c_1.$$

When computing the first variation one has to realize that the perturbation in case 8 is of the form $y' + th'$ and that it is not sufficient to simply set $y' = u$. A local minimizer has to fulfill

$$\delta J(u)h' = 0 \quad \text{for all } h \in C_0^1[a,b],$$

which by Lemma 1.3.2 implies $F_u(\cdot, u) = c_1$. On the other hand, we see in case 8 that $c_1 = 0$ if $D = C^{1,pw}[a,b]$.

Exercises

1.5.1. Assume that the Lagrange function of the functional $J(y) = \int_a^b F(x,y,y')dx$ is two times continuously partially differentiable with respect to all three variables and that $y \in C^{1,pw}[a,b] \cap C^1[x_{i-1},x_i]$ is a solution of the Euler-Lagrange equation

$$\frac{d}{dx}F_{y'}(\cdot,y,y') = F_y(\cdot,y,y') \quad \text{on } [x_{i-1},x_i] \subset [a,b].$$

If $F_{y'y'}(x_0,y(x_0),y'(x_0)) \neq 0$ for some $x_0 \in (x_{i-1},x_i)$, prove that y is two times continuously differentiable in a neighborhood of x_0 in (x_{i-1},x_i). Hence, **local ellipticity** implies **local regularity.**

Hint: Apply the implicit function theorem.

1.5.2. Find extremals of $J(y) = \int_0^1 F(x,y,y')dx$ in $D = C^{1,pw}[0,1] \cap \{y(0) = 0,$ $y(1) = 1\}$ where

a) $F(x,y,y') = y'$,   b) $F(x,y,y') = yy'$,   c) $F(x,y,y') = xyy'$.

Compute the supremum of J in case c).

1.5.3. Find solutions of the Euler-Lagrange equations in $C^2[0,1]$ for

a) $J(y) = \int_0^1 ((y')^2 + 2y)dx, \quad y(0) = 0, \, y(1) = 1,$

b) $J(y) = \int_{-1}^2 ((y')^2 + 2yy')dx, \quad y(-1) = 1, \, y(2) = 0,$

c) $J(y) = \int_0^1 ((y')^2 + 2xy' + x^2)dx, \quad y(0) = 0, \, y(1) = 0,$

d) $J(y) = \int_0^2 ((y')^2 + 2yy' + y^2)dx, \quad y(0) = 0, \, y(2) = 1.$

Compute the second variations of J and find out whether the solutions of the Euler-Lagrange equations are local or global extremals among all admitted solutions fulfilling the same boundary conditions.

1.6 Minimal Surfaces of Revolution

When the graph of a continuous positive function y is revolved around the x-axis it generates a surface of revolution as sketched in Figure 1.9.

The area of a surface of revolution generated by a function $y \in C^1[a,b]$ is given by

$$J(y) = 2\pi \int_a^b y\sqrt{1+(y')^2}dx. \tag{1.6.1}$$

The variational problem reads as follows: Which curve connecting (a,A) and (b,B) generates a surface of revolution having the smallest surface area? Such a surface is called minimal surface of revolution.

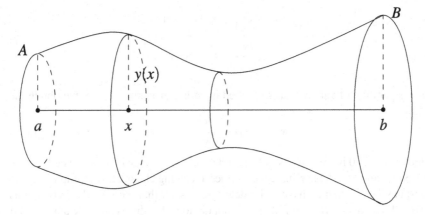

Fig. 1.9 A Surface of Revolution

The functional (1.6.1) is defined on $D = C^1[a,b] \cap \{y(a) = A,\ y(b) = B\}$. By a rescaling we normalize the problem as follows:

$$\tilde{y}(\tilde{x}) = y(a + A\tilde{x})/A \quad \text{for } \tilde{x} \in \left[0, \frac{b-a}{A}\right] = [0, \tilde{b}] \tag{1.6.2}$$

gives

$$J(y) = 2\pi A^2 \int_0^{\tilde{b}} \tilde{y}\sqrt{1 + (\tilde{y}')^2}d\tilde{x}, \quad \tilde{y}(0) = 1,\ \tilde{y}(\tilde{b}) = \frac{B}{A} = \tilde{B}. \tag{1.6.3}$$

We omit the factor $2\pi A^2$ as well as the tilde, and we study the normalized problem

$$J(y) = \int_0^b y\sqrt{1 + (y')^2}dx \quad \text{on} \tag{1.6.4}$$
$$D = C^1[0,b] \cap \{y(0) = 1,\ y(b) = B\}.$$

The functional (1.6.4) is treated as the special case 7 in Paragraph 1.5. The required additional regularity of a local minimizer is guaranteed by Exercise 1.5.1: In view of

$$F_{y'y'}(y,y') = \frac{y}{(1+(y')^2)^{3/2}} > 0, \quad \text{provided } y > 0, \tag{1.6.5}$$

on $[a,b]$, any positive solution of the Euler-Lagrange equation in D is in $C^2(a,b)$, and as discussed in case 7 of Paragraph 1.5, it solves the differential equation

$$F(y,y') - y'F_{y'}(y,y') = c_1 \quad \text{on} \quad [a,b]. \tag{1.6.6}$$

For the Lagrange function $F(y,y') = y\sqrt{1+(y')^2}$, equation (1.6.6) yields

$$y = c_1 \sqrt{1 + (y')^2} \quad \text{and}$$

$$y' = \sqrt{\frac{y^2 - c_1^2}{c_1^2}} = f(y; c_1). \tag{1.6.7}$$

By a separation of variables as described in case 7 of Paragraph 1.5, we obtain the solution

$$y(x) = c_1 \cosh\left(\frac{x + c_2}{c_1}\right), \tag{1.6.8}$$

which for $c_1 > 0$ is a positive solution of the Euler-Lagrange equation. The function (1.6.8) is called a catenary since it describes a hanging chain, cf. Paragraph 2.3. The constants $c_1 > 0$ and c_2 have to be determined such that the boundary conditions $y(0) = 1$ and $y(b) = B$ are fulfilled. It is not obvious that this is possible at all, and if so, that it is possible in a unique way. An experiment with a soap film between two rings shows that an increase of the distance b between the rings causes a contraction of the surface followed by a ripping of the film such that it forms finally two discs. That scenario is also described by the mathematics above.

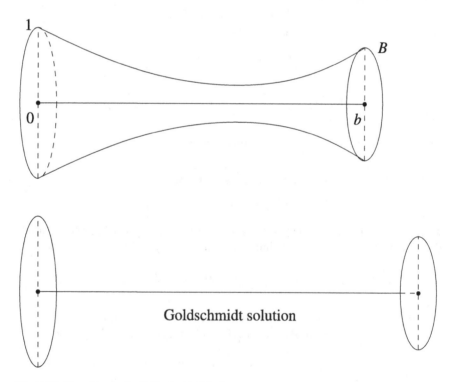

Goldschmidt solution

Fig. 1.10 Transition to the Goldschmidt Solution

If b is large compared to B, the area of any surface of revolution is bigger than the area $\pi(1 + B^2)$ of the two discs. This "Goldschmidt solution," named after B. Goldschmidt (1807–1851) and sketched in Figure 1.10, is not an admitted solution of the original variational problem. The latter has no solution for large b, since (1.6.8) cannot fulfill the boundary conditions. If b decreases there exist two solutions (1.6.8), whose areas are still bigger than that of the Goldschmidt solution, but the upper solution generates a locally minimizing surface. If b decreases even more then the upper solution generates a globally minimizing surface with an area that is smaller than that of the Goldschmidt solution.

A precise analysis of this problem can be found in [3], p. 80, p. 436, [2], p. 82, [13], p. 298.

1.7 Dido's Problem

As explained in the Introduction, the variational problem can be reduced to the following: Which curve with prescribed length L in the upper half-plane having its endpoints on the x-axis encloses together with the x-axis a region of maximal area? Admitted, however, are only curves that are graphs of functions of x. In Paragraph 2.2 we revisit this problem in more generality.

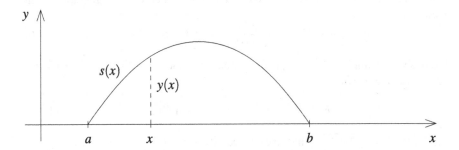

Fig. 1.11 Dido's Problem for Graphs of Functions

Since the endpoints are not fixed it is convenient to parametrize the curve by its arc length. If $y \in C^1[a,b] \cap \{y(a) = 0,\ y(b) = 0\}$ then the arc length from $(a,0)$ to $(x, y(x))$ is given by

$$s(x) = \int_a^x \sqrt{1 + (y'(\xi))^2}\,d\xi \quad \text{where } s(a) = 0,\ s(b) = L. \tag{1.7.1}$$

Since $s(x)$ is strictly monotone the inverse function exists and is a continuously differentiable function $x : [0, L] \to [a, b]$. Then

$$\{(x, y(x)) | x \in [a, b]\} = \{(x(s), y(x(s))) | s \in [0, L]\} \tag{1.7.2}$$

is the parametrization of the curve by its arc length, cf. Figure 1.11.

Next the area $\int_a^b y\,dx$ has to be expressed by the arc length:

$$\frac{dx}{ds}(s) = \frac{1}{\frac{ds}{dx}(x(s))} = \frac{1}{\sqrt{1+(y'(x(s)))^2}},$$

$$\frac{d}{ds}y(x(s)) = y'(x(s))\frac{dx}{ds}(s), \quad \text{and thus} \tag{1.7.3}$$

$$\frac{dx}{ds}(s) = \sqrt{1 - \left(\frac{d}{ds}y(x(s))\right)^2}.$$

The area transforms according to the substitution formula:

$$\int_a^b y(x)dx = \int_0^L y(x(s))\frac{dx}{ds}(s)ds = \int_0^L y(x(s))\sqrt{1 - \left(\frac{d}{ds}y(x(s))\right)^2}\,ds. \tag{1.7.4}$$

Denoting $\tilde{y}(s) = y(x(s))$ we obtain the functional

$$J(\tilde{y}) = \int_0^L \tilde{y}\sqrt{1-(\tilde{y}')^2}\,ds, \quad \text{defined on}$$
$$\tilde{D} = C^1[0,L] \cap \{\tilde{y}(0) = 0,\ \tilde{y}(L) = 0\}, \tag{1.7.5}$$

which is to be maximized. We proceed as in the special case 7 of Paragraph 1.5. In view of

$$F_{\tilde{y}'\tilde{y}'}(\tilde{y},\tilde{y}') = -\frac{\tilde{y}}{(1-(\tilde{y}')^2)^{3/2}} < 0 \quad \text{for } \tilde{y} > 0, \tag{1.7.6}$$

Exercise 1.5.1 implies that any positive solution of the Euler-Lagrange equation is in $C^2(0,L)$ and solves

$$F(\tilde{y},\tilde{y}') - \tilde{y}'F_{\tilde{y}'}(\tilde{y},\tilde{y}') = c_1 \quad \text{on } [0,L]. \tag{1.7.7}$$

For $F(\tilde{y},\tilde{y}') = \tilde{y}\sqrt{1-(\tilde{y}')^2}$ this gives the equations

$$\tilde{y} = c_1\sqrt{1-(\tilde{y}')^2} \quad \text{and}$$
$$\tilde{y}' = \sqrt{1 - \left(\frac{\tilde{y}}{c_1}\right)^2} = f(\tilde{y};c_1), \tag{1.7.8}$$

which admit solutions (obtained by separation of variables)

$$\tilde{y}(s) = c_1\sin\left(\frac{s+c_2}{c_1}\right). \tag{1.7.9}$$

For $c_1 > 0$ and $0 < (s + c_2)/c_1 < \pi$, these are positive solutions of the Euler-Lagrange equation. The boundary conditions determine the constants as follows:

$$\tilde{y}(0) = c_1 \sin \frac{c_2}{c_1} = 0, \quad c_2 = 0,$$
$$\tilde{y}(L) = c_1 \sin \frac{L}{c_1} = 0, \quad c_1 = \frac{L}{\pi}, \tag{1.7.10}$$

since the solution must be positive. Finally, by (1.7.3),

$$x(s) = a + \int_0^s \frac{dx}{ds}(\sigma)d\sigma = a + \int_0^s \sqrt{1 - (\tilde{y}'(\sigma))^2}\,d\sigma$$
$$= a + \int_0^s \sin \frac{\pi}{L}\sigma d\sigma = a + \frac{L}{\pi} - \frac{L}{\pi} \cos \frac{\pi}{L}s, \tag{1.7.11}$$

and the curve

$$\left\{ \left(a + \frac{L}{\pi} - \frac{L}{\pi} \cos \frac{\pi}{L}s, \frac{L}{\pi} \sin \frac{\pi}{L}s \right) \Big| s \in [0, L] \right\} \tag{1.7.12}$$

is a semi-circle having center $(a + \frac{L}{\pi}, 0)$, radius $\frac{L}{\pi}$, and length L, cf. Figure 1.12.

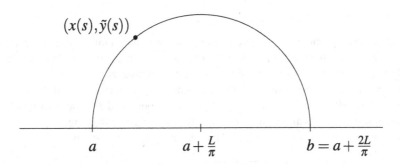

Fig. 1.12 Solution of Dido's Problem

1.8 The Brachistochrone Problem of Johann Bernoulli

The problem was posed by Johann Bernoulli in 1696 as follows: Given two points A and B in a vertical plane, what is the curve traced out by a point mass M acted on only by gravity, which starts at A and reaches B in shortest time?

That curve is called **brachistochrone**.

We treat the problem following Euler and Lagrange, and we point out that Bernoulli could not know their results published more than 50 years later. Below we sketch Bernoulli's arguments based on Fermat's principle of least time for a beam of light and on Snell's refraction law.

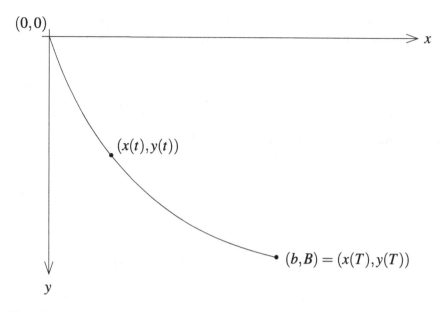

Fig. 1.13 An Admitted Curve for Bernoulli's Problem

The physical problem to be solved first is to express the running time of the point mass in terms of its trajectory. Let the starting point of the trajectory be the origin $(0,0)$ of a coordinate system whose y-axis points downwards.

We parametrize the trajectory by the time t: $\{(x(t),y(t))|t \in [0,T]\}$. Here $(x(0),y(0)) = (0,0)$, $(x(T),y(T)) = (b,B)$ is the endpoint and T is the running time. We assume continuous differentiability such that the length of the tangent $(\dot{x}(t),\dot{y}(t))$ is the velocity $v(t) = \sqrt{\dot{x}(t)^2 + \dot{y}(t)^2}$. Then the arc length is given by

$$s(t) = \int_0^t \sqrt{\dot{x}(\tau)^2 + \dot{y}(\tau)^2}d\tau = \int_0^t v(\tau)d\tau \quad \text{and}$$
$$\frac{ds}{dt}(t) = v(t). \tag{1.8.1}$$

By conservation of energy, the sum of kinetic and potential energy is constant along the curve:

$$\frac{1}{2}mv^2 + mg(h_0 - y) = mgh_0 \quad \text{and}$$
$$v = \sqrt{2gy}, \tag{1.8.2}$$

where m is the mass of M, g is the gravitational acceleration, and the constant h_0 is a fictitious height. For $t = 0$ we have $y(0) = 0$ and $v(0) = 0$.

By "physical evidence" we assume that the trajectory can also be parametrized by x: $\{(x(t),y(t))|t \in [0,T]\} = \{(x,\tilde{y}(x))|x \in [0,b]\}$. The arc length is now

$$s(t) = \int_a^{x(t)} \sqrt{1 + (\tilde{y}'(\xi))^2} d\xi \quad \text{and the velocity is}$$

$$\frac{ds}{dt}(t) = \sqrt{1 + (\tilde{y}'(x(t)))^2} \dot{x}(t) = v(t) = \sqrt{2g\tilde{y}(x(t))},$$

(1.8.3)

by $(1.8.2)_2$. This gives finally the running time

$$T = \int_0^T 1 dt = \int_0^T \sqrt{\frac{1 + (\tilde{y}'(x(t)))^2}{2g\tilde{y}(x(t))}} \dot{x}(t) dt$$

$$= \int_0^b \sqrt{\frac{1 + (\tilde{y}'(x))^2}{2g\tilde{y}(x)}} dx,$$

(1.8.4)

by the substitution formula. Omitting the factor $\frac{1}{\sqrt{2g}}$, the functional to be minimized is

$$J(\tilde{y}) = \int_0^b \sqrt{\frac{1 + (\tilde{y}')^2}{\tilde{y}}} dx,$$

(1.8.5)

which has the following anomalies: For physical reasons we expect $\tilde{y}'(0) = +\infty$, and since $\tilde{y}(0) = 0$, (1.8.5) is an improper integral. Finally, only positive functions are admitted such that the domain of definition of J is

$$\tilde{D} = C[0,b] \cap C^{1,pw}(0,b] \cap \{\tilde{y}(0) = 0, \tilde{y}(b) = B\}$$

$$\cap \{\tilde{y} > 0 \text{ in } (0,b]\} \cap \{J(\tilde{y}) < \infty\}.$$

(1.8.6)

We compute the first variation on an interval $[\delta, b]$ where δ is small: For any $\delta > 0$ there is a $d > 0$ such that $\tilde{y}(x) \geq d > 0$ for $\tilde{y} \in \tilde{D}$ and $x \in [\delta, b]$. Choose $h \in C_0^{1,pw}[\delta, b]$ with a support $supp(h) \subset [\delta, b]$. For any h there exists an $\varepsilon > 0$ such that $\tilde{y} + th \in \tilde{D}$ for $t \in (-\varepsilon, \varepsilon)$. For a local minimizer y of J on D we obtain

$$\delta J(\tilde{y})h = \int_\delta^b F_{\tilde{y}}(\tilde{y}, \tilde{y}')h + F_{\tilde{y}'}(\tilde{y}, \tilde{y}')h' dx = 0$$

$$\text{for all } h \in C_0^{1,pw}[\delta, b].$$

(1.8.7)

Relation (1.8.7), in turn, implies that y solves the Euler-Lagrange equation piecewise on $[\delta, b]$. In particular, by $(1.4.3)_1$,

$$f := F_{\tilde{y}'}(\tilde{y}, \tilde{y}') \in C[\delta, b] \quad \text{or}$$

$$\frac{\tilde{y}'}{\sqrt{1 + (\tilde{y}')^2}} = f\sqrt{\tilde{y}} \in C[\delta, b], \ |f\sqrt{\tilde{y}}| < 1 \quad \text{or}$$

$$\tilde{y}' = \sqrt{\frac{f^2\tilde{y}}{1 - f^2\tilde{y}}} \in C[\delta, b] \quad \text{and} \quad \tilde{y} \in C^1[\delta, b],$$

(1.8.8)

since $\tilde{y} \in C[\delta, b]$. Therefore the supplement of Proposition 1.4.1 applies, and since $\delta > 0$ is arbitrarily small, $\tilde{y}$ solves the Euler-Lagrange equation in $(0, b]$. In view of

$$F_{\tilde{y}'\tilde{y}'}(\tilde{y}, \tilde{y}') = \frac{1}{\sqrt{\tilde{y}}} \frac{1}{(1 + (\tilde{y}')^2)^{3/2}} > 0 \quad \text{on } (0, b], \tag{1.8.9}$$

Exercise 1.5.1 guarantees that $\tilde{y} \in C^2(0, b)$ and the method of the special case 7 of Paragraph 1.5 is applicable, yielding

$$F(\tilde{y}, \tilde{y}') - \tilde{y}' F_{\tilde{y}'}(\tilde{y}, \tilde{y}') = c_1 \quad \text{on} \quad (0, b] \quad \text{and finally}$$

$$\tilde{y}' = \sqrt{\frac{2r - \tilde{y}}{\tilde{y}}} \quad \text{where} \quad 2r = \frac{1}{c_1^2} > 0. \tag{1.8.10}$$

This differential equation is not solved by known special functions. We proceed in the same way as Bernoulli, i.e., we transform (1.8.10) into a parametric differential equation. To this purpose the parameter x is replaced by a new parameter τ which is not the physical time (Figure 1.13):

$$(x, \tilde{y}(x)) = (\hat{x}(\tau), \hat{y}(\tau)), \quad x \in [0, b], \quad \tau \in [\tau_0, \tau_b]. \tag{1.8.11}$$

The ansatz $\hat{y}(\tau) = r(1 - \cos \tau)$ yields by $(1.8.10)_2$

$$\hat{y}(\tau) = \tilde{y}(\hat{x}(\tau)), \quad \frac{d\hat{y}}{d\tau}(\tau) = \tilde{y}'(\hat{x}(\tau)) \frac{d\hat{x}}{d\tau}(\tau),$$

$$\frac{d\hat{x}}{d\tau}(\tau) = \frac{d\hat{y}}{d\tau}(\tau)/\tilde{y}'(\hat{x}(\tau)) = \frac{d\hat{y}}{d\tau}(\tau)\sqrt{\frac{\hat{y}(\tau)}{2r - \hat{y}(\tau)}} \tag{1.8.12}$$

$$= r\sin\tau \sqrt{\frac{1 - \cos\tau}{1 + \cos\tau}} = r(1 - \cos\tau),$$

where we use $\sin\tau = \sqrt{1 - \cos^2\tau}$. Integration gives

$$\hat{x}(\tau) = r(\tau - \sin\tau) + c_2,$$
$$\hat{y}(\tau) = r(1 - \cos\tau) \quad \text{for } \tau \in [\tau_0, \tau_b]. \tag{1.8.13}$$

The four constants are determined by the boundary conditions:

$$\hat{y}(\tau_0) = r(1 - \cos\tau_0) = 0 \quad \Rightarrow \tau_0 = 0,$$
$$\hat{x}(\tau_0) = \hat{x}(0) = c_2 = 0$$
$$\hat{x}(\tau_b) = r(\tau_b - \sin\tau_b) = b \tag{1.8.14}$$
$$\hat{y}(\tau_b) = r(1 - \cos\tau_b) = B.$$

Accordingly $\dfrac{b}{B} = \dfrac{\tau_b - \sin\tau_b}{1 - \cos\tau_b} =: f(\tau_b)$. The function $f(\tau_b)$ for $\tau_b \in (0, 2\pi)$ is sketched in Figure 1.14, cf. also Exercise 1.8.1:

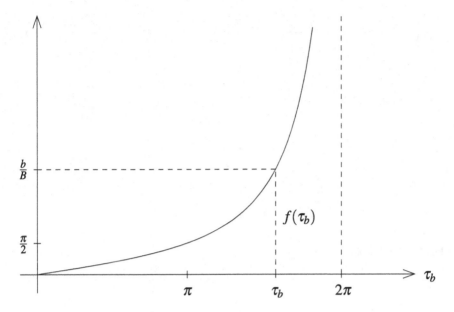

Fig. 1.14 Graph of the Function f

The function $f(\tau_b)$ is strictly monotonically increasing, $f(0) = 0$ and $\lim\limits_{\tau_b \to 2\pi} f(\tau_b) = +\infty$. This implies:

There is precisely one $\tau_b \in (0, 2\pi)$ such that $f(\tau_b) = \dfrac{b}{B} > 0$,

$$\frac{b}{B} \le \frac{\pi}{2} \Rightarrow \tau_b \le \pi, \quad \frac{b}{B} > \frac{\pi}{2} \Rightarrow \tau_b \in (\pi, 2\pi).$$

(1.8.15)

The last constant r is determined by

$$r = \frac{B}{1 - \cos \tau_b}.$$

(1.8.16)

The resulting brachistochrone,

$$\hat{x}(\tau) = r(\tau - \sin \tau),$$
$$\hat{y}(\tau) = r(1 - \cos \tau) \quad \text{for } \tau \in [0, \tau_b],$$

(1.8.17)

is a cycloid, depicted in Figure 1.15.

A fixed point on the perimeter of a rolling wheel with radius r traces out a cycloid. It is remarkable that for $\frac{b}{B} \le \frac{\pi}{2}$, i.e., for $\tau_b \le \pi$, the trajectory descends, whereas for $\frac{b}{B} > \frac{\pi}{2}$, i.e., for $\tau_b \in (\pi, 2\pi)$, it ascends.

The parameter τ gives the physical running time t on a cycloid with parameter r as follows. By analogy to (1.8.4),

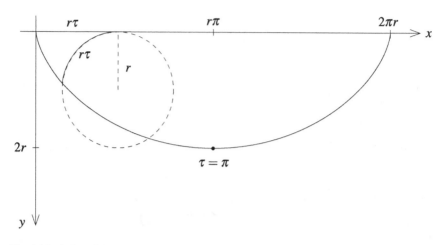

Fig. 1.15 A Cycloid

$$t_0 = \int_0^{x(t_0)} \sqrt{\frac{1+(\tilde{y}')^2}{2g\tilde{y}}} \, dx = \frac{1}{\sqrt{2g}} \int_0^{\tau_0} \sqrt{\frac{1+(\tilde{y}'(\hat{x}(\tau))^2}{\tilde{y}(\hat{x}(\tau))}} \frac{d\hat{x}}{d\tau}(\tau) d\tau, \qquad (1.8.18)$$

where we apply the substitution formula with $x(t_0) = \hat{x}(\tau_0)$. Using $\tilde{y}(\hat{x}(\tau)) = \hat{y}(\tau)$, the formulas (1.8.12), (1.8.17), and (1.8.18) give $\sqrt{\frac{r}{g}}\tau_0$. Thus the relation between the parameter τ and the physical time t is

$$t = \sqrt{\frac{r}{g}}\tau, \quad \text{and in particular, } T = \sqrt{\frac{r}{g}}\tau_b. \qquad (1.8.19)$$

We have to leave open whether the cycloid gives indeed the minimal value. For the functional (1.8.5) we have $J(\tilde{y}) = \sqrt{2g}T < \infty$, whence $\tilde{y} \in \tilde{D}$, cf. (1.8.6). Furthermore its tangent at $(0,0)$ is vertical:

$$\frac{d\tilde{y}}{dx}(x) = \frac{d\hat{y}}{d\tau}(\tau) \bigg/ \frac{d\hat{x}}{d\tau}(\tau) \to +\infty \quad \text{for} \quad \begin{cases} x \searrow 0, \\ \tau \searrow 0. \end{cases} \qquad (1.8.20)$$

Now we sketch how Bernoulli solved his variational problem. He didn't have the results of Euler and Lagrange but he knew Fermat's principle of least time. That principle, in turn, implies Snell's refraction law (5) as described in the Introduction. Bernoulli discretized the continuous problem. In thin layers he assumed a constant velocity along straight line segments. Increasing velocities decrease the slopes according to Snell's refraction law as sketched in Figure 1.16.

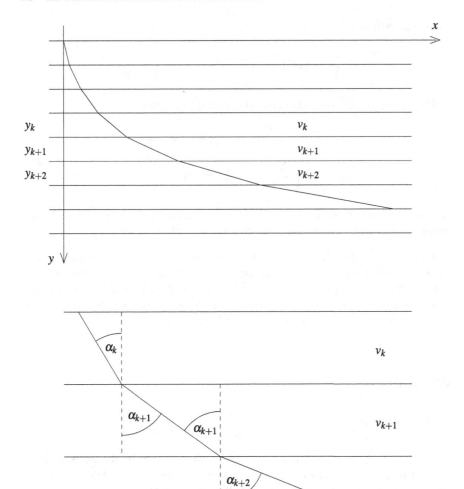

Fig. 1.16 On Bernoulli's Solution of the Brachistochrone Problem

Snell's law gives

$$\frac{\sin \alpha_k}{\sin \alpha_{k+1}} = \frac{v_k}{v_{k+1}}, \quad \frac{\sin \alpha_{k+1}}{\sin \alpha_{k+2}} = \frac{v_{k+1}}{v_{k+2}} \quad \text{etc.} \quad \text{or}$$

$$\frac{v_k}{\sin \alpha_k} = c \quad \text{for all } k.$$

(1.8.21)

The slope of the line segment in the kth layer is $y'_k = \tan\left(\frac{\pi}{2} - \alpha_k\right)$, where it has to be taken into account that the y-axis points downwards. Well-known formulas of trigonometry imply

$$\sqrt{1+(y_k')^2} = \sqrt{1+\tan^2\left(\frac{\pi}{2}-\alpha_k\right)} = \frac{1}{\cos(\frac{\pi}{2}-\alpha_k)} = \frac{1}{\sin\alpha_k}, \tag{1.8.22}$$

and using $v_k = \sqrt{2gy_k}$ from $(1.8.2)_2$ and $(1.8.21)_2$ one obtains

$$\sqrt{2gy_k(1+(y_k')^2)} = c \quad \text{or}$$

$$y_k' = \sqrt{\frac{2r-y_k}{y_k}} \quad \text{where} \quad 2r = \frac{c^2}{2g}. \tag{1.8.23}$$

A transition from the discretized to the continuous problem yields finally the differential equation $(1.8.10)_2$, whose parametric form is solved by a cycloid.

Remark. *Bernoulli did not prove that the cycloid is indeed the curve that a point mass traces out to reach a given endpoint in shortest time. For him and his contemporaries this was evident. As a matter of fact, his proof using Snell's refraction law gives stronger evidence than a proof using the Euler-Lagrange equation. Weierstraß proved the minimizing property of the cycloid much later, cf. [2], p. 46.*

Exercises

1.8.1. Analyze the function $f(\tau) = \dfrac{\tau - \sin\tau}{1-\cos\tau}$ for $\tau \in [0, 2\pi)$.

1.8.2. Let $(0,0)$ be the starting point and $(b,B) = (b, \frac{2}{\pi}b)$ be the endpoint in a coordinate system $\{(x,y)\}$ where the y-axis points downwards. Compare the running time of a point mass acted on by gravity on the line segment and on the cycloid from $(0,0)$ to (b,B). Compute the ratio of the running times.

1.9 Natural Boundary Conditions

Let the functional

$$J(y) = \int_a^b F(x,y,y')dx. \tag{1.9.1}$$

be defined on $D = C^{1,pw}[a,b]$. Then for any $y \in D$, $y + th \in D$ for all $h \in C^{1,pw}[a,b]$ and for all $t \in \mathbb{R}$. Under the hypotheses of Proposition 1.2.2 the first variation exists in all y and in all directions h in $D = C^{1,pw}[a,b]$ and is given by

$$\delta J(y)h = \int_a^b F_y(x,y,y')h + F_{y'}(x,y,y')h'dx. \tag{1.9.2}$$

For a local minimizer $\delta J(y)h = 0$ for all $h \in D$ and therefore also for all $h \in C_0^{1,pw}[a,b] \subset D$. By Proposition 1.4.1 y solves the Euler-Lagrange equation (1.4.3), and, in addition, y fulfills the natural boundary conditions:

Proposition 1.9.1. *Under the hypotheses of Proposition 1.4.1 a local minimizer $y \in C^{1,pw}[a,b]$ of the functional (1.9.1) solves the Euler-Lagrange equation (1.4.3) and fulfills the natural boundary conditions*

$$F_{y'}(a,y(a),y'(a)) = 0 \quad and$$
$$F_{y'}(b,y(b),y'(b)) = 0. \qquad (1.9.3)$$

Proof. For $y \in D$, the statements (1.4.3) hold, and integration by parts, cf. Lemma 1.3.3, yields

$$\delta J(y)h = \int_a^b F_y h + F_{y'} h' dx$$

$$= \int_a^b (F_y - \frac{d}{dx} F_{y'}) h dx + F_{y'} h \Big|_a^b, \qquad (1.9.4)$$

where we use the abbreviations $F_y = F_y(\cdot,y,y')$ and $F_{y'} = F_{y'}(\cdot,y,y')$. By the Euler-Lagrange equation in its weak form (1.4.4) and its strong form (1.4.3), relation (1.9.4) implies

$$F_{y'}(b,y(b),y'(b))h(b) - F_{y'}(a,y(a),y'(a))h(a) = 0. \qquad (1.9.5)$$

Choosing $h(b) = 0$ and $h(a) \neq 0$ proves the natural boundary condition at $x = a$ and $h(a) = 0$, $h(b) \neq 0$ proves it at $x = b$. □

Remark. *If for a local minimizer $y \in D$, one boundary condition $y(a) = A$ or $y(b) = B$ is prescribed, then at the respective free boundary the natural boundary is fulfilled.*

Examples:

1. The functional $J(y) = \int_a^b \sqrt{1+(y')^2} dx$ defined on $D = C^{1,pw}[a,b]$ describes the length of a graph $\{(x,y(x))|x \in [a,b]\}$ between the vertical lines $x = a$ and $x = b$. In Example 4 of Paragraph 1.5 we have seen that a minimizer of J is among the lines $y(x) = c_1 x + c_2$. The natural boundary conditions are

$$F_{y'}(a,y(a),y'(a)) = \frac{y'(a)}{\sqrt{1+(y'(a))^2}} = 0 \quad whence \quad y'(a) = 0,$$

   and analogously $y'(b) = 0$. Minimizers are therefore $y(x) = c_2$.
2. What is the curve traced out by a point mass acted on only by gravity which starts at $(0,0)$ and reaches the vertical line $x = b$ in shortest time?

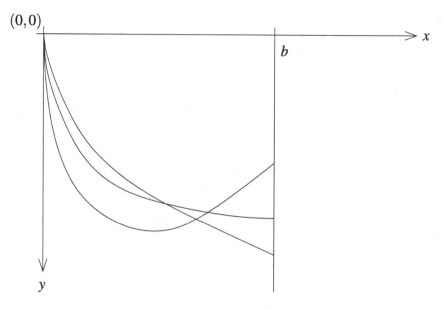

Fig. 1.17   Admitted Curves for Example 2

The functional (1.8.5) is defined on (1.8.6) without a boundary condition at $x = b$. A minimizer fulfills the Euler-Lagrange equation (1.8.10) with the boundary condition $\tilde{y}(0) = 0$. The natural boundary condition at $x = b$ is

$$\frac{\tilde{y}'(b)}{\sqrt{\tilde{y}(b)(1 + (\tilde{y}'(b))^2}} = 0 \quad \text{whence} \quad \tilde{y}'(b) = 0.$$

After transformation into a parametric differential equation, solutions are again cycloids (1.8.17), and the natural boundary condition for $\hat{y}(\tau) = \tilde{y}(\hat{x}(\tau))$ implies

$$\frac{d\hat{y}}{d\tau}(\tau_b) = \tilde{y}'(x(\tau_b))\frac{d\hat{x}}{d\tau}(\tau_b) = \tilde{y}'(b)\frac{d\hat{x}}{d\tau}(\tau_b) = 0, \quad \text{and whence,}$$
$$r \sin \tau_b = 0.$$

This gives $\tau_b = \pi$, and by $\hat{x}(\tau_b) = r(\pi - \sin \pi) = r\pi = b$, we obtain the cycloid

$$\hat{x}(\tau) = \frac{b}{\pi}(\tau - \sin \tau),$$
$$\hat{y}(\tau) = \frac{b}{\pi}(1 - \cos \tau) \quad \text{for} \quad \tau \in [0, \pi],$$

meeting the vertical line at $(b, \frac{2}{\pi}b)$ orthogonally (Figure 1.17).

Exercises

1.9.1.   a) Find all solutions of the Euler-Lagrange equation of the functional

$$J(y) = \int_0^1 (y')^2 + y \, dx$$

on $C^{1,pw}[0,1]$.
b) Which solutions fulfill the natural boundary conditions?
c) Which solutions fulfill the boundary conditions $y(0) = 0$, $y(1) = 1$?
d) Which solutions are local extremals without and with boundary conditions? Which are globally extremal?

1.9.2. Does the functional

$$J(y) = \int_a^b (y')^2 + \arctan y \, dx$$

have local or global extremals in $C^{1,pw}[a,b]$? Is the functional bounded below?

1.10 Functionals in Parametric Form

In some cases we have parametrized graphs $\{(x,y(x))|x \in [a,b]\}$ of functions like $\{(\tilde{x}(t), \tilde{y}(t))|t \in [t_a,t_b]\}$. Apparently the class of parametrized curves is bigger than the class of graphs, cf. Figure 1.18.

But even if the admitted curves are graphs, a parametrization can give more information about minimizing curves. This will be demonstrated in the next paragraph.

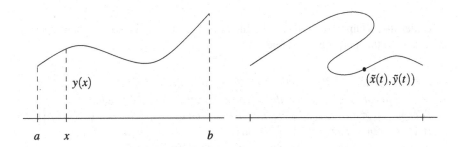

Fig. 1.18   A Graph of a Function and a Parametrized Curve

Definition 1.10.1. *A functional*

$$J(x,y) = \int_{t_a}^{t_b} \Phi(x,y,\dot{x},\dot{y})dt, \quad \text{defined on}$$

$$D \subset (C^{1,pw}[t_a,t_b])^2 \quad \text{with a continuous function} \tag{1.10.1}$$

$$\Phi : \mathbb{R}^4 \to \mathbb{R},$$

is called a functional in parametric form or shortly a parametric functional. Curves in D may have to fulfill boundary conditions at $t = t_a$ or at $t = t_b$.

Without loss of generality we assume that each component of $(x,y) \in (C^{1,pw}[t_a,t_b])^2 = C^{1,pw}[t_a,t_b] \times C^{1,pw}[t_a,t_b]$ has the same partition $t_a = t_0 < t_1 < \cdots < t_m = t_b$ and both components x and y are in $C^1[t_{i-1},t_i], i = 1,\ldots,m$. The integral (1.10.1) is defined as a sum of integrals over $[t_{i-1},t_i]$, cf. (1.1.3).

Two examples: $\int_{t_a}^{t_b} \sqrt{\dot{x}^2 + \dot{y}^2}dt$ is the length of a parametrized curve $(x,y) \in D$. The Lagrange function $L(x,y,\dot{x},\dot{y}) = \frac{1}{2}m(\dot{x}^2 + \dot{y}^2) - V(x,y)$, with a potential $V : \mathbb{R}^2 \to \mathbb{R}$, is called the free energy and $\int_{t_a}^{t_b} L(x,y,\dot{x},\dot{y})dt$ is called the action of the point mass m along the curve $\{(x(t),y(t))|t \in [t_a,t_b]\}$.

There is an important difference between these two examples: Whereas the length of a curve does not depend on its parametrization, the physical quantities like the kinetic energy and the action along a curve require its parametrization by physical time. The following definition takes this difference into account:

Definition 1.10.2. *A parametric functional (1.10.1) is invariant if*

$$\Phi(x,y,\alpha\dot{x},\alpha\dot{y}) = \alpha\Phi(x,y,\dot{x},\dot{y}) \tag{1.10.2}$$

for all $\alpha > 0$ and for all $(x,y,\dot{x},\dot{y}) \in \mathbb{R}^4$.

Under the assumption (1.10.2), the functional (1.10.1) is invariant when reparametrized. To this purpose we define:

Definition 1.10.3. *Let the function $\varphi \in C^1[\tau_a,\tau_b]$ fulfill $\varphi(\tau_a) = t_a$, $\varphi(\tau_b) = t_b$, and $\frac{d\varphi}{d\tau}(\tau) > 0$ for all $\tau \in [\tau_a,\tau_b]$ (with one-sided derivatives at the boundary).*
Then $\varphi : [\tau_a,\tau_b] \to [t_a,t_b]$ is bijective and $\{(x(\varphi(\tau)),y(\varphi(\tau))) = (\tilde{x}(\tau),\tilde{y}(\tau))|\tau \in [\tau_a,\tau_b]\}$ is a reparametrization of the curve $\{(x(t),y(t))|t \in [t_a,t_b]\}$.
If $(x,y) \in (C^{1,pw}[t_a,t_b])^2$ (with boundary conditions at $t = t_a$ or $t = t_b$) then $(\tilde{x},\tilde{y}) \in (C^{1,pw}[\tau_a,\tau_b])^2$ (with boundary conditions at $\tau = \tau_a$ or $\tau = \tau_b$).

Proposition 1.10.1. *If the parametric functional is invariant according to Definition 1.10.2, then for any reparametrization, it follows that*

$$J(\tilde{x}, \tilde{y}) = \int_{\tau_a}^{\tau_b} \Phi\left(\tilde{x}, \tilde{y}, \frac{d\tilde{x}}{d\tau}, \frac{d\tilde{y}}{d\tau}\right) d\tau = J(x, y). \tag{1.10.3}$$

Proof. By analogy with (1.1.3) the integral (1.10.1) is the sum of integrals over $[t_{i-1}, t_i]$ where $(x, y) \in (C^1[t_{i-1}, t_i])^2$. Then $(\tilde{x}, \tilde{y}) \in (C^1[\tau_{i-1}, \tau_i])^2$ where $\varphi(\tau_i) = t_i, i = 0, \ldots, m$ and $J(\tilde{x}, \tilde{y})$ is the sum of integrals over $[\tau_{i-1}, \tau_i]$. For any summand, we apply (1.10.2), and the substitution formula yields

$$\int_{\tau_{i-1}}^{\tau_i} \Phi(\tilde{x}, \tilde{y}, \frac{d\tilde{x}}{d\tau}, \frac{d\tilde{y}}{d\tau}) d\tau$$

$$= \int_{\tau_{i-1}}^{\tau_i} \Phi(x(\varphi(\tau)), y(\varphi(\tau)), \dot{x}(\varphi(\tau))\frac{d\varphi}{d\tau}(\tau), \dot{y}(\varphi(\tau))\frac{d\varphi}{d\tau}(\tau)) d\tau$$

$$= \int_{\tau_{i-1}}^{\tau_i} \Phi(x(\varphi(\tau)), y(\varphi(\tau)), \dot{x}(\varphi(\tau)), \dot{y}(\varphi(\tau)))\frac{d\varphi}{d\tau}(\tau) d\tau \tag{1.10.4}$$

$$= \int_{t_{i-1}}^{t_i} \Phi(x, y, \dot{x}, \dot{y}) dt. \qquad \square$$

Proposition 1.10.1 is applicable to $J(x, y) = \int_{t_a}^{t_b} \sqrt{\dot{x}^2 + \dot{y}^2} dt$, and therefore the length of a curve (x, y) does not depend on its parametrization.

Next we define the first variation for parametric functionals. Like in Paragraph 1.2 we assume that for $(x, y) \in D \subset (C^{1,pw}[t_a, t_b])^2$, the perturbation $(x, y) + s(h_1, h_2) \in D$, where $h = (h_1, h_2) \in (C_0^{1,pw}[t_a, t_b])^2$ and $s \in (-\varepsilon, \varepsilon)$. The Gâteaux differential $dJ(x, y, h_1, h_2)$ is given by $g'(0)$ for $g(s) = J((x, y) + s(h_1, h_2))$, provided the derivative exists. This is the case under the assumptions of the following Proposition:

Proposition 1.10.2. *Let the parametric functional (1.10.1) be defined on $D \subset (C^{1,pw}[t_a, t_b])^2$ and let its Lagrange function $\Phi : \mathbb{R}^4 \to \mathbb{R}$ be continuously partially differentiable with respect to all four variables (or continuously totally differentiable). Then for all $(x, y) \in D$ and all $h = (h_1, h_2) \in (C_0^{1,pw}[t_a, t_b])^2$, the Gâteaux differential exists and it is represented by*

$$dJ(x, y, h_1, h_2) = \int_{t_a}^{t_b} \Phi_x h_1 + \Phi_y h_2 + \Phi_{\dot{x}} \dot{h}_1 + \Phi_{\dot{y}} \dot{h}_2 dt. \tag{1.10.5}$$

Here $\Phi_x, \Phi_y, \Phi_{\dot{x}}, \Phi_{\dot{y}}$ denote the partial derivatives with respect to the four variables and in (1.10.5) we use the abbreviations $\Phi_x = \Phi_x(x, y, \dot{x}, \dot{y})$ etc.

Proof. The proof is a minor modification of the proof of Proposition 1.2.1. $\quad \square$

Since the Gâteaux differential (1.10.5) is linear in $h = (h_1, h_2)$, we call it as in Definition 1.2.2 the first variation of J in (x, y) in direction $h = (h_1, h_2)$, denoted by

$$dJ(x, y, h_1, h_2) = \delta J(x, y)h. \tag{1.10.6}$$

Finally, if J is defined on all of $(C^{1,pw}[t_a, t_b])^2$ then

$$J : (C^{1,pw}[t_a, t_b])^2 \to \mathbb{R} \quad \text{is continuous, and}$$
$$\delta J(x, y) : (C^{1,pw}[t_a, t_b])^2 \to \mathbb{R} \quad \text{is linear and continuous.} \tag{1.10.7}$$

If a curve $(x, y) \in D \subset (C^{1,pw}[t_a, t_b])^2$ is a local minimizer of J according to Definition 1.4.1, meaning

$$J(x, y) \le J(\hat{x}, \hat{y}) \quad \text{for all } (\hat{x}, \hat{y}) \in D, \text{ where}$$
$$\|x - \hat{x}\|_{1, pw, [t_a, t_b]} < d \quad \text{and} \quad \|y - \hat{y}\|_{1, pw, [t_a, t_b]} < d, \tag{1.10.8}$$

then the function $g(s) = J((x, y) + s(h_1, h_2))$ has a local minimum at $s = 0$, whence $g'(0) = 0$. This implies, cf. Proposition 1.4.1:

Proposition 1.10.3. *Let the curve $(x, y) \in D \subset (C^{1,pw}[t_a, t_b])^2$ be a local minimizer of the parametric functional (1.10.1) and let the Lagrange function $\Phi : \mathbb{R}^4 \to \mathbb{R}$ be continuously totally differentiable. Then*

$$(\Phi_{\dot{x}}(x, y, \dot{x}, \dot{y}), \Phi_{\dot{y}}(x, y, \dot{x}, \dot{y})) \in (C^{1,pw}[t_a, t_b])^2 \quad and$$
$$\frac{d}{dt} \Phi_{\dot{x}}(x, y, \dot{x}, \dot{y}) = \Phi_x(x, y, \dot{x}, \dot{y}) \quad piecewise\ on\ [t_a, t_b], \tag{1.10.9}$$
$$\frac{d}{dt} \Phi_{\dot{y}}(x, y, \dot{x}, \dot{y}) = \Phi_y(x, y, \dot{x}, \dot{y}) \quad piecewise\ on\ [t_a, t_b].$$

Proof. We follow the proof of Proposition 1.4.1. In view of $g'(0) = 0$, (1.10.5) yields

$$\delta J(x, y)h = \int_{t_a}^{t_b} \Phi_x h_1 + \Phi_y h_2 + \Phi_{\dot{x}} \dot{h}_1 + \Phi_{\dot{y}} \dot{h}_2 \, dt = 0 \tag{1.10.10}$$

for all $h = (h_1, h_2) \in (C_0^{1,pw}[t_a, t_b])^2$. Choosing $h_2 \equiv 0$ and $h_1 \equiv 0$, respectively, we obtain

$$\int_{t_a}^{t_b} \Phi_x h_1 + \Phi_{\dot{x}} \dot{h}_1 \, dt = 0 \quad \text{for all } h_1 \in C_0^{1,pw}[t_a, t_b],$$
$$\int_{t_a}^{t_b} \Phi_y h_2 + \Phi_{\dot{y}} \dot{h}_2 \, dt = 0 \quad \text{for all } h_2 \in C_0^{1,pw}[t_a, t_b]. \tag{1.10.11}$$

The claim (1.10.9) is a consequence of Lemma 1.3.4. □

Equations (1.10.9) are the Euler-Lagrange equations for a locally minimizing curve of the parametric functional (1.10.1). Our remarks following Proposition 1.4.1 are valid here as well.

For an invariant parametric functional (1.10.1), any local minimizer $(x,y) \in D \subset (C^{1,pw}[t_a,t_b])^2$ remains a local minimizer under reparametrization $(\tilde{x},\tilde{y}) \in \tilde{D} \subset (C^{1,pw}[\tau_a,\tau_b])^2$, cf. Proposition 1.10.1. Consequently, $(\tilde{x},\tilde{y})$ fulfills the same Euler-Lagrange equations as does (x,y).

The following proposition states the invariance of the Euler-Lagrange equations.

Proposition 1.10.4. *If the parametric functional (1.10.1) is invariant according to Definition 1.10.2, then for any reparametrization $(\tilde{x}(\tau),\tilde{y}(\tau)) = (x(\varphi(\tau)),y(\varphi(\tau)))$ of a curve $(x,y) \in (C^{1,pw}[t_a,t_b])^2$ according to Definition 1.10.3, the following equivalence holds for a continuously totally differentiable Lagrange function Φ:*

$$(\Phi_{\dot{x}}(\tilde{x},\tilde{y},\frac{d}{d\tau}\tilde{x},\frac{d}{d\tau}\tilde{y}),\Phi_{\dot{y}}(\tilde{x},\tilde{y},\frac{d}{d\tau}\tilde{x},\frac{d}{d\tau}\tilde{y})) \in (C^{1,pw}[\tau_a,\tau_b])^2,$$

$$\frac{d}{d\tau}\Phi_{\dot{x}}(\tilde{x},\tilde{y},\frac{d}{d\tau}\tilde{x},\frac{d}{d\tau}\tilde{y}) = \Phi_x(\tilde{x},\tilde{y},\frac{d}{d\tau}\tilde{x},\frac{d}{d\tau}\tilde{y}), \qquad (1.10.12)$$

$$\frac{d}{d\tau}\Phi_{\dot{y}}(\tilde{x},\tilde{y},\frac{d}{d\tau}\tilde{x},\frac{d}{d\tau}\tilde{y}) = \Phi_y(\tilde{x},\tilde{y},\frac{d}{d\tau}\tilde{x},\frac{d}{d\tau}\tilde{y}),$$

holds piecewise on $[\tau_a,\tau_b]$, respectively, if and only if

$$(\Phi_{\dot{x}}(x,y,\dot{x},\dot{y}),\Phi_{\dot{y}}(x,y,\dot{x},\dot{y})) \in (C^{1,pw}[t_a,t_b])^2,$$

$$\frac{d}{dt}\Phi_{\dot{x}}(x,y,\dot{x},\dot{y}) = \Phi_x(x,y,\dot{x},\dot{y}), \qquad (1.10.13)$$

$$\frac{d}{dt}\Phi_{\dot{y}}(x,y,\dot{x},\dot{y}) = \Phi_y(x,y,\dot{x},\dot{y}),$$

holds piecewise on $[t_a,t_b]$, respectively.

Proof. By differentiation, relation (1.10.2) gives

$$\Phi_x(x,y,\alpha\dot{x},\alpha\dot{y}) = \alpha\Phi_x(x,y,\dot{x},\dot{y}),$$
$$\Phi_y(x,y,\alpha\dot{x},\alpha\dot{y}) = \alpha\Phi_y(x,y,\dot{x},\dot{y}),$$
$$\Phi_{\dot{x}}(x,y,\alpha\dot{x},\alpha\dot{y}) = \Phi_{\dot{x}}(x,y,\dot{x},\dot{y}), \qquad (1.10.14)$$
$$\Phi_{\dot{y}}(x,y,\alpha\dot{x},\alpha\dot{y}) = \Phi_{\dot{y}}(x,y,\dot{x},\dot{y}),$$

where the last two equations are divided by $\alpha > 0$. Inserting $\tilde{x}(\tau) = x(\varphi(\tau)), \frac{d}{d\tau}\tilde{x}(\tau) = \dot{x}(\varphi(\tau))\frac{d\varphi}{d\tau}(\tau)$, $\tilde{y}(\tau) = y(\varphi(\tau))$, and $\frac{d}{d\tau}\tilde{y}(\tau) = \dot{y}(\varphi(\tau))\frac{d\varphi}{d\tau}(\tau)$ into $(1.10.14)_{3,4}$ yields

$$\Phi_{\dot{x}}(\tilde{x},\tilde{y},\frac{d}{d\tau}\tilde{x},\frac{d}{d\tau}\tilde{y})(\tau) = \Phi_{\dot{x}}(x,y,\dot{x},\dot{y})(\varphi(\tau)),$$
$$\qquad (1.10.15)$$
$$\Phi_{\dot{y}}(\tilde{x},\tilde{y},\frac{d}{d\tau}\tilde{x},\frac{d}{d\tau}\tilde{y})(\tau) = \Phi_{\dot{y}}(x,y,\dot{x},\dot{y})(\varphi(\tau)),$$

which proves the equivalence of $(1.10.12)_1$ and $(1.10.13)_1$. By $(1.10.15)_1$ and $(1.10.14)_1$ we obtain

$$
\begin{aligned}
\frac{d}{d\tau}\Phi_{\dot{x}}(\tilde{x},\tilde{y},\frac{d}{d\tau}\tilde{x},\frac{d}{d\tau}\tilde{y})(\tau) &= \frac{d}{dt}\Phi_{\dot{x}}(x,y,\dot{x},\dot{y})(\varphi(\tau))\frac{d\varphi}{d\tau}(\tau), \\
\Phi_x(\tilde{x},\tilde{y},\frac{d}{d\tau}\tilde{x},\frac{d}{d\tau}\tilde{y})(\tau) &= \Phi_x(x,y,\dot{x},\dot{y})(\varphi(\tau))\frac{d\varphi}{d\tau}(\tau).
\end{aligned}
\tag{1.10.16}
$$

Since $\frac{d\varphi}{d\tau}(\tau) > 0$ for all $\tau \in [\tau_a,\tau_b]$, the relations (1.10.16) prove the equivalence of $(1.10.12)_2$ and $(1.10.13)_2$. Analogously the equivalence of $(1.10.12)_3$ and $(1.10.13)_3$ follows from $(1.10.15)_2$ and $(1.10.14)_2$.   □

Remark. *Be careful with reparametrizations of parametric functionals. If, for instance, a curve is parametrized by its arc length, then the Lagrange function of the parametric functional describing the length of the curve simplifies to the constant 1, cf. (2.6.6). But now the parameter interval depends on the individual curve.*

The natural boundary conditions on local minimizers read as follows:

Proposition 1.10.5. *Let $(x,y) \in D \subset (C^{1,pw}[t_a,t_b])^2$ be a local minimizer of the parametric functional (1.10.1) with continuously totally differentiable Lagrange function Φ. If the components x and/or y are free at $t = t_a$ and/or $t = t_b$, then they fulfill the natural boundary conditions*

$$
\begin{aligned}
\Phi_{\dot{x}}(x(t_a),y(t_a),\dot{x}(t_a),\dot{y}(t_a)) &= 0 \quad \text{and/or} \\
\Phi_{\dot{y}}(x(t_a),y(t_a),\dot{x}(t_a),\dot{y}(t_a)) &= 0 \quad \text{and/or} \\
\Phi_{\dot{x}}(x(t_b),y(t_b),\dot{x}(t_b),\dot{y}(t_b)) &= 0 \quad \text{and/or} \\
\Phi_{\dot{y}}(x(t_b),y(t_b),\dot{x}(t_b),\dot{y}(t_b)) &= 0, \quad \text{respectively.}
\end{aligned}
\tag{1.10.17}
$$

Proof. We consider only the case when x is free at $t = t_a$. Then $(x,y) + s(h_1,h_2) \in D$ for any $h = (h_1,h_2) \in C^{1,pw}[t_a,t_b]$ with arbitrary $h_1(t_a)$ but satisfying $h_1(t_b) = 0$, $h_2(t_a) = 0$, $h_2(t_b) = 0$. Since $\delta J(x,y)h = 0$ we obtain after integration by parts, allowed by $(1.10.9)_1$,

$$
\begin{aligned}
\int_{t_a}^{t_b} &\Phi_x h_1 + \Phi_y h_2 + \Phi_{\dot{x}} \dot{h}_1 + \Phi_{\dot{y}} \dot{h}_2 dt \\
&= \int_{t_a}^{t_b} \left(\Phi_x - \frac{d}{dt}\Phi_{\dot{x}}\right) h_1 + \left(\Phi_y - \frac{d}{dt}\Phi_{\dot{y}}\right) h_2 dt - (\Phi_{\dot{x}} h_1)(t_a) = 0.
\end{aligned}
\tag{1.10.18}
$$

All other boundary terms vanish by the choice of h_1 and h_2. Since the local minimizer fulfills the Euler-Lagrange equations $(1.10.9)_{2,3}$, only the boundary term in (1.10.18) remains. If $h_1(t_a) \neq 0$, we obtain $(1.10.17)_1$.   □

It is not necessary to confine ourselves to plane curves. Furthermore the Lagrange function can depend explicitly on the parameter. We generalize:

A curve in $\mathbb{R}^n$, $n \in \mathbb{N}$, is given by $\{x(t) = (x_1(t), \ldots, x_n(t)) | t \in [t_a, t_b]\}$. A functional

$$J(x) = \int_{t_a}^{t_b} \Phi(t, x, \dot{x}) dt, \tag{1.10.19}$$

with a continuous Lagrange function $\Phi : [t_a, t_b] \times \mathbb{R}^n \times \mathbb{R}^n \to \mathbb{R}$ can be defined on $D \subset (C^{1,pw}[t_a, t_b])^n$. We call it a **parametric functional for curves in $\mathbb{R}^n$**. Boundary conditions can be imposed for all or only some components of $x \in D$.

If Φ does not depend explicitly on the parameter and if

$$\Phi(x, \alpha \dot{x}) = \alpha \Phi(x, \dot{x}) \tag{1.10.20}$$

for all $\alpha > 0$ and for all $(x, \dot{x}) \in \mathbb{R}^n \times \mathbb{R}^n$, then the parametric functional (1.10.19) is called **invariant**. The proof of Proposition 1.10.1 for $n > 2$ again shows that any reparametrization of the curve x according to Definition (1.10.3) leaves the functional invariant.

If $\Phi : [t_a, t_b] \times \mathbb{R}^n \times \mathbb{R}^n \to \mathbb{R}$ is continuously partially differentiable with respect to the last $2n$ variables, then the **first variation** of J in x and in direction $h = (h_1, \ldots, h_n) \in (C_0^{1,pw}[t_a, t_b])^n$ exists and it is represented by

$$\delta J(x)h = \int_{t_a}^{t_b} \sum_{k=1}^{n} (\Phi_{x_k} h_k + \Phi_{\dot{x}_k} \dot{h}_k) dt$$

$$= \int_{t_a}^{t_b} (\Phi_x, h) + (\Phi_{\dot{x}}, \dot{h}) dt. \tag{1.10.21}$$

In (1.10.21) we use the abbreviations $\Phi_x = (\Phi_{x_1}, \ldots, \Phi_{x_n})$, $\Phi_{\dot{x}} = (\Phi_{\dot{x}_1}, \ldots, \Phi_{\dot{x}_n})$, and also employ the Euclidean scalar product $(\ ,\ )$ in $\mathbb{R}^n$. The argument of Φ_{x_k} and of $\Phi_{\dot{x}_k}$ is the vector $(t, x(t), \dot{x}(t)) \in [t_a, t_b] \times \mathbb{R}^n \times \mathbb{R}^n$.

The distance between two curves $x, \hat{x} \in (C^{1,pw}[t_a, t_b])^n$ is defined by $\max_{k \in \{1, \ldots, n\}} \|x_k - \hat{x}_k\|_{1,pw,[t_a,t_b]}$ which, in turn, allows us to define a local minimizer of J as in Definition 1.4.1. For a local minimizer $x \in D \subset (C^{1,pw}[t_a, t_b])^n$, the first variation vanishes, i.e., $\delta J(x)h = 0$ for all $h \in (C_0^{1,pw}[t_a, t_b])^n$, and as expounded in Proposition 1.10.3 for $n = 2$, this implies **the system of Euler-Lagrange equations**

$$\Phi_{\dot{x}}(\cdot, x, \dot{x}) \in (C^{1,pw}[t_a, t_b])^n, \quad \text{and}$$

$$\frac{d}{dt} \Phi_{\dot{x}}(\cdot, x, \dot{x}) = \Phi_x(\cdot, x, \dot{x}) \quad \text{piecewise on } [t_a, t_b]. \tag{1.10.22}$$

The proof of Proposition 1.10.4 can be extended to this system: If the parametric functional does not depend explicitly on the parameter, and if it is invariant in the sense of (1.10.20), then the system of Euler-Lagrange equations is invariant under reparametrizations.

Finally, there is the analogue of Proposition 1.10.5: If the kth component x_k of a local minimizer x is free at $t = t_a$ and/or at $t = t_b$, then it fulfills the **natural boundary conditions**

$$\Phi_{\dot{x}_k}(t_a,x(t_a),\dot{x}(t_a)) = 0 \text{ and/or } \Phi_{\dot{x}_k}(t_b,x(t_b),\dot{x}(t_b)) = 0, \text{ respectively. } \quad (1.10.23)$$

Among the historically first variational problems, we find physical examples, in particular in **Lagrangian mechanics**:

Let $\{x(t) = (x_1(t),x_2(t),x_3(t))|t \in [t_a,t_b]\}$ be the trajectory of a point mass m in the space $\mathbb{R}^3$ parametrized by the time t. For $x \in (C^1[t_a,t_b])^3$

$$T = \frac{1}{2}m(\dot{x}_1^2 + \dot{x}_2^2 + \dot{x}_3^2) = \frac{1}{2}m\|\dot{x}\|^2 = T(\dot{x}) \quad (1.10.24)$$

is the kinetic energy, and

$$V = V(x_1,x_2,x_3) = V(x) \quad (1.10.25)$$

is the potential energy. ($\|\ \|$ is the Euclidean norm in $\mathbb{R}^3$). We assume continuous total differentiability of V. Then

$$\begin{aligned}
E &= T + V &&\text{ is the total energy,}\\
L &= T - V &&\text{ is the free energy, and}\\
J(x) &= \int_{t_a}^{t_b} L(x,\dot{x})dt &&\text{ is the action}
\end{aligned} \quad (1.10.26)$$

of the point mass m along the trajectory $\{x(t)|t \in [t_a,t_b]\}$. The function $L : \mathbb{R}^3 \times \mathbb{R}^3 \to \mathbb{R}$ is the Lagrangian, a nomenclature that is transferred to all variational functionals.

According to the "principle of least action" the minimization of the action leads to the system of Euler-Lagrange equations, which a trajectory has to fulfill:

$$\begin{aligned}
&\frac{d}{dt}L_{\dot{x}}(x,\dot{x}) = L_x(x,\dot{x}), \quad \text{or}\\
&m\ddot{x}_1 = -V_{x_1}(x_1,x_2,x_3)\\
&m\ddot{x}_2 = -V_{x_2}(x_1,x_2,x_3)\\
&m\ddot{x}_3 = -V_{x_3}(x_1,x_2,x_3), \quad \text{or}\\
&m\ddot{x} = -\nabla V(x) = -\text{grad}V(x).
\end{aligned} \quad (1.10.27)$$

The equations (1.10.27) are the equations of motions for a point mass m. Since the total energy is conserved along a trajectory that fulfills (1.10.27), cf. Exercise 1.10.4, the system (1.10.27) is called a conservative system. Uniqueness of a solution is only possible if (natural) boundary conditions are imposed. However, it is not obvious that a solution minimizes the action. A sufficient condition is given in Exercise 1.10.6.

We remark that the action is not invariant in the sense of (1.10.20). Therefore the physical time cannot be replaced by a different parameter without changing the physical meaning of equations (1.10.27).

Remark. *In a different notation, (1.10.19) can be rewritten as*

$$J(y) = \int_a^b F(x,y,y')dx \qquad (1.10.28)$$

for $y(x) = (y_1(x),\ldots,y_n(x))$, $y'(x) = (y_1'(x),\ldots,y_n'(x))$. *The functional (1.10.28) generalizes the functional (1.1.1) in a natural way. The Lagrange function of (1.10.28)* $F : [a,b] \times \mathbb{R}^n \times \mathbb{R}^n \to \mathbb{R}$ *is a continuous and a continuously partially differentiable function with respect to the last 2n variables by assumption. The Euler-Lagrange equations are similar to (1.4.3), differing only by* $F_y = (F_{y_1},\ldots,F_{y_n})$, $F_{y'} = (F_{y_1'},\ldots,F_{y_n'})$, *and by the fact that (1.4.3) now becomes a system of n differential equations.*

Exercises

1.10.1. Prove that a curve $(x,y) \in (C^{1,pw}[t_a,t_b])^2$ is the graph of a function $\tilde{y} \in C^{1,pw}[a,b]$ if $x(t_a) = a$, $x(t_b) = b$ and $\dot{x}(t) > 0$ piecewise for $t \in [t_{i-1},t_i], i = 1,\ldots,m$. Here $t_a = t_0 < t_1 < \cdots < t_{m-1} < t_m = t_b$.

1.10.2. The parametric functional,

$$J(x) = \int_{t_a}^{t_b} (F(x),\dot{x})dt,$$

defined for a totally differentiable vector field $F : \mathbb{R}^n \to \mathbb{R}^n$ and for a curve $x \in D = (C^{1,pw}[t_a,t_b])^n \cap \{x(t_a) = A, x(t_b) = B\}$, is called the integral of the field F along the curve x. Here $(\ ,\ )$ is the Euclidean scalar product in $\mathbb{R}^n$.

a) Compute the first variation $\delta J(x) : (C_0^{1,pw}[t_a,t_b])^n \to \mathbb{R}^n$.
b) Give the system of Euler-Lagrange equations. Does it have solutions in any D?
c) Assume

$$\frac{\partial F_i}{\partial x_k}(x) = \frac{\partial F_k}{\partial x_i}(x) \qquad \text{for all } x \in \mathbb{R}^n \quad \text{and} \quad i,k = 1,\ldots,n.$$

Show that $\delta J(x) = 0$ for all $x \in D$ and that in this case any $x \in D$ is a solution of the Euler-Lagrange system piecewise. What does this mean for the functional J?

A Lagrange function (a Lagrangian) with identically vanishing first variation is called a "Null Lagrangian."

1.10.3. Let the curve $x \in (C^2[t_a,t_b])^n$ be a local minimizer of the parametric functional

$$J(x) = \int_{t_a}^{t_b} \Phi(x, \dot{x}) dt,$$

where the Lagrange function $\Phi : \mathbb{R}^n \times \mathbb{R}^n \to \mathbb{R}$ is continuously totally differentiable. Prove that

$$\Phi(x, \dot{x}) - (\dot{x}, \Phi_{\dot{x}}(x, \dot{x})) = c_1 \quad \text{on} \ [t_a, t_b],$$

where $(\ , \ )$ is the scalar product in $\mathbb{R}^n$.

1.10.4. Adopt the definitions (1.10.24)–(1.10.27). Show that a local minimizer $x \in (C^1[t_a, t_b])^3$ of the action is in $(C^2[t_a, t_b])^3$ and that the total energy $E = E(x, \dot{x}) = const$ for $t \in [t_a, t_b]$.

1.10.5. Compute the second variation of the parametric functional

$$J(x) = \int_{t_a}^{t_b} \Phi(t, x, \dot{x}) dt$$

in x in direction h, where the Lagrange function $\Phi : \mathbb{R} \times \mathbb{R}^n \times \mathbb{R}^n \to \mathbb{R}$ is continuous and two times continuously partially differentiable with respect to the last $2n$ variables. That is, compute

$$\frac{d^2}{ds^2} J(x + sh)|_{s=0} = \delta^2 J(x)(h, h),$$

where $x, h \in (C^{1, pw}[t_a, t_b])^n$, cf. Exercise 1.2.2.

1.10.6. a) Adopt the definitions (1.10.24)–(1.10.27) and assume that the potential energy $V : \mathbb{R}^3 \to \mathbb{R}$ is two times continuously differentiable. Compute the second variation $\delta^2 J(x)(h, h)$ in $x, h \in (C^1[t_a, t_b])^3$, cf. Exercise 1.10.5.
b) Let the Hessian of the potential energy

$$D^2 V(x) = \left(\frac{\partial^2 V}{\partial x_i \partial x_j}(x) \right)_{\substack{i=1,2,3 \\ j=1,2,3}}$$

fulfill

$$(D^2 V(x) h, h) \leq 0 \quad \text{for all} \ x, h \in \mathbb{R}^3.$$

Show that any solution of the equations of motions (1.10.27) is a global minimizer of the action among all trajectories $x \in (C^2[t_a, t_b])^3$ fulfilling the same boundary conditions at $t = t_a$ and at $t = t_b$.
Hint: Exercise 1.4.4.

1.11 The Weierstraß-Erdmann Corner Conditions

We mentioned before in Paragraph 1.10 that a functional in parametric form can give more information about a minimizer of a functional in nonparametric form (1.1.1). Therefore we consider (1.1.1) as a special case of (1.10.1).

A reparametrization according to Definition 1.10.3 is given by

$$\varphi : [\tau_a, \tau_b] \to [a,b], \qquad \varphi \in C^1[\tau_a, \tau_b],$$
$$\frac{d\varphi}{d\tau}(\tau) > 0 \quad \text{for all} \quad \tau \in [\tau_a, \tau_b]. \tag{1.11.1}$$

Let $y \in C^{1,pw}[a,b]$ and set

$$x = \varphi(\tau) = \tilde{x}(\tau) \in [a,b],$$
$$y(x) = y(\varphi(\tau)) = y(\tilde{x}(\tau)) = \tilde{y}(\tau).$$
$$\text{Then} \quad (\tilde{x}, \tilde{y}) \in C^1[\tau_a, \tau_b] \times C^{1,pw}[\tau_a, \tau_b], \quad \text{and}$$
$$\{(x, y(x)) | x \in [a,b]\} = \{(\tilde{x}(\tau), \tilde{y}(\tau)) | \tau \in [\tau_a, \tau_b]\} \tag{1.11.2}$$

is a reparametrization of the graph of y. Then on $[\tau_a, \tau_b]$, we have piecewise

$$\frac{d\tilde{y}}{d\tau}(\tau) = y'(\varphi(\tau)) \frac{d\varphi}{d\tau}(\tau) = y'(x) \frac{d\tilde{x}}{d\tau}(\tau), \quad \text{or}$$
$$y'(x) = \frac{\dot{\tilde{y}}}{\dot{\tilde{x}}}(\tau) \quad \text{where } x = \tilde{x}(\tau) \text{ and } (\dot{\,}) = \frac{d(\,)}{d\tau}. \tag{1.11.3}$$

The substitution formula transforms the functional (1.1.1) into a parametric one:

$$J(y) = \int_a^b F(x,y,y')dx = \int_{\tau_a}^{\tau_b} F\left(\tilde{x}, \tilde{y}, \frac{\dot{\tilde{y}}}{\dot{\tilde{x}}}\right) \dot{\tilde{x}} d\tau = J(\tilde{x}, \tilde{y}). \tag{1.11.4}$$

The Lagrange function of the parametric functional is invariant in the sense of Definition 1.10.2:

$$\Phi(\tilde{x}, \tilde{y}, \dot{\tilde{x}}, \dot{\tilde{y}}) = F\left(\tilde{x}, \tilde{y}, \frac{\dot{\tilde{y}}}{\dot{\tilde{x}}}\right) \dot{\tilde{x}}, \tag{1.11.5}$$

Any reparametrization (1.11.2) preserves the value of $J(y)$.

Proposition 1.11.1. *Let $y \in D \subset C^{1,pw}[a,b]$ be a global minimizer of*

$$J(y) = \int_a^b F(x,y,y')dx, \tag{1.11.6}$$

where admitted functions in D have possibly to fulfill boundary conditions at $x = a$ and/or at $x = b$. Then any reparametrization (1.11.2) of the graph $\{(x, y(x)) | x \in [a,b]\}$ is a local minimizer of the corresponding parametric functional

$$J(\tilde{x}, \tilde{y}) = \int_{\tau_a}^{\tau_b} F\left(\tilde{x}, \tilde{y}, \frac{\dot{\tilde{y}}}{\dot{\tilde{x}}}\right) \dot{\tilde{x}} d\tau, \tag{1.11.7}$$

defined on $\tilde{D} \subset C^1[\tau_a, \tau_b] \times C^{1,pw}[\tau_a, \tau_b]$. Admissible curves in $\tilde{D}$ fulfill $\tilde{x}(\tau_a) = a$, $\tilde{x}(\tau_b) = b$, and $\tilde{y}$ satisfies the boundary conditions that are possibly prescribed by $y \in D$. Furthermore $\dot{\tilde{x}}(\tau) > 0$ for $\tau \in [\tau_a, \tau_b]$.

Proof. Since $[\tau_a, \tau_b]$ is compact, $\dot{\tilde{x}}(\tau) \geq d > 0$ for $\tau \in [\tau_a, \tau_b]$. Any admissible curve near $(\tilde{x}, \tilde{y})$ in the sense of $(1.10.8)_2$ is given by

$$\{(\tilde{x}(\tau) + h_1(\tau), \tilde{y}(\tau) + h_2(\tau)) | \tau \in [\tau_a, \tau_b]\}, \quad \text{where}$$
$$h_1 \in C_0^1[\tau_a, \tau_b], \ h_2 \in C^{1,pw}[\tau_a, \tau_b] \quad \text{and} \tag{1.11.8}$$
$$\|h_1\|_{1,[\tau_a,\tau_b]} < d, \quad \|h_2\|_{1,pw;[\tau_a,\tau_b]} < d.$$

If $\tilde{y}$ must satisfy prescribed boundary conditions at $\tau = \tau_a$ and/or $\tau = \tau_b$, then necessarily $h_2(\tau_a) = 0$ and/or $h_2(\tau_b) = 0$. By $\dot{\tilde{x}}(\tau) + \dot{h}_1(\tau) > 0$ for $\tau \in [\tau_a, \tau_b]$, the perturbation (1.11.8) is the graph of a function $\hat{y} \in D \subset C^{1,pw}[a,b]$, cf. Exercise 1.10.1. Therefore, in view of (1.11.4),

$$J(\tilde{x}, \tilde{y}) = J(y) \leq J(\hat{y}) = J(\tilde{x} + h_1, \ \tilde{y} + h_2), \tag{1.11.9}$$

which proves the claim.   □

Why must $y \in D$ be a global minimizer of the nonparametric functional in order to guarantee that the reparametrized curve $(\tilde{x}, \tilde{y})$ is a local minimizer of the corresponding parametric functional?

The answer is given in Figure 1.19:

A perturbation of the graph $\{(x, y(x)) | x \in [a,b]\}$ by $\{(x + h_1(x), y(x)) | x \in [a,b]\} = \{(x, \hat{y}(x)) | x \in [a,b]\}$, where $h_1 \in C_0^1[a,b]$ is sketched in Figure 1.19, has a distance $\|y - \hat{y}\|_{1,pw,[a,b]} \geq |y'(x) - \hat{y}'(x)|$ for any point $x \in [a,b]$. However small $\|h_1\|_{1,[a,b]}$ might be, which measures the distance between curves, the distance between y and $\hat{y}$ is still big.

However, $\|y - \hat{y}\|_{0,[a,b]}$ gets small for small $\|h_1\|_1$ and $\|h_2\|_1$ in (1.11.8). Therefore Proposition 1.11.1 holds for a "strong local minimizer" of (1.11.6), cf. the remarks after Definition 1.4.1.

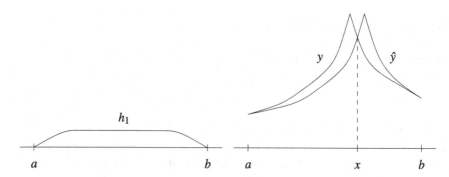

Fig. 1.19   Perturbation of a Graph considered as a Curve

Remark. *A global minimizer of the nonparametric functional is not necessarily a global minimizer of the corresponding parametric functional if the constraint $\dot{\tilde{x}}(\tau) > 0$ is given up. Here is an example:*

$$J(y) = \int_0^1 (y')^2 dx \quad \text{defined on}$$

$$D = C^{1,pw}[0,1] \cap \{y(0) = 0, y(1) = 1\}.$$

By example 2 in Paragraph 1.5, the global minimizer is the line segment given by
$y(x) = x$. The corresponding parametric functional is given by

$$J(\tilde{x}, \tilde{y}) = \int_{\tau_0}^{\tau_1} \left(\frac{\dot{\tilde{y}}}{\dot{\tilde{x}}}\right)^2 \dot{\tilde{x}} d\tau = \int_{\tau_0}^{\tau_1} \frac{\dot{\tilde{y}}^2}{\dot{\tilde{x}}} d\tau \quad \text{defined on}$$

$$\tilde{D} = (C^1[\tau_0, \tau_1] \times C^{1,pw}[\tau_0, \tau_1]) \cap \{(\tilde{x}(\tau_0), (\tilde{y}(\tau_0)) = (0,0), (\tilde{x}(\tau_1), (\tilde{y}(\tau_1)) = (1,1)\},$$
$$\dot{\tilde{x}}(\tau) > 0 \quad on \quad [\tau_0, \tau_1].$$

Giving up the constraint that $\dot{\tilde{x}}$ has to be positive, we now require only that $\tilde{x} \in$
$C^{1,pw}[\tau_0, \tau_1]$ with $\dot{\tilde{x}} \neq 0$ piecewise.
     Choosing the curves

$$\left.\begin{array}{l} \tilde{x}(\tau) = p\tau \\ \tilde{y}(\tau) = q\tau \end{array}\right\} \quad for \quad \tau \in [0,1],$$

$$\left.\begin{array}{l} \tilde{x}(\tau) = p + (1-p)(\tau-1) \\ \tilde{y}(\tau) = q + (1-q)(\tau-1) \end{array}\right\} \quad for \quad \tau \in [1,2],$$

where $p \neq 0$ and $p \neq 1$, then the curves connect $(\tilde{x}(0), \tilde{y}(0)) = (0,0)$ and $(\tilde{x}(2),$
$\tilde{y}(2)) = (1,1)$. Furthermore,

$$J(\tilde{x}, \tilde{y}) = 1 + \frac{(p-q)^2}{p(1-p)},$$

and we see that J can have any value in $\mathbb{R}$.
     The given curve can be composed by arbitrarily many pieces as sketched in
Figure 1.20 for $0 < q < 1 < p$. As long as p and q are fixed, all such composi-
tions give the same value of J. This shows that the parametric functional has any
real value in any neighborhood of $(\frac{1}{2}\tau, \frac{1}{2}\tau)$ in $(C[0,2])^2$.

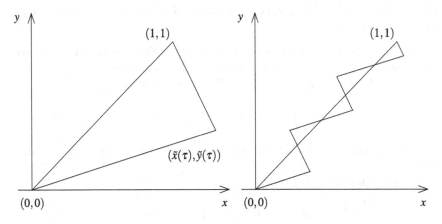

Fig. 1.20   A Parametric Versus a Nonparametric Functional

Next we make use of the regularity $(1.10.9)_1$ of a local minimizer.

Proposition 1.11.2. *Let* $y \in D \subset C^{1,pw}[a,b]$ *be a global (or a strong local) minimizer of the functional*

$$J(y) = \int_a^b F(x,y,y')dx \qquad (1.11.10)$$

satisfying boundary conditions or not. Assume that the Lagrange function $F : \mathbb{R}^3 \to \mathbb{R}$ *is continuously totally differentiable. Then*

$$
\begin{aligned}
&F_{y'}(\cdot,y,y') \in C[a,b], \quad \text{and} \\
&F(\cdot,y,y') - y'F_{y'}(\cdot,y,y') \in C[a,b].
\end{aligned} \qquad (1.11.11)
$$

In addition, if F does not depend explicitly on x, then $F(y,y') - y'F_{y'}(y,y') = c_1$ *on* $[a,b]$.

Proof. According to Proposition 1.11.1 the curve $\{(x,y(x))|x \in [a,b]\}$ is a local minimizer of the corresponding parametric functional (1.11.7) with Lagrange function $\Phi(\tilde{x},\tilde{y},\dot{\tilde{x}},\dot{\tilde{y}}) = F(\tilde{x},\tilde{y},\frac{\dot{\tilde{y}}}{\dot{\tilde{x}}})\dot{\tilde{x}}$. Then $\Phi : \mathbb{R}^4 \cap \{\dot{\tilde{x}} > 0\} \to \mathbb{R}$ is continuously totally differentiable, and according to Proposition 1.10.3, the curve fulfills the Euler-Lagrange equations. Due to the invariance of the Lagrange function, any reparametrized curve has that same property, cf. Proposition 1.10.4. We obtain for (1.11.5)

$$\Phi_{\dot{\tilde{x}}}(\tilde{x},\tilde{y},\dot{\tilde{x}},\dot{\tilde{y}}) = F\left(\tilde{x},\tilde{y},\frac{\dot{\tilde{y}}}{\dot{\tilde{x}}}\right) - F_{y'}\left(\tilde{x},\tilde{y},\frac{\dot{\tilde{y}}}{\dot{\tilde{x}}}\right)\frac{\dot{\tilde{y}}}{\dot{\tilde{x}}} \in C^{1,pw}[\tau_a,\tau_b], \qquad (1.11.12)$$

and for the reparametrization $\{(x,y(x))|x \in [a,b]\}$, this gives by (1.11.2) and (1.11.3)

$$
\begin{aligned}
&\tilde{x} = x, \quad \dot{\tilde{x}} = 1, \quad \tilde{y} = y, \quad \dot{\tilde{y}} = y', \quad \tau_a = a, \quad \tau_b = b \quad \text{and} \\
&F(\cdot,y,y') - y'F_{y'}(\cdot,y,y') \in C^{1,pw}[a,b] \subset C[a,b].
\end{aligned} \qquad (1.11.13)
$$

The first claim of (1.11.11) is part of Proposition 1.4.1. The last addition is given as Exercise 1.11.1. $\qquad \square$

The continuity conditions (1.11.11) are the so-called **Weierstraß-Erdmann corner conditions**. They admit only specific corners for minimizers.

We recall Example 6 of Paragraph 1.5. We generalize it to

$$J(y) = \int_a^b W(y')dx \quad \text{on} \quad D \subset C^{1,pw}[a,b] \qquad (1.11.14)$$

where the potential $W : \mathbb{R} \to \mathbb{R}$ sketched in Figure 1.21 is continuously differentiable.

Any local minimizer y satisfies

$$\frac{d}{dx}W'(y') = 0 \quad \text{or} \quad W'(y') = c_1 \quad \text{on} \quad [a,b], \qquad (1.11.15)$$

since, due to the first corner condition, the function $W'(y')$ is continuous on $[a,b]$. The second corner condition for global minimizers implies

$$W(y') - y'W'(y') = c_2 \quad \text{on} \quad [a,b] \tag{1.11.16}$$

by the supplement of Proposition 1.11.2. This can also be seen directly from (1.11.15): The derivative y' and therefore $W'(y') - y'W'(y')$ is piecewise constant and by continuity constant $[a,b]$. Relation (1.11.15) provides three constants for y', namely

$$y' = c_1^1, c_1^2, c_1^3 , \tag{1.11.17}$$

cf. Figures 1.21 and 1.5. Upon substitution into (1.11.16), we obtain

$$W(c_1^i) - c_1^i W'(c_1^i) = c_2 \quad \text{for } i = 1, 2, 3 \quad \text{and therefore}$$
$$W(c_1^1) - W(c_1^2) = c_1^1 W'(c_1^1) - c_1^2 W'(c_1^2)$$
$$= (c_1^1 - c_1^2)W'(c_1^1) = (c_1^1 - c_1^2)W'(c_1^2), \tag{1.11.18}$$
$$\frac{W(c_1^2) - W(c_1^1)}{c_1^2 - c_1^1} = W'(c_1^1) = W'(c_1^2).$$

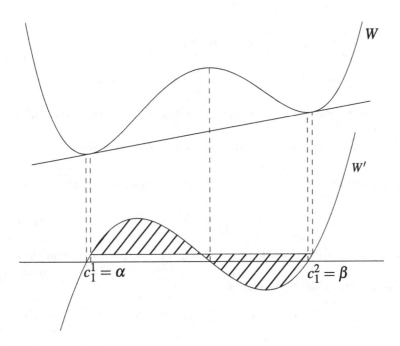

Fig. 1.21 A W-Potential

The geometric interpretation is that the slope of the secant through $W(c_1^1)$ and $W(c_1^2)$ equals the slope of the tangent in $W(c_1^1)$ as well as in $W(c_1^2)$. A second interpretation is the following:

$$W(c_1^2) - W(c_1^1) = \int_{c_1^1}^{c_1^2} W'(z)dz = (c_1^2 - c_1^1)W'(c_1^i), \quad i = 1,2. \tag{1.11.19}$$

Both conditions allow only two specific constants $c_1^1 = \alpha$ and $c_1^2 = \beta$ sketched in Figure 1.21. The horizontal line limiting with the graph of W' the hatched areas of equal size is called "Maxwell line."

The Weierstraß-Erdmann corner conditions give necessary conditions on the slopes of a global minimizer at a corner. However, any straight line segment over $[a,b]$ fulfills the Euler-Lagrange equation (1.11.15) and the corner condition (1.11.16) as well. Therefore a separate discussion depending on the potential W and on possible boundary conditions has to decide on global minimizers of (1.11.14).

We do this now for the Example 6 of Paragraph 1.5 where $W(z) = (z^2 - 1)^2$. Here $c_1^1 = \alpha = -1$, $c_1^2 = \beta = 1$ and we find infinitely many global minimizers without boundary conditions as sketched in Figure 1.4. However, for the boundary conditions $y(0) = 0$ and $y(1) = 2$ corners with slopes ± 1 are not possible. The only candidate for a minimizer is the line $y = 2x$. We show that it is indeed a global minimizer.

The tangent to the graph of W in $(2, W(2)) = (2,9)$ is the line $W(2) + W'(2)(z - 2) = 9 + 24(z - 2)$, and Figure 1.5 shows that $W(z) \geq W(2) + W'(2)(z - 2)$ for all $z \in \mathbb{R}$. For $\hat{y} \in C^{1,pw}[0,1] \cap \{y(0) = 0, \ y(1) = 2\}$, we obtain

$$\begin{aligned}
J(\hat{y}) &= \int_0^1 W(\hat{y}')dx \geq \int_0^1 W(2) + W'(2)(\hat{y}' - 2)dx \\
&= \int_0^1 W(2)dx = J(y), \quad \text{since } \int_0^1 \hat{y}'dx = \hat{y}(1) - \hat{y}(0) = 2.
\end{aligned} \tag{1.11.20}$$

This proves that the line segment between $(0,0)$ and $(1,2)$ is the global minimizer of J in $D = C^{1,pw}[0,1] \cap \{y(0) = 0, \ y(1) = 2\}$.

Next we give a sufficient condition to exclude broken minimizers.

Proposition 1.11.3. *Let $y \in D \subset C^{1,pw}[a,b]$ be a local minimizer of the functional*

$$J(y) = \int_a^b F(x,y,y')dx, \tag{1.11.21}$$

where the Lagrange function F is continuous, is continuously differentiable, once with respect to the second and twice with respect to the third variable. Boundary conditions can possibly be imposed. If

$$F_{y'y'}(x,y(x),z) \neq 0 \quad \text{for all } x \in [a,b], \ z \in \mathbb{R}, \tag{1.11.22}$$

then $y \in C^1[a,b]$.

Proof. Assume that $y \in C^{1,pw}[a,b] \setminus C^1[a,b]$. Then there is a $x_i \in (a,b)$ such that

$$y'_-(x_i) = \lim_{x \nearrow x_i} y'(x) \neq \lim_{x \searrow x_i} y'(x) = y'_+(x_i). \qquad (1.11.23)$$

For $f(z) = F(x_i, y(x_i), z)$, we have

$$\begin{aligned} f'(z) &= F_{y'}(x_i, y(x_i), z), \quad \text{and} \\ f'(y'_-(x_i)) &= f'(y'_+(x_i)), \end{aligned} \qquad (1.11.24)$$

due to the first Weierstraß-Erdmann corner condition $(1.11.11)_1$, or due to $(1.4.3)_1$, which holds also for local minimizers. The existence of some z between $y'_-(x_i)$ and $y'_+(x_i)$ such that

$$f''(z) = F_{y'y'}(x_i, y(x_i), z) = 0, \qquad (1.11.25)$$

guaranteed by Rolle's theorem, contradicts the assumption (1.11.22).   □

Proposition 1.11.3 and Exercise 1.5.1 imply the following regularity theorem:

Proposition 1.11.4. *Let* $y \in D \subset C^{1,pw}[a,b]$ *be a local minimizer of the functional*

$$J(y) = \int_a^b F(x,y,y')dx, \qquad (1.11.26)$$

where the Lagrange function F is twice continuously differentiable with respect to all three variables. If

$$F_{y'y'}(x,y(x),z) \neq 0 \quad \text{for all } x \in [a,b],\ z \in \mathbb{R}, \qquad (1.11.27)$$

then $y \in C^2[a,b]$.

Condition (1.11.27) means **"ellipticity"** of the Euler-Lagrange equation (1.4.5) along a minimizer. We return to this in Chapter 3.

Exercises

1.11.1. Let $y \in D \subset C^{1,pw}[a,b]$ be a global minimizer of the functional

$$J(y) = \int_a^b F(y,y')dx,$$

where the Lagrange function $F : \mathbb{R}^2 \to \mathbb{R}$ is continuously totally differentiable. Show that

$$F(y,y') - y'F_{y'}(y,y') = c_1 \quad \text{on } [a,b].$$

Compare that result with the special case 7 of Paragraph 1.5.

1.11.2. Compute global minimizers of

$$J(y) = \int_a^b W(y')dx \quad \text{in } D = C^{1,pw}[a,b], \quad \text{where } W(z) = \frac{1}{2}z^4 + \frac{1}{3}z^3 - \frac{1}{2}z^2.$$

Are they unique?

1.11.3. Compute and sketch global minimizers of the functional of Exercise 1.11.2 if $D = C^{1,pw}[a,b] \cap \{y(a) = 0, \ y(b) = 0\}$.

Hint: Compute $J(y)$ for a sawtooth function having slopes $c_1^1 = \alpha < 0$ and $c_1^2 = \beta > 0$ according to Figure 1.21 and fulfilling the boundary conditions. Show that $J(\hat{y}) \geq J(y)$ for all $\hat{y} \in D$.

Chapter 2
Variational Problems with Constraints

2.1 Isoperimetric Constraints

Many early variational problems like Dido's problem or the problem of the hanging
chain have constraints in a natural way: Maximize the area with given perimeter
or minimize the potential energy of a hanging chain with given length. These constraints belong to the class of isoperimetric constraints, and they are of the same
type as the functional to be maximized or minimized.

> **Definition 2.1.1.** *Consider a functional*
>
> $$J(y) = \int_a^b F(x,y,y')dx, \qquad (2.1.1)$$
>
> *defined on $D \subset C^{1,pw}[a,b]$. A constraint of the type*
>
> $$K(y) = \int_a^b G(x,y,y')dx = c \qquad (2.1.2)$$
>
> *is called an isoperimetric constraint. The Lagrange functions F and G are continu-
> ous.*

The goal of this paragraph is a necessary condition on a local minimizer $y \in D$
of J under an isoperimetric constraint.

We recall the following theorem from calculus:

> Let $f,g : \mathbb{R}^2 \to \mathbb{R}$ be continuously totally differentiable.
> If f is locally extremal at $x_0 \in \mathbb{R}^2$ subject to $g(x) = c$, and
> if $\nabla g(x_0) \neq 0$ holds, then
> $\nabla f(x_0) + \lambda \nabla g(x_0) = 0$ for some $\lambda \in \mathbb{R}$. $\qquad (2.1.3)$

© Springer International Publishing AG 2018
H. Kielhöfer, *Calculus of Variations*, Texts in Applied Mathematics 67,
https://doi.org/10.1007/978-3-319-71123-2_2

This is the method of Lagrange multipliers, which is proved in its general version in the Appendix. In Figure 2.1, we visualize the case (2.1.3) by a hiking map with the contour lines of a landscape described by f and a trail to a summit described by $g(x) = c$.

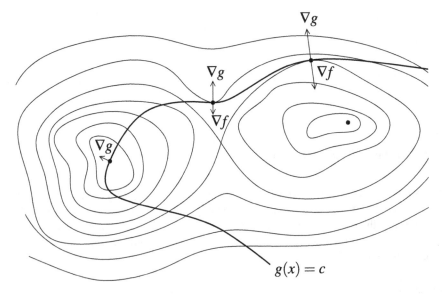

Fig. 2.1 Local Extremals under a Constraint

Hikers, even when they are not mathematicians, realize that a point where the trail and a contour line are tangential is locally extremal on their trail. Since the gradients are orthogonal to level curves, the Lagrange multiplier rule (2.1.3) gives precisely the points where the contour lines and the trail are tangent to each other. On the summit, the gradient of f vanishes and the multiplier rule is satisfied with $\lambda = 0$.

Proposition 2.1.1. *Let $y \in D \subset C^{1,pw}[a,b]$ be a local minimizer of the functional (2.1.1) under the isoperimetric constraint (2.1.2), i.e.,*

$$J(y) \leq J(\tilde{y}) \quad \text{for all } \tilde{y} \in D \cap \{K(y) = c\}$$
$$\text{satisfying } \|y - \tilde{y}\|_{1,pw} < d. \tag{2.1.4}$$

The domain D involves possibly boundary conditions. The Lagrange functions $F, G : [a,b] \times \mathbb{R} \times \mathbb{R} \to \mathbb{R}$ are assumed to be continuous and continuously partially differentiable with respect to the last two variables.
Furthermore assume that y is not critical for the constraint K, which means that $\delta K(y) : C_0^{1,pw}[a,b] \to \mathbb{R}$ is linear, continuous, and surjective (or not identically zero).

Then, there is some $\lambda \in \mathbb{R}$ such that

$$F_{y'}(\cdot,y,y') + \lambda G_{y'}(\cdot,y,y') \in C^{1,pw}[a,b] \quad and$$

$$\frac{d}{dx}(F_{y'}(\cdot,y,y') + \lambda G_{y'}(\cdot,y,y')) = F_y(\cdot,y,y') + \lambda G_y(\cdot,y,y') \tag{2.1.5}$$

piecewise on $[a,b]$.

Proof. By assumption on G, the first variation $\delta K(y)$ exists and is a linear and continuous mapping, and also there is a function $h_0 \in C_0^{1,pw}[a,b]$ with $\delta K(y)h_0 = 1$. For an arbitrary $h \in C_0^{1,pw}[a,b]$, we define

$$f(x_1,x_2) = J(y+x_1h+x_2h_0) \quad and$$
$$g(x_1,x_2) = K(y+x_1h+x_2h_0). \tag{2.1.6}$$

We remark that $y+x_1h+x_2h_0 \in D$ for all $x = (x_1,x_2) \in \mathbb{R}^2$, since $D \subset C^{1,pw}[a,b]$ is possibly constrained by boundary conditions only. As expounded in Paragraph 1.2,

$$\frac{\partial f}{\partial x_1}(0,0) = \lim_{x_1 \to 0} \frac{J(y+x_1h) - J(y)}{x_1} = \delta J(y)h,$$

$$\frac{\partial f}{\partial x_2}(0,0) = \delta J(y)h_0, \tag{2.1.7}$$

$$\frac{\partial g}{\partial x_1}(0,0) = \delta K(y)h, \quad \frac{\partial g}{\partial x_2}(0,0) = \delta K(y)h_0 = 1.$$

The arguments in the proof of Proposition 1.2.1 yield also that

$$\frac{\partial f}{\partial x_1}(x) = \delta J(y+x_1h+x_2h_0)h, \quad \frac{\partial f}{\partial x_2}(x) = \delta J(y+x_1h+x_2h_0)h_0,$$

$$\frac{\partial g}{\partial x_1}(x) = \delta K(y+x_1h+x_2h_0)h, \quad \frac{\partial g}{\partial x_2}(x) = \delta K(y+x_1h+x_2h_0)h_0. \tag{2.1.8}$$

The representation (1.2.8) of the first variation shows that for fixed y,h, and h_0, the partial derivatives (2.1.8) are continuous with respect to $x = (x_1,x_2)$. This follows because the partial derivatives $F_y, F_{y'}$ and $G_y, G_{y'}$ are all continuous on $[a,b] \times \mathbb{R} \times \mathbb{R}$ and uniformly so on compact subsets, cf. the arguments in (1.2.10)–(1.2.12). Consequently, the functions $f,g : \mathbb{R}^2 \to \mathbb{R}$ defined by (2.1.6) are continuously totally differentiable.

By assumption on y, the point $x = (0,0) \in \mathbb{R}^2$ is a local minimizer of f subject to $g(x) = c$, and since the gradient $\nabla g(0) \neq 0$ by $(2.1.7)_3$, the multiplier rule (2.1.3) yields

$$\nabla f(0) + \lambda \nabla g(0) = 0 \quad \text{for some } \lambda \in \mathbb{R} \quad \text{or}$$

$$\delta J(y)h + \lambda \delta K(y)h = 0, \tag{2.1.9}$$

$$\delta J(y)h_0 + \lambda \delta K(y)h_0 = 0 \quad \text{due to (2.1.7).}$$

The definitions of f and g in (2.1.6) depend on y, h, and h_0. Whereas y and h_0 are fixed, and $h \in C_0^{1,pw}[a,b]$ is arbitrary. It is crucial that the multiplier λ does not depend on h. As a matter of fact, in view of $\delta K(y)h_0 = 1$ and of $(2.1.9)_3$,

$$\lambda = -\delta J(y)h_0 \quad \text{for all } h \in C_0^{1,pw}[a,b]. \tag{2.1.10}$$

Then, $(2.1.9)_2$ reads:

$$\int_a^b (F_y + \lambda G_y)h + (F_{y'} + \lambda G_{y'})h' dx = 0 \quad \text{for all } h \in C_0^{1,pw}[a,b], \tag{2.1.11}$$

where the argument of all functions is $(x, y(x), y'(x))$. Lemma 1.3.4 finally implies the claim (2.1.5). $\qquad\qquad\square$

If the functional (2.1.1) is defined on all of $D = C^{1,pw}[a,b]$, which means that the boundary of $y \in D$ is free at $x = a$ and at $x = b$, then a local minimizer under an isoperimetric constraint fulfills natural boundary conditions.

Proposition 2.1.2. *Let $y \in D = C^{1,pw}[a,b]$ be a local minimizer of the functional (2.1.1) under the isoperimetric constraint (2.1.2) in the sense of (2.1.4). Assume the hypotheses of Proposition 2.1.1.*

Then, y fulfills (2.1.5), and with the same multiplier $\lambda \in \mathbb{R}$, it fulfills the natural boundary conditions

$$\begin{aligned} F_{y'}(a, y(a), y'(a)) + \lambda G_{y'}(a, y(a), y'(a)) = 0 \quad \text{and} \\ F_{y'}(b, y(b), y'(b)) + \lambda G_{y'}(b, y(b), y'(b)) = 0. \end{aligned} \tag{2.1.12}$$

Proof. Following the proof of Proposition 2.1.1, we choose $h_0 \in C^{1,pw}[a,b]$ such that $\delta K(y)h_0 = 1$. Then with arbitrary $h \in C_0^{1,pw}[a,b]$, we obtain (2.1.9), (2.1.10), (2.1.11), and finally (2.1.5). But now an arbitrary $h \in C^{1,pw}[a,b]$ is admitted in (2.1.6), and thus, (2.1.9), (2.1.10), and (2.1.11) hold for any $h \in C^{1,pw}[a,b]$. Integration by parts, cf. Lemma 1.3.3, yields

$$\begin{aligned} &\delta J(y)h + \lambda \delta K(y)h \\ &= \int_a^b (F_y + \lambda G_y)h + (F_{y'} + \lambda G_{y'})h' dx \\ &= \int_a^b (F_y + \lambda G_y - \frac{d}{dx}(F_{y'} + \lambda G_{y'}))h dx + (F_{y'} + \lambda G_{y'})h \big|_a^b = 0, \end{aligned} \tag{2.1.13}$$

where the argument of all functions is $(x, y(x), y'(x))$. Due to (2.1.5), the only term left in (2.1.13) is

$$(F_{y'}(x, y(x), y'(x)) + \lambda G_{y'}(x, y(x), y'(x)))h(x) \big|_a^b = 0, \tag{2.1.14}$$

proving the claim (2.1.12) by an arbitrary choice of $h(a)$ and $h(b)$. $\qquad\qquad\square$

Remark. *If only one boundary value is prescribed in* $D \subset C^{1,pw}[a,b]$, *then any local minimizer subject to the isoperimetric constraint fulfills one natural boundary condition (2.1.12) at the free end.*

Next we generalize Propositions 2.1.1 and 2.1.2 to functionals and isoperimetric constraints that are defined for vector-valued functions. Also the constraint consists of several components.

Definition 2.1.2. *The functional*

$$J(y) = \int_a^b F(x,y,y')dx, \quad y(x) = (y_1(x),\ldots,y_n(x)), \tag{2.1.15}$$

as well as the isoperimetric constraints

$$K_i(y) = \int_a^b G_i(x,y,y')dx = c_i, \ i = 1,\ldots,m, \tag{2.1.16}$$

are defined for $y \in D \subset (C^{1,pw}[a,b])^n$. *The Lagrange functions* $F,G_i : [a,b] \times \mathbb{R}^n \times \mathbb{R}^n \to \mathbb{R}$ *are continuous. We express the m constraints as a single vector-valued constraint as follows:*

$$K(y) = (K_1(y),\ldots,K_m(y)) = (c_1,\ldots,c_m) = c. \tag{2.1.17}$$

We point out that the dimensions n and m are completely independent.
The domain of definition D of J determines the space of admissible perturbations as follows: If

$$D = \{y \in (C^{1,pw}[a,b])^n \mid y_k(a) = A_k \quad \text{and/or} \quad y_k(b) = B_k\}$$
$$\text{for certain } k \in \{1,\ldots,n\}, \text{then} \tag{2.1.18}$$
$$D_0 = \{h \in (C^{1,pw}[a,b])^n \mid h_k(a) = 0 \quad \text{and/or} \quad h_k(b) = 0\}$$
$$\text{for the same choice of indices.}$$

All remaining components have a free boundary at $x = a$ and/or $x = b$.

Proposition 2.1.3. *Let* $y \in D \subset (C^{1,pw}[a,b])^n$ *be a local minimizer of the functional (2.1.15) under the isoperimetric constraints (2.1.17), i.e.,*

$$J(y) \leq J(\tilde{y}) \quad \text{for all } \tilde{y} \in D \cap \{K(y) = c\}$$
$$\text{where } \max_{k=1,\ldots,n} \|y_k - \tilde{y}_k\|_{1,pw} < d. \tag{2.1.19}$$

The Lagrange functions $F,G_i : [a,b] \times \mathbb{R}^n \times \mathbb{R}^n \to \mathbb{R}, i = 1,\ldots,m$, *are assumed to be continuous and continuously partially differentiable with respect to the last 2n variables.*

If y is not critical for the constraints, i.e., if $\delta K(y) = (\delta K_1(y), \ldots, \delta K_m(y))$: $D_0 \to \mathbb{R}^m$ is linear, continuous, and surjective (or if the m linear functionals $\delta K_1(y), \ldots, \delta K_m(y)$ are linearly independent in the sense of Exercise 2.1.1), then holds:

$$F_{y'}(\cdot, y, y') + \sum_{i=1}^{m} \lambda_i G_{i,y'}(\cdot, y, y') \in (C^{1,pw}[a,b])^n,$$

$$\frac{d}{dx}(F + \sum_{i=1}^{m} \lambda_i G_i)_{y'} = (F + \sum_{i=1}^{m} \lambda_i G_i)_y \quad \textit{piecewise on } [a,b] \tag{2.1.20}$$

for some $\lambda = (\lambda_1, \ldots, \lambda_m) \in \mathbb{R}^m$.
Here, we agree upon the notation $F_{y'} = (F_{y_1'}, \ldots, F_{y_n'})$, $F_y = (F_{y_1}, \ldots, F_{y_n})$, and analogously for $G_{i,y'}$, $G_{i,y}$. The argument of all functions in (2.1.20) is $(x, y(x), y'(x)) = (x, y_1(x), \ldots, y_n(x), y_1'(x), \ldots, y_n'(x))$, and (2.1.20)$_2$ is a system of n differential equations.

Proof. By assumption on G_i, the first variations $\delta K_i(y)$ exist, they are linear and continuous. Due to the surjectivity of $\delta K(y)$, there exist functions $h_1, \ldots, h_m \in D_0$ such that

$$\delta K_i(y)h_j = \delta_{ij} = \begin{cases} 1 & \text{for } i = j, \\ 0 & \text{for } i \neq j, \end{cases} \quad i, j = 1, \ldots, m. \tag{2.1.21}$$

We define for arbitrary $h \in (C_0^{1,pw}[a,b])^n$

$$\begin{aligned} f(s, t_1, \ldots, t_m) &= J(y + sh + t_1 h_1 + \cdots + t_m h_m), \\ \Psi_i(s, t_1, \ldots, t_m) &= K_i(y + sh + t_1 h_1 + \cdots + t_m h_m), \ i = 1, \ldots, m. \end{aligned} \tag{2.1.22}$$

In view of (2.1.18)$_1$, the functions of the arguments in (2.1.22) are in D, and as expounded in the proof of Proposition 2.1.1, the functions

$$\begin{aligned} f &: \mathbb{R}^{m+1} \to \mathbb{R} \quad \text{and} \\ \Psi &= (\Psi_1, \ldots, \Psi_m) : \mathbb{R}^{m+1} \to \mathbb{R}^m \end{aligned} \tag{2.1.23}$$

are continuously totally differentiable.

The choice of the functions $h_1, \ldots, h_m$ in (2.1.21) yields the following structure of the Jacobian matrix:

$$D\Psi(0) = \begin{pmatrix} \delta K_1(y)h & 1 & \cdots & 0 \\ \vdots & \vdots & \ddots & \vdots \\ \delta K_m(y)h & 0 & \cdots & 1 \end{pmatrix}. \tag{2.1.24}$$

Since the rank of $D\Psi(0)$ is maximal, the set $\{(s, t_1, \ldots, t_m) | \Psi(s, t_1, \ldots, t_m) = c \in \mathbb{R}^m\}$ is locally a one-dimensional manifold near $0 \in \mathbb{R}^{m+1}$, i.e., a curve, cf. the

Appendix. By assumption on y, the function f at $0 \in \mathbb{R}^{m+1}$ is locally minimal subject to the constraint $\Psi(s, t_1, \ldots, t_m) = c$. Therefore, the vector-valued Lagrange multiplier rule (proved in the Appendix (A.20)–(A.23)) is applicable and reads:

$$\nabla f(0) + \sum_{i=1}^{m} \lambda_i \nabla \Psi_i(0) = 0 \quad \text{for } \lambda = (\lambda_1, \ldots, \lambda_m) \in \mathbb{R}^m. \tag{2.1.25}$$

The partial derivatives of the gradients are computed as in (2.1.7), whence the $m+1$ equations (2.1.25) give

$$\delta J(y)h + \sum_{i=1}^{m} \lambda_i \delta K_i(y)h = 0,$$

$$\delta J(y)h_j + \sum_{i=1}^{m} \lambda_i \delta K_i(y)h_j = 0, \ j = 1, \ldots, m. \tag{2.1.26}$$

In view of (2.1.21), the last m equations yield

$$\lambda_j = -\delta J(y)h_j \quad \text{for } j = 1, \ldots, m, \tag{2.1.27}$$

which means that the Lagrange multipliers λ_j do not depend on $h \in (C_0^{1,pw}[a,b])^n$, despite the fact that the definitions of the functions f and Ψ_i involve h. The first equation (2.1.26) reads

$$\int_a^b ((F + \sum_{i=1}^{m} \lambda_i G_i)_y, h) + ((F + \sum_{i=1}^{m} \lambda_i G_i)_{y'}, h')dx = 0, \tag{2.1.28}$$

where (,) denotes the Euclidean scalar product in $\mathbb{R}^n$.

Choosing $h = (0, \ldots, \tilde{h}, \ldots 0)$, where $\tilde{h} \in C_0^{1,pw}[a,b]$ in the k-th component is arbitrary, (2.1.28) gives

$$\int_a^b (F + \sum_{i=1}^{m} \lambda_i G_i)_{y_k} \tilde{h} + (F + \sum_{i=1}^{m} \lambda_i G_i)_{y_k'} \tilde{h}' dx = 0, \tag{2.1.29}$$

and thus, Lemma 1.3.4 proves the claim (2.1.20) for the k-th component, $k = 1, 2, \ldots, n$. $\qquad \square$

By (2.1.18), some boundaries of some components of admitted functions in $D \subset (C^{1,pw}[a,b])^n$ are free. Local minimizers under isoperimetric constraints fulfill natural boundary conditions there. We give here only the result, since a proof following Proposition 2.1.2 along with the result of Proposition 2.1.3 is straight forward.

Proposition 2.1.4. *Under the hypotheses of Proposition 2.1.3, a local minimizer y under isoperimetric constraints whose k-th component y_k has a free boundary at*

$x = a$ and/or $x = b$ fulfills (2.1.20), and with the same multipliers $\lambda_i \in \mathbb{R}$, it fulfills the natural boundary conditions

$$\left(F + \sum_{i=1}^{m} \lambda_i G_i\right)_{y'_k}(a, y(a), y'(a)) = 0 \quad and/or$$

$$\left(F + \sum_{i=1}^{m} \lambda_i G_i\right)_{y'_k}(b, y(b), y'(b)) = 0. \tag{2.1.30}$$

Next we investigate parametric functionals introduced in Paragraph 1.10, which are now subject to isoperimetric constraints.

Definition 2.1.3. *Consider a parametric functional*

$$J(x) = \int_{t_a}^{t_b} \Phi(t, x, \dot{x}) dt, \quad x(t) = (x_1(t), \ldots, x_n(t)), \tag{2.1.31}$$

defined on $D \subset (C^{1,pw}[t_a, t_b])^n$. Constraints of the type

$$K_i(x) = \int_{t_a}^{t_b} \Psi_i(t, x, \dot{x}) dt = c_i, \; i = 1, \ldots, m, \tag{2.1.32}$$

are called isoperimetric constraints. The Lagrange functions $\Phi, \Psi_i : [t_a, t_b] \times \mathbb{R}^n \times \mathbb{R}^n \to \mathbb{R}$ are continuous, and we express the m constraints as a single vector constraint as follows:

$$K(x) = (K_1(x), \ldots, K_m(x)) = (c_1, \ldots, c_m) = c. \tag{2.1.33}$$

We give the main proposition on the Euler-Lagrange equations and on the natural boundary conditions for a local minimizer subject to isoperimetric constraints. The proof is completely analogous to those of the preceding propositions. Also the domain of definition $D_0 \subset (C^{1,pw}[t_a, t_b])^n$ corresponds to that of (2.1.18).

Proposition 2.1.5. *Let the curve $x \in D \subset (C^{1,pw}[t_a, t_b])^n$ be a local minimizer of the parametric functional (2.1.31) under the isoperimetric constraints (2.1.33), i.e.,*

$$J(x) \leq J(\hat{x}) \quad for \; all \; \hat{x} \in D \cap \{K(x) = c\}$$
$$where \; \max_{k=1,\ldots,n} \|x_k - \hat{x}_k\|_{1,pw} < d. \tag{2.1.34}$$

The Lagrange functions $\Phi, \Psi_i : [t_a, t_b] \times \mathbb{R}^n \times \mathbb{R}^n \to \mathbb{R}$ are assumed to be continuous and continuously partially differentiable with respect to the last $2n$ variables.

If x is not critical for the constraints, i.e., if $\delta K(x) = (\delta K_1(x),\ldots,\delta K_m(x))$: $D_0 \to \mathbb{R}^m$ is linear, continuous, and surjective (or if the m linear functionals $\delta K_1(x),\ldots,\delta K_m(x)$ are linearly independent in the sense of Exercise 2.1.1), then there holds:

$$\Phi_{\dot{x}}(\cdot,x,\dot{x}) + \sum_{i=1}^{m} \lambda_i \Psi_{i,\dot{x}}(\cdot,x,\dot{x}) \in (C^{1,pw}[t_a,t_b])^n,$$

$$\frac{d}{dt}\left(\Phi + \sum_{i=1}^{m} \lambda_i \Psi_i\right)_{\dot{x}} = \left(\Phi + \sum_{i=1}^{m} \lambda_i \Psi_i\right)_x \quad \text{piecewise on} \quad [t_a,t_b].$$

(2.1.35)

Here, we agree upon the notation $\Phi_{\dot{x}} = (\Phi_{\dot{x}_1},\ldots,\Phi_{\dot{x}_n})$, $\Phi_x = (\Phi_{x_1},\ldots,\Phi_{x_n})$, and analogously upon $\Psi_{i,\dot{x}}$, $\Psi_{i,x}$. The argument of all functions in (2.1.35) is $(t,x(t),\dot{x}(t)) = (t,x_1(t),\ldots,x_n(t),\dot{x}_1(t),\ldots,\dot{x}_n(t))$, and (2.1.35)$_2$ is a system of differential equations.

If the k-th component x_k of the local minimizer x is free at $t = t_a$ and/or $t = t_b$, then x fulfills the natural boundary conditions

$$\left(\Phi + \sum_{i=1}^{m} \lambda_i \Psi_i\right)_{\dot{x}_k}(t_a,x(t_a),\dot{x}(t_a)) = 0 \quad \text{and/or}$$

$$\left(\Phi + \sum_{i=1}^{m} \lambda_i \Psi_i\right)_{\dot{x}_k}(t_b,x(t_b),\dot{x}(t_b)) = 0,$$

(2.1.36)

respectively, with the same multipliers $\lambda_i \in \mathbb{R}$.

If the Lagrange functions Φ and Ψ_i do not depend explicitly on t, their invariance

$$\begin{aligned} \Phi(x,\alpha\dot{x}) &= \alpha\Phi(x,\dot{x}), \\ \Psi_i(x,\alpha\dot{x}) &= \alpha\Psi_i(x,\dot{x}) \quad \text{for } i=1,\ldots,m \quad \text{and all } \alpha > 0, \end{aligned}$$

(2.1.37)

implies that the integrals (2.1.31), (2.1.32), as well as the Euler-Lagrange equations are invariant with respect to reparametrizations of the admitted curves, cf. Definition 1.10.3 and Proposition 1.10.4. The extension to n components and to a system of n differential equations is apparent.

Remark. *The noncriticality of a local minimizer for the isoperimetric constraints depends on the domain of definition D_0. Here is an example:*

$$\text{For} \quad K(y) = \int_a^b y'\,dx = c, \quad \text{we obtain}$$

$$\delta K(y)h = \int_a^b h'\,dx = h(b) - h(a) \quad \text{and}$$

$$\delta K(y) : C_0^{1,pw}[a,b] \to \mathbb{R} \quad \text{vanishes identically, whereas}$$

$$\delta K(y) : C^{1,pw}[a,b] \to \mathbb{R} \quad \text{is surjective.}$$

Exercises

2.1.1. Prove the equivalence of the following statements, where $D_0 \subset (C^{1,pw}[a,b])^n$ is a subspace:

i) $\delta K(y) = (\delta K_1(y), \dots, \delta K_m(y)) : D_0 \to \mathbb{R}^m$ is surjective.
ii) $\delta K_1(y), \dots, \delta K_m(y)$ are linearly independent in the following sense:
   $\sum_{i=1}^{m} \lambda_i \delta K_i(y)h = 0$ for all $h \in D_0$ implies $\lambda_1 = \cdots = \lambda_m = 0$.

2.1.2. Prove for

$$J(y) = \int_0^1 (y')^2 dx \quad \text{and} \quad K(y) = \int_0^1 y^2 dx,$$

the equivalence of the following statements:

i) y is a global minimizer of J on $D = C^{1,pw}[0,1] \cap \{y(0) = 0, y(1) = 0\}$
   under the isoperimetric constraint $K(y) = 1$.
ii) $y \in C^2[0,1]$, $\quad K(y) = 1$, $\quad y'' = \lambda y$ on $[0,1]$, $\quad y(0) = 0$, $y(1) = 0$,
   $-\lambda \int_0^1 h^2 dx \leq \int_0^1 (h')^2 dx \quad$ for all $h \in C_0^{1,pw}[0,1]$ and some $\lambda \in \mathbb{R}$.

   The inequality in ii) is called **Poincaré inequality** (Poincaré, 1854–1912).
   Give y and $\lambda < 0$ explicitly assuming i) or ii).
   The existence of a global minimizer is proved in Paragraph 3.3.

2.1.3. Compute a global minimizer $y \in D = C^{1,pw}[0,1]$ of

$$J(y) = \int_0^1 (y')^2 dx,$$

under the isoperimetric constraints

$$K_1(y) = \int_0^1 y^2 dx = 1,$$

$$K_2(y) = \int_0^1 y\, dx = m,$$

provided it exists.
   Distinguish the cases $m^2 = 1$ and $m^2 < 1$.
   If it exists for $m = 0$, prove a Poincaré inequality for all $h \in C^{1,pw}[0,1] \cap \{\int_0^1 h\, dx = 0\}$.

2.2 Dido's Problem as a Variational Problem with Isoperimetric Constraint

We treat now the variational problem introduced in Paragraph 1.7 in a more "natural way": Which closed curve with a given perimeter bounds a maximal area?

Piecewise continuously differentiable closed curves are admitted:

$$D = (C^{1,pw}[t_a, t_b])^2 \cap \{x(t_a) = x(t_b), \ \|\dot{x}(t)\| > 0 \text{ for } t \in [t_{i-1}, t_i], \ i = 1, \ldots, m\}. \tag{2.2.1}$$

The condition $\|\dot{x}(t)\| > 0$ excludes cusps. The isoperimetric constraint for curves in D reads

$$K(x) = \int_{t_a}^{t_b} \|\dot{x}\| dt = \sum_{i=1}^{m} \int_{t_{i-1}}^{t_i} \sqrt{\dot{x}_1^2 + \dot{x}_2^2} \, dt = L. \tag{2.2.2}$$

What is the area bounded by a curve in D? We use Green's formula for a plane continuously totally differentiable vector field $f = (f_1, f_2) : \mathbb{R}^2 \to \mathbb{R}^2$ over a bounded domain $\Omega \subset \mathbb{R}^2$, whose boundary $\partial\Omega$ is a piecewise continuously differentiable positively oriented curve, cf. Figure 2.2. Green's formula reads

$$\int_{\Omega} \left(\frac{\partial f_1}{\partial x_1} + \frac{\partial f_2}{\partial x_2} \right) dx = \int_{\partial\Omega} f_1 dx_2 - f_2 dx_1. \tag{2.2.3}$$

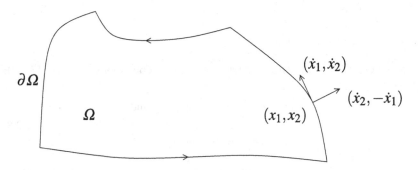

Fig. 2.2 On Green's Formula

We choose the vector field $f(x_1, x_2) = (x_1, x_2)$, and we obtain

$$\int_{\Omega} 2 dx = \int_{\partial\Omega} x_1 dx_2 - x_2 dx_1 = \int_{t_a}^{t_b} x_1 \dot{x}_2 - \dot{x}_1 x_2 dt, \tag{2.2.4}$$

and hence, the parametric functional to be maximized is

$$J(x) = \frac{1}{2} \int_{t_a}^{t_b} x_1 \dot{x}_2 - \dot{x}_1 x_2 dt. \tag{2.2.5}$$

We show that the hypotheses of Proposition 2.1.5 are fulfilled. Excluding the origin in $\mathbb{R}^2$, the conditions on differentiability are fulfilled. The constraint $K(x) = L$ is critical for the following curves:

$$\delta K(x)h = 0 \quad \text{for all } h \in (C_0^{1,pw}[t_a,t_b])^2 \Leftrightarrow$$

$$\frac{d}{dt}\frac{\dot{x}_k}{\sqrt{\dot{x}_1^2 + \dot{x}_2^2}} = 0 \quad \text{piecewise on } [t_a,t_b] \text{ for } k = 1,2 \Leftrightarrow$$

$$\frac{\dot{x}_k}{\sqrt{\dot{x}_1^2 + \dot{x}_2^2}} = c_k \quad \text{on } [t_a,t_b] \text{ for } k = 1,2, \quad \text{since} \qquad (2.2.6)$$

$$\frac{\dot{x}_k}{\sqrt{\dot{x}_1^2 + \dot{x}_2^2}} \in C^{1,pw}[t_a,t_b] \subset C[t_a,t_b] \quad \text{for } k = 1,2,$$

cf. Proposition 1.10.3. A closed curve cannot have tangents in one constant direction, and therefore it is not critical. By (2.1.35)

$$\left(-x_2 + \lambda\frac{\dot{x}_1}{\sqrt{\dot{x}_1^2 + \dot{x}_2^2}}, \; x_1 + \lambda\frac{\dot{x}_2}{\sqrt{\dot{x}_1^2 + \dot{x}_2^2}}\right) \in (C^{1,pw}[t_a,t_b])^2,$$

$$\frac{d}{dt}\left(-x_2 + \lambda\frac{\dot{x}_1}{\|\dot{x}\|}\right) = \dot{x}_2, \quad \frac{d}{dt}\left(x_1 + \lambda\frac{\dot{x}_2}{\|\dot{x}\|}\right) = -\dot{x}_1 \qquad (2.2.7)$$

piecewise on $[t_a,t_b]$.

Since the expression $(2.2.7)_1$ as well as $(x_2,-x_1)$ are continuous on $[t_a,t_b]$, $(2.2.7)_2$ gives

$$2x_2 - \lambda\frac{\dot{x}_1}{\|\dot{x}\|} = c_1 \quad \text{on } [t_a,t_b] \text{ and}$$

$$2x_1 + \lambda\frac{\dot{x}_2}{\|\dot{x}\|} = c_2 \quad \text{on } [t_a,t_b]. \qquad (2.2.8)$$

Multiplication of $(2.2.8)_1$ by $\dot{x}_2$ and of $(2.2.8)_2$ by $\dot{x}_1$ and addition of the resulting equations yield

$$2(x_1\dot{x}_1 + x_2\dot{x}_2) = c_2\dot{x}_1 + c_1\dot{x}_2,$$

$$\text{or } \frac{d}{dt}(x_1^2 + x_2^2) = \frac{d}{dt}(c_2x_1 + c_1x_2) \quad \text{piecewise on } [t_a,t_b]. \qquad (2.2.9)$$

Equation $(2.2.9)_2$ implies due to continuity

$$x_1^2 + x_2^2 - c_2x_1 - c_1x_2 = c_3 \quad \text{on } [t_a,t_b], \text{ and finally}$$

$$(x_1 - \frac{c_2}{2})^2 + (x_2 - \frac{c_1}{2})^2 = c_3 + \frac{c_1^2}{4} + \frac{c_2^2}{4} = R^2 \quad \text{on } [t_a,t_b]. \qquad (2.2.10)$$

Any circle with arbitrary center $(\frac{c_2}{2}, \frac{c_1}{2})$, radius R, and perimeter L bounds a maximal area of size $\frac{L^2}{4\pi}$.

Exercise

2.2.1. Compute the continuously differentiable curve starting in $(0,A)$ on the positive y-axis and ending on the positive x-axis that encloses with the positive axes a given fixed area S that creates a minimal surface of revolution when rotating around the x-axis.

2.3 The Hanging Chain

At first sight, a hanging chain acted on only by gravity looks like a parabola, cf. Figure 2.3.

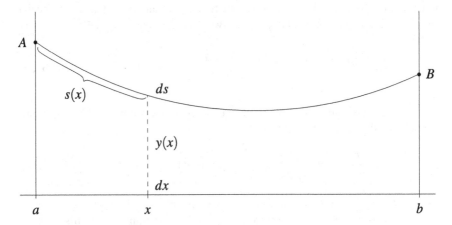

Fig. 2.3   A Hanging Chain

The variational principles of mechanics require that the potential energy of the hanging chain is minimal. Let the chain hang between (a,A) and (b,B) and denote $y(x)$ the height of the chain over $x \in [a,b]$. Then, its potential energy is given by

$$g\rho \int_a^b y\sqrt{1+(y')^2}dx \tag{2.3.1}$$

where g is the gravitational acceleration and ρ is the density of the homogeneous mass of the chain. (The mass at height $y(x)$ is $\rho ds = \rho\sqrt{1+(y'(x))^2}dx$.) The variational problem then is the following:

$$\text{Minimize} \quad J(y) = \int_a^b y\sqrt{1+(y')^2}dx$$

$$\text{in} \quad D = C^1[a,b] \cap \{y(a) = A, \ y(b) = B\}$$

under the isoperimetric constraint   (2.3.2)

$$K(y) = \int_a^b \sqrt{1+(y')^2}dx = L,$$

where L is the length of the chain.

A necessary condition is obviously $(b-a)^2 + (B-A)^2 \le L^2$, and in case of equality only the straight line between (a,A) and (b,B) fulfills the constraint.

The hypotheses of Proposition 2.1.1 are fulfilled if y is not critical for the constraint K. As shown in Example 4 of Paragraph 1.5, the only critical function for the functional K is the straight line which is excluded in case $(b-a)^2 + (B-A)^2 < L^2$.

Therefore, the Euler-Lagrange equation (2.1.5) holds. Neither the Lagrange function F nor G depend explicitly on x, and therefore, we can proceed as in the special case 7 of Paragraph 1.5. To that purpose, we need the regularity $y \in C^2(a,b)$. By virtue of

$$(F_{y'y'} + \lambda G_{y'y'})(y,y') = (y+\lambda)\frac{1}{(1+(y')^2)^{3/2}} \ne 0$$

$$\text{for} \quad y+\lambda \ne 0,$$ (2.3.3)

Exercise 1.5.1 guarantees that regularity in any open interval where $(2.3.3)_2$ is satisfied. Since the Euler-Lagrange equation,

$$\frac{d}{dx}\left((y+\lambda)\frac{y'}{\sqrt{1+(y')^2}}\right) = \sqrt{1+(y')^2} > 0 \quad \text{on } [a,b],$$ (2.3.4)

has no piecewise constant solutions, there is an interval $(x_1,x_2) \subset [a,b]$ where $(2.3.3)_2$ holds, and by the regularity $y \in C^2(x_1,x_2)$, the minimizer y fulfills

$$F(y,y') + \lambda G(y,y') - y'(F_{y'}(y,y') + \lambda G_{y'}(y,y')) = c_1,$$ (2.3.5)

cf. case 7 of Paragraph 1.5. Here, (2.3.5) means

$$y+\lambda = c_1\sqrt{1+(y')^2} \quad \text{on } (x_1,x_2).$$ (2.3.6)

By separation of variables, cf. case 7 of Paragraph 1.5, we obtain as in (1.6.8)

$$y(x)+\lambda = c_1 \cosh\left(\frac{x+c_2}{c_1}\right) \quad \text{for } x \in (x_1,x_2).$$ (2.3.7)

Clearly $c_1 \ne 0$. Hence, $y(x)+\lambda \ne 0$ not only for $x \in (x_1,x_2)$ but for all $x \in [a,b]$. Check that (2.3.7) is indeed a solution of the Euler-Lagrange equation on the entire interval $[a,b]$.

The constants $c_1 \ne 0$, c_2 and λ are determined by the boundary conditions $y(a) = A$, $y(b) = B$ and $K(y) = L$. The computation is not trivial, and it gives for $(b-a)^2 +$

$(B-A)^2 < L^2$ two solutions $c_1 > 0$ and $c_1 < 0$. The hanging chain with minimal potential energy is described by $c_1 > 0$, and the solution with $c_1 < 0$ describes the shape of a chain of length L with maximal potential energy.

Since the hyperbolic cosine describes the hanging chain, it is also called a catenary.

2.4 The Weierstraß-Erdmann Corner Conditions under Isoperimetric Constraints

Using the parametrization (1.11.2) for the functional (2.1.1), as well as for the constraint (2.1.2), we obtain

$$J(y) = J(\tilde{x}, \tilde{y}) = \int_{\tau_a}^{\tau_b} F(\tilde{x}, \tilde{y}, \frac{\dot{\tilde{y}}}{\dot{\tilde{x}}}) \dot{\tilde{x}} d\tau \quad \text{and}$$

$$K(y) = K(\tilde{x}, \tilde{y}) = \int_{\tau_a}^{\tau_b} G(\tilde{x}, \tilde{y}, \frac{\dot{\tilde{y}}}{\dot{\tilde{x}}}) \dot{\tilde{x}} d\tau = c. \tag{2.4.1}$$

We assume that the Lagrange functions $F, G : \mathbb{R}^3 \to \mathbb{R}$ are continuously totally differentiable. The domain of definition $D \subset C^{1,pw}[a,b]$ of the functional J is possibly characterized by boundary conditions at $x = a$ and/or $x = b$. Arguing as in the proof of Proposition 1.11.1, we realize that a global minimizer $y \in D \subset C^{1,pw}[a,b]$ of J under the isoperimetric constraint $K(y) = c$ is a local minimizer of $(2.4.1)_1$ under the isoperimetric constraint $(2.4.1)_2$. This is true for any parametrization that is admitted. The domain of definition $\tilde{D} \subset (C^{1,pw}[\tau_a, \tau_b])^2$ is determined by the domain of definition D of J, but in any case, $\tilde{x}(\tau_a) = a$ and $\tilde{x}(\tau_b) = b$.

If y is not critical for the constraint K, Proposition 2.1.3 holds with $n = 1$ and $m = 1$. In order to apply Proposition 2.1.5 with $n = 2$ and $m = 1$, the curve $\{(x, y(x)) | x \in [a,b]\}$ has to be noncritical for the parametric constraint K $(2.4.1)_2$.

By Proposition 1.10.2, we obtain for $\Psi(\tilde{x}, \tilde{y}, \dot{\tilde{x}}, \dot{\tilde{y}}) = G(\tilde{x}, \tilde{y}, \frac{\dot{\tilde{y}}}{\dot{\tilde{x}}}) \dot{\tilde{x}}$ and $\tilde{h} = (0, h_2)$,

$$\delta K(\tilde{x}, \tilde{y})\tilde{h} = \int_{\tau_a}^{\tau_b} G_y(\tilde{x}, \tilde{y}, \frac{\dot{\tilde{y}}}{\dot{\tilde{x}}}) \dot{\tilde{x}} h_2 + G_{y'}(\tilde{x}, \tilde{y}, \frac{\dot{\tilde{y}}}{\dot{\tilde{x}}}) \dot{h}_2 d\tau$$

$$= \int_a^b G_y(x, y, y')h + G_{y'}(x, y, y')h' dx \tag{2.4.2}$$

$$= \delta K(y)h,$$

where we use the substitutions $x = \varphi(\tau) = \tilde{x}(\tau)$ and $h(x) = h_2(\varphi^{-1}(x)) = h_2(\tau)$, cf. (1.11.2), (1.11.3).

Relation (2.4.2) proves: If $\delta K(y) : D \subset C^{1,pw}[a,b] \to \mathbb{R}$ is not identically zero, i.e., if y is not critical for K, then $\delta K(\tilde{x}, \tilde{y}) : \tilde{D} \subset (C^{1,pw}[\tau_a, \tau_b])^2 \to \mathbb{R}$ is not identically zero either, which means that the curve $\{(x, y(x)) | x \in [a,b]\} = \{(\tilde{x}(\tau), \tilde{y}(\tau)) | \tau \in [\tau_a, \tau_b]\}$ is not critical for the parametric constraint as well.

Proposition 2.4.1. *Let $y \in D \subset C^{1,pw}[a,b]$ be a global minimizer of the functional*

$$J(y) = \int_a^b F(x,y,y')dx \qquad (2.4.3)$$

under the isoperimetric constraint

$$K(y) = \int_a^b G(x,y,y')dx = c \qquad (2.4.4)$$

where the Lagrange functions $F,G : \mathbb{R}^3 \to \mathbb{R}$ are continuously totally differentiable. Assume that y is not critical for K. Then, the Lagrange equation (2.1.5) and

$$\begin{aligned}
&F_{y'}(\cdot,y,y') + \lambda G_{y'}(\cdot,y,y') \in C[a,b], \\
&F(\cdot,y,y') + \lambda G(\cdot,y,y') - y'(F_{y'}(\cdot,y,y') + \lambda G_{y'}(\cdot,y,y')) \in C[a,b],
\end{aligned} \qquad (2.4.5)$$

hold for the same $\lambda \in \mathbb{R}$.
If F and G do not depend explicitly on x, then

$$F(y,y') + \lambda G(y,y') - y'(F_{y'}(y,y') + \lambda G_{y'}(y,y')) = c_1. \qquad (2.4.6)$$

holds on $[a,b]$.

Proof. We know that the parametrized curve $\{(x,y(x))|x \in [a,b]\}$ is a local minimizer of $(2.4.1)_1$ under the constraint $(2.4.1)_2$. Furthermore the curve is not critical for $(2.4.1)_2$. By Proposition 2.1.5, the regularity $(2.1.35)_1$ then holds for the Lagrange functions $\Phi(\tilde{x},\tilde{y},\dot{\tilde{x}},\dot{\tilde{y}}) = F(\tilde{x},\tilde{y},\frac{\dot{\tilde{y}}}{\dot{\tilde{x}}})\dot{\tilde{x}}$ and $\Psi(\tilde{x},\tilde{y},\dot{\tilde{x}},\dot{\tilde{y}}) = G(\tilde{x},\tilde{y},\frac{\dot{\tilde{y}}}{\dot{\tilde{x}}})\dot{\tilde{x}}$, and in particular,

$$\begin{aligned}
&\Phi_{\dot{\tilde{x}}}(\tilde{x},\tilde{y},\dot{\tilde{x}},\dot{\tilde{y}}) + \lambda \Psi_{\dot{\tilde{x}}}(\tilde{x},\tilde{y},\dot{\tilde{x}},\dot{\tilde{y}}) \in C^{1,pw}[\tau_a,\tau_b] \subset C[\tau_a,\tau_b], \\
&\Phi_{\dot{\tilde{x}}}(\tilde{x},\tilde{y},\dot{\tilde{x}},\dot{\tilde{y}}) = F(\tilde{x},\tilde{y},\frac{\dot{\tilde{y}}}{\dot{\tilde{x}}}) - F_{y'}(\tilde{x},\tilde{y},\frac{\dot{\tilde{y}}}{\dot{\tilde{x}}})\frac{\dot{\tilde{y}}}{\dot{\tilde{x}}}, \\
&\Psi_{\dot{\tilde{x}}}(\tilde{x},\tilde{y},\dot{\tilde{x}},\dot{\tilde{y}}) = G(\tilde{x},\tilde{y},\frac{\dot{\tilde{y}}}{\dot{\tilde{x}}}) - G_{y'}(\tilde{x},\tilde{y},\frac{\dot{\tilde{y}}}{\dot{\tilde{x}}})\frac{\dot{\tilde{y}}}{\dot{\tilde{x}}}.
\end{aligned} \qquad (2.4.7)$$

Due to the invariance of the functionals according to Definition 1.10.2, Proposition 2.1.5 holds for any reparametrization of the curve $(\tilde{x},\tilde{y})$. Choosing the parameter x, we have $\tilde{x} = x$, $\dot{\tilde{x}} = 1$, $\tilde{y} = y$, $\dot{\tilde{y}} = y'$, $\tau_a = a$, $\tau_b = b$, which proves $(2.4.5)_2$ by (2.4.7). The continuity $(2.4.5)_1$ is proved in Proposition 2.1.1. The supplement (2.4.6) is Exercise 2.4.1. $\qquad\square$

We apply Proposition 2.4.1 to the following **Example**: Minimize

$$J(y) = \int_a^b W(y')dx \quad \text{in} \quad D = C^{1,pw}[a,b] \qquad (2.4.8)$$

under the isoperimetric constraint

$$K(y) = \int_a^b y' dx = m. \tag{2.4.9}$$

The potential $W : \mathbb{R} \to \mathbb{R}$ is continuously differentiable and sketched in Figure 1.21. Since no boundary conditions are imposed, we have $\delta K(y)h = \int_a^b h' dx = h(b) - h(a)$ for all $y \in D$, which are therefore not critical for the constraint (2.4.9).

Proposition 2.4.1 states the necessary conditions, which a global minimizer of (2.4.8) under (2.4.9) has to fulfill:

$$W'(y') + \lambda \in C^{1,pw}[a,b] \subset C[a,b],$$
$$\frac{d}{dx}(W'(y') + \lambda) = 0 \quad \text{or} \quad W'(y') + \lambda = c_1 \quad \text{on} \quad [a,b], \tag{2.4.10}$$
$$W(y') + \lambda y' - y'(W'(y') + \lambda) = W(y') - y'W'(y') = c_2 \quad \text{on} \quad [a,b].$$

Equations (2.4.10) have the same solutions as the variational problem (2.4.8) without a constraint, cf. (1.11.14)–(1.11.19):
A global minimizer is a line or a sawtooth function having two specific slopes $c_1^1 = \alpha$ and $c_1^2 = \beta$ that we sketch in Figure 1.21 and that fulfill (1.11.18) or (1.11.19). The number of corners, however, is not determined. For $\alpha < \beta$, let

$$y' = \alpha \quad \text{on intervals } I_i^\alpha \subset [a,b], \ i = 1,\ldots,m_1,$$
$$y' = \beta \quad \text{on intervals } I_i^\beta \subset [a,b], \ i = 1,\ldots,m_2, \tag{2.4.11}$$
$$\text{where} \quad \bigcup_{i=1}^{m_1} I_i^\alpha = I_\alpha, \quad \bigcup_{i=1}^{m_2} I_i^\beta = I_\beta, \quad I_\alpha \cup I_\beta = [a,b].$$

One possible function y is sketched in Figure 2.4, and we remark that any $y + c$ fulfills the necessary conditions as well.

A function (2.4.11) fulfills the constraint (2.4.9) if

$$\alpha|I_\alpha| + \beta|I_\beta| = m,$$
$$|I_\alpha| + |I_\beta| = b - a \tag{2.4.12}$$

holds. $|I_\alpha|, |I_\beta|$ denote the total length of all intervals I_i^α and I_i^β, respectively. Relations (2.4.12) imply

$$\alpha < \frac{m}{b-a} < \beta \quad \text{and}$$
$$|I_\alpha| = \frac{(b-a)\beta - m}{\beta - \alpha}, \quad |I_\beta| = \frac{m - (b-a)\alpha}{\beta - \alpha}. \tag{2.4.13}$$

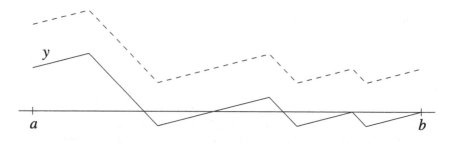

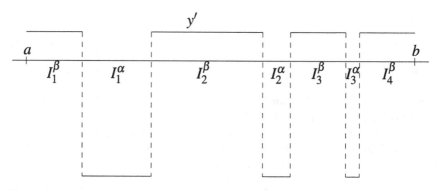

Fig. 2.4 A Global Minimizer

For

$$\frac{m}{b-a} \leq \alpha \quad \text{or} \quad \beta \leq \frac{m}{b-a}$$

$$y' = \frac{m}{b-a} \quad \text{on } [a,b]$$

(2.4.14)

gives the only function that fulfills the constraint (2.4.9).

Are the functions given in (2.4.11)–(2.4.14) indeed global minimizers? Figure 1.21 shows that

$$W(\alpha) + \frac{W(\beta) - W(\alpha)}{\beta - \alpha}(\tilde{y}' - \alpha) \leq W(\tilde{y}')$$

(2.4.15)

holds for all $\tilde{y}' \in \mathbb{R}$, and integration of (2.4.15) yields, using $\int_a^b \tilde{y}' dx = m$,

$$W(\alpha)\frac{(b-a)\beta - m}{\beta - \alpha} + W(\beta)\frac{m - (b-a)\alpha}{\beta - \alpha} \leq \int_a^b W(\tilde{y}')dx.$$

(2.4.16)

Therefore, the sawtooth function (2.4.11) with (2.4.13) fulfills

$$J(y) = \int_a^b W(y')dx \leq \int_a^b W(\tilde{y}')dx = J(\tilde{y})$$

(2.4.17)

for all $\tilde{y} \in C^{1,pw}[a,b]$ fulfilling the constraint (2.4.9).

Again as shown by Figure 1.21 in case (2.4.14),

$$W\left(\frac{m}{b-a}\right) + W'\left(\frac{m}{b-a}\right)\left(\tilde{y}' - \frac{m}{b-a}\right) \leq W(\tilde{y}') \qquad (2.4.18)$$

holds for all $\tilde{y}' \in \mathbb{R}$. Integration of (2.4.18) yields, using $\int_a^b \tilde{y}' dx = m$,

$$W\left(\frac{m}{b-a}\right)(b-a) \leq \int_a^b W(\tilde{y}')dx. \qquad (2.4.19)$$

Therefore, the line given by (2.4.14) fulfills

$$J(y) = \int_a^b W(y')dx \leq \int_a^b W(\tilde{y}')dx = J(\tilde{y}) \qquad (2.4.20)$$

for all $\tilde{y} \in C^{1,pw}[a,b]$ fulfilling the constraint (2.4.9).

All functions that fulfill the necessary conditions given in Proposition 2.4.1 are global minimizers of the functional (2.4.8) under the constraint (2.4.9).

Substituting $y' = u$ in (2.4.8) and in (2.4.9), the functions given above are global minimizers of

$$J(u) = \int_a^b W(u)dx \quad \text{in} \quad D = C^{pw}[a,b], \qquad (2.4.21)$$

under the isoperimetric constraint

$$K(u) = \int_a^b u \, dx = m. \qquad (2.4.22)$$

Indeed, any $y \in C^{1,pw}[a,b]$ gives via $y' = u$ a function in $C^{pw}[a,b]$, and conversely, any $u \in C^{pw}[a,b]$ gives by integration a function $y \in C^{1,pw}[a,b]$. Note that no boundary conditions are involved.

An application of problem (2.4.21), (2.4.22) is given by J. Carr, M.E. Gurtin, and M. Slemrod in "Structured Phase Transitions on a Finite Interval," Arch. Rat. Mech. Anal. 86, 317–351 (1984). They investigate a so-called spinodal decomposition of a binary alloy. Stable states are described by minimizers of the total energy (2.4.21) under preservation of mass (2.4.22). Here, the values of the minima of the potential W at $u = \alpha$ and at $u = \beta$ are identical. The function u measures the relative density (or concentration) of one component, and it is scaled as follows: If $u(x) = \alpha$ or $u(x) = \beta$, then only one component is present at $x \in [a,b]$, respectively. Thus, the piecewise constant minimizers (2.4.11) describe complete decompositions of the two components of the alloy. However, the distribution of the two components, also called phases, is completely arbitrary.

Which distribution of the two phases or which pattern of the interfaces is preferred, cannot be determined by the model (2.4.21), (2.4.22).

More information is given by the energy

$$J_\varepsilon(u) = \int_a^b \varepsilon(u')^2 + W(u)dx, \qquad (2.4.23)$$

where for small $\varepsilon > 0$, the additional term models the energy of interfaces between the phases. This functional in one or higher dimensions is the so-called Allen-Cahn or Modica-Mortola functional. One expects that global minimizers u_ε of $J_\varepsilon(u)$ under preservation of mass (2.4.22) converge to a global minimizer u_0 of $J_0(u) = J(u)$ as ε tends to zero and that this limit describes the physically relevant distribution of the two phases. This limiting behavior is proved in the above-mentioned paper. Moreover, the minimizers u_ε are monotonic and converge pointwise to u_0 which has a single interface. The situation is sketched in Figure 2.5.

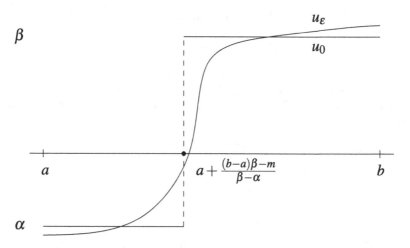

Fig. 2.5 Special Global Minimizers

The one-dimensional model (2.4.23) is interesting from a mathematical point of view but it does not describe a real situation. More realistic models in two or higher dimensions are studied in the literature. We mention a result for a two-dimensional model showing the analog limiting behavior, which is presented by H. Kielhöfer in "Minimizing Sequences Selected via Singular Perturbations and their Pattern Formation," Arch. Rat. Mech. Anal. 155, 261–276 (2000). For the piecewise constant limit describing a complete decomposition of the two components of the alloy, the "minimal interface criterion" holds, cf. L. Modica, "The Gradient Theory of Phase Transitions and the Minimal Interface Criterion," Arch. Rat. Mech. Anal. 98, 123–142 (1987).

Exercise

2.4.1. Prove the supplement (2.4.6) of Proposition 2.4.1.

2.5 Holonomic Constraints

The notion of a holonomic constraint - an expression commonly accredited to H. Hertz - was introduced in classical mechanics. The point masses are not free in space but are forced to be on some manifold. For instance, a pendulum moves on a circle and airplanes or deep-sea vessels run on a sphere.

We present the mathematical definition:

Definition 2.5.1. *Let a parametric functional*

$$J(x) = \int_{t_a}^{t_b} \Phi(x, \dot{x}) dt, \tag{2.5.1}$$

be defined on $D \subset (C^1[t_a, t_b])^n$. Then,

$$\Psi(x) = 0, \tag{2.5.2}$$

is called a holonomic constraint. Here,

$$\Psi : \mathbb{R}^n \to \mathbb{R}^m \quad where\ n > m, \tag{2.5.3}$$

is a continuously totally differentiable function, and we assume that the Jacobian matrix

$$D\Psi(x) = \left(\frac{\partial \Psi_i}{\partial x_j}(x) \right)_{\substack{i=1,\dots,m \\ j=1,\dots,n}} \in \mathbb{R}^{m \times n} \tag{2.5.4}$$

has the maximal rank m for all x satisfying $\Psi(x) = 0$.

In the Appendix we prove that the set,

$$M = \{x \in \mathbb{R}^n | \Psi(x) = 0\}, \quad \text{is a } (n-m)\text{-dimensional} \tag{2.5.5}$$
continuously differentiable manifold.

The constraint (2.5.2) forces admitted curves in $D \subset (C^{1,pw}[t_a, t_b])^n$ to be on the manifold M. The domain of definition $D \subset (C^{1,pw}[t_a, t_b])^n$ may be characterized by boundary conditions on some components at $t = t_a$ and/or $t = t_b$, which have clearly to be compatible with the constraint.

Again Lagrange's multiplier rule is the crucial tool to deduct necessary conditions on local constrained minimizers, but the derivation is not as simple as for isoperimetric constraints. The main difference is that the Lagrange multipliers are no longer constant. For technical reasons, we need more regularity of the Lagrange function Φ, of the constraint Ψ, and of the local minimizer x.

Proposition 2.5.1. *Let the curve $x \in D \subset (C^2[t_a, t_b])^n$ be a local minimizer of the parametric functional (2.5.1) under the holonomic constraint (2.5.2). In particular,*

$$J(x) \leq J(\hat{x}) \quad \text{for all } \hat{x} \in D \cap \{\Psi(x) = 0\}$$
$$\text{satisfying} \quad \max_{k=1,\ldots,n} \|x_k - \hat{x}_k\|_1 < d. \tag{2.5.6}$$

We assume that the functions $\Phi : \mathbb{R}^n \times \mathbb{R}^n \to \mathbb{R}$ and $\Psi : \mathbb{R}^n \to \mathbb{R}^m$ are twice and three times continuously partially differentiable, respectively. Then, there exists a continuous function

$$\lambda = (\lambda_1, \ldots, \lambda_m) : [t_a, t_b] \to \mathbb{R}^m, \tag{2.5.7}$$

such that x is a solution of the following system of n differential equations:

$$\frac{d}{dt} \Phi_{\dot{x}}(x, \dot{x}) = \Phi_x(x, \dot{x}) + \sum_{i=1}^{m} \lambda_i \nabla \Psi_i(x) \quad \text{on } [t_a, t_b]. \tag{2.5.8}$$

Here, we agree upon $\Phi_{\dot{x}} = (\Phi_{\dot{x}_1} \ldots, \Phi_{\dot{x}_n})$, $\Phi_x = (\Phi_{x_1} \ldots, \Phi_{x_n})$, $x = (x_1, \ldots, x_n)$, $\dot{x} = (\dot{x}_1, \ldots, \dot{x}_n)$, and $\nabla \Psi_i = (\Psi_{i,x_1}, \ldots, \Psi_{i,x_n})$. The functions $\lambda_1, \ldots, \lambda_m$ are called Lagrange multipliers.

Proof. As a first step, we construct an admitted perturbation of the minimizing curve $x \in D \cap \{\Psi(x) = 0\}$. That perturbation is no longer of the simple form $x + sh$, since we need to guarantee that the perturbation $x + h(s, \cdot)$ stays on the manifold M.

For that purpose, we use the nomenclature of the Appendix. Let

$$h \in (C_0^2[t_a, t_b])^n \quad \text{be a curve in } \mathbb{R}^n, \tag{2.5.9}$$

and for $x \in M$, let $P(x) : \mathbb{R}^n \to T_x M$ be the orthogonal projection on the tangent space $T_x M$ in x. This operator is twice continuously differentiable with respect to $x \in M$, cf. (A.17), (A.19). Then,

$$a(t) = P(x(t))h(t) \in T_{x(t)}M \quad \text{for } [t_a, t_b],$$
$$a \in (C^2[t_a, t_b])^n \quad \text{and} \quad a(t_a) = 0, \ a(t_b) = 0. \tag{2.5.10}$$

For $x \in M$, let $Q(x) : \mathbb{R}^n \to N_x M$ be the orthogonal projection on the normal space $N_x M$. Again this operator is twice continuously differentiable with respect to $x \in M$, cf. (A.16), and for any fixed $t_0 \in [t_a, t_b]$,

$$Q(x(t)) : N_{x(t_0)}M \to N_{x(t)}M \quad \text{is bijective, provided } |t - t_0| < \delta, \tag{2.5.11}$$

where $0 < \delta$ is sufficiently small, cf. (A.18). We define

$$H : \mathbb{R} \times [t_a, t_b] \times N_{x(t_0)}M \to \mathbb{R}^m \quad \text{by}$$
$$H(s, t, z) = \Psi(x(t) + sa(t) + Q(x(t))z), \tag{2.5.12}$$

and by the assumptions on x and Ψ, the properties of the curve a, and the projection Q, the function H satisfies:

H is twice continuously differentiable with respect to all three variables,

$H(0,t_0,0) = \Psi(x(t_0)) = 0$,

$D_z H(0,t_0,0) = D\Psi(x(t_0))Q(x(t_0)) : N_{x(t_0)}M \to \mathbb{R}^m$ is bijective,

since $Q(x(t_0))|_{N_{x(t_0)}M} = I$, and

$D\Psi(x(t_0)) : N_{x(t_0)}M \to \mathbb{R}^m$ is bijective, cf. (A.9).

$$(2.5.13)$$

By the theorem on implicit functions, there exists a unique function

$$\beta : (-\varepsilon_0, \varepsilon_0) \times (t_0 - \delta_0, t_0 + \delta_0) \to N_{x(t_0)}M, \quad \text{satisfying}$$
$$\beta(0,t_0) = 0 \text{ and } H(s,t,\beta(s,t)) = 0 \tag{2.5.14}$$
$$\text{for } (s,t) \in (-\varepsilon_0, \varepsilon_0) \times (t_0 - \delta_0, t_0 + \delta_0).$$

Moreover that function β is twice continuously differentiable with respect to its two variables. (If $t_0 = t_a$ or $t_0 = t_b$, choose the domain of definitions $(-\varepsilon_0, \varepsilon_0) \times [t_a, t_a + \delta_0)$ and $(-\varepsilon_0, \varepsilon_0) \times (t_b - \delta_0, t_b]$, respectively.) Defining

$$h(s,t) = sa(t) + Q(x(t))\beta(s,t) = sa(t) + b(s,t), \quad \text{we obtain}$$
$$\Psi(x(t) + h(s,t)) = 0, \quad \text{meaning} \quad x(t) + h(s,t) \in M \tag{2.5.15}$$
$$\text{for } (s,t) \in (-\varepsilon_0, \varepsilon_0) \times (t_0 - \delta_0, t_0 + \delta_0).$$

Moreover, by construction,

$$a(t) \in T_{x(t)}M \quad \text{and} \quad b(s,t) \in N_{x(t)}M. \tag{2.5.16}$$

Due to the local uniqueness of β in (2.5.14) and by the injectivity of $Q(x(t))$ on $N_{x(t_0)}M$, cf. (2.5.11) for $0 < \delta \le \delta_0$, the perturbation h has the following properties:

$$sa(t) = 0 \quad \Rightarrow \quad \beta(s,t) = 0. \quad \text{In particular}$$
$$\beta(0,t) = 0, \quad \beta(s,t_a) = 0, \quad \beta(s,t_b) = 0, \quad \text{meaning}$$
$$h(0,t) = 0 \quad \text{for} \quad t \in (t_0 - \delta_0, t_0 + \delta_0) \quad \text{and} \tag{2.5.17}$$
$$h(s,t_a) = 0, \quad h(s,t_b) = 0.$$

By $\Psi(x(t) + sa(t) + b(s,t)) = 0$ for all $s \in (-\varepsilon_0, \varepsilon_0)$ and for any fixed $t \in (t_0 - \delta_0, t_0 + \delta_0)$, the derivative with respect to s vanishes, i.e.,

$$\frac{\partial}{\partial s}\Psi(x(t) + sa(t) + b(s,t))|_{s=0}$$
$$= D\Psi(x(t))(a(t) + \frac{\partial}{\partial s}b(0,t)) = D\Psi(x(t))\frac{\partial}{\partial s}b(0,t) = 0, \tag{2.5.18}$$
$$\text{since} \quad a(t) \in T_{x(t)}M = \text{Kern } D\Psi(x(t)) , \text{cf. (A.7).}$$

By construction $b(s,t) \in N_{x(t)}M$, cf. (2.5.16), and hence, $\frac{\partial}{\partial s}b(0,t) \in N_{x(t)}M$. Then, injectivity of $D\Psi(x(t))$ on $N_{x(t)}M$, cf. (A.9), implies

$$\frac{\partial}{\partial s}b(0,t) = 0, \quad \text{and by definition of } h \text{ in (2.5.15)},$$

$$\frac{\partial}{\partial s}h(0,t) = a(t) \quad \text{for all} \quad t \in (t_0 - \delta_0, t_0 + \delta_0). \tag{2.5.19}$$

Finally, due to the continuity of the second derivatives, we note

$$\frac{\partial}{\partial s}\frac{\partial}{\partial t}h(0,t) = \frac{\partial}{\partial t}\frac{\partial}{\partial s}h(0,t) = \frac{\partial}{\partial t}a(t) = \dot{a}(t) \tag{2.5.20}$$
$$\text{for all } t \in (t_0 - \delta_0, t_0 + \delta_0).$$

Thus far, h is constructed only on $(-\varepsilon_0, \varepsilon_0) \times (t_0 - \delta_0, t_0 + \delta_0)$, where δ_0 and ε_0 depend on t_0. Since $[t_a, t_b]$ is compact, finitely many open intervals $(t_0 - \delta_0, t_0 + \delta_0)$ cover $[t_a, t_b]$, and for simplicity we denote the smallest among the finitely many $\varepsilon_0 > 0$ again by ε_0. For t in an intersection of two intervals $(t_0 - \delta_0, t_0 + \delta_0)$, the two functions $\beta(s,t)$ coincide on the intersection of the two rectangles $(-\varepsilon_0, \varepsilon_0) \times (t_0 - \delta_0, t_0 + \delta_0)$ by uniqueness, cf. (2.5.14). We summarize:

There exists a perturbation

$h : (-\varepsilon_0, \varepsilon_0) \times [t_a, t_b] \to \mathbb{R}^n$ which is twice continuously differentiable

with respect to both variables and which has the property that

$x(t) + h(s,t) \in M$ for $(s,t) \in (-\varepsilon_0, \varepsilon_0) \times [t_a, t_b]$.

The perturbation is composed of two terms, $h(s,t) = sa(t) + b(s,t)$,

where $a(t) \in T_{x(t)}M$, $b(s,t) \in N_{x(t)}M$,   satisfying

$h(s, t_a) = 0$, $h(s, t_b) = 0$ for $s \in (-\varepsilon_0, \varepsilon_0)$,

$h(0,t) = 0$, $\frac{\partial}{\partial s}h(0,t) = a(t)$ and $\frac{\partial^2}{\partial s \partial t}h(0,t) = \dot{a}(t)$ for $t \in [t_a, t_b]$.
$$\tag{2.5.21}$$

In a second step of the proof, we make use of the minimizing property of x under the holonomic constraint. Therefore, in view of the properties of the perturbation h, the real-valued function

$J(x + h(s, \cdot))$ is locally minimal at $s = 0$, whence

$$\frac{d}{ds}J(x + h(s, \cdot))|_{s=0} = 0. \tag{2.5.22}$$

Following the arguments in the proof of Proposition 1.2.1, the derivative with respect to s is computed as follows:

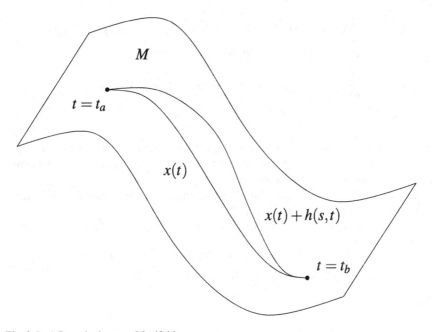

Fig. 2.6 A Perturbation on a Manifold

$$\frac{d}{ds}\int_{t_a}^{t_b}\Phi(x+h(s,\cdot),\dot{x}+\frac{\partial}{\partial t}h(s,\cdot))dt|_{s=0}$$

$$=\int_{t_a}^{t_b}\sum_{k=1}^{n}(\Phi_{x_k}(x,\dot{x})\frac{\partial}{\partial s}h_k(0,\cdot)+\Phi_{\dot{x}_k}(x,\dot{x})\frac{\partial^2}{\partial s\partial t}h_k(0,\cdot))dt \qquad (2.5.23)$$

$$=\int_{t_a}^{t_b}(\Phi_x(x,\dot{x}),a)+(\Phi_{\dot{x}}(x,\dot{x}),\dot{a})dt=0,$$

where, as usual, $\Phi_x=(\Phi_{x_1},\ldots,\Phi_{x_n})$, $\Phi_{\dot{x}}=(\Phi_{\dot{x}_1},\ldots,\Phi_{\dot{x}_n})$, and $(\quad,\quad)$ is the scalar product in $\mathbb{R}^n$. We also use $(2.5.21)_8$.

Due to the assumed regularity of x and Φ, integration by parts is possible, which yields

$$\int_{t_a}^{t_b}(\Phi_x(x,\dot{x})-\frac{d}{dt}\Phi_{\dot{x}}(x,\dot{x}),a)dt=0. \qquad (2.5.24)$$

Note that due to $a(t_a)=0$ and $a(t_b)=0$, the boundary terms vanish.

The following computations make use of both the properties of the projections summarized in (A.14), and of the Definition (A.7) of $T_{x(t)}M$ and $N_{x(t)}M$:

$$Q(x(t))(\frac{d}{dt}\Phi_{\dot{x}}(x(t),\dot{x}(t)) - \Phi_x(x(t),\dot{x}(t))) \in N_{x(t)}M = \text{ range } D\Psi(x(t))^*,$$

$$D\Psi(x(t))^* = (\nabla\Psi_1(x(t))\cdots\nabla\Psi_m(x(t))) \in \mathbb{R}^{n\times m},$$

$$Q(x(t))(\frac{d}{dt}\Phi_{\dot{x}}(x(t),\dot{x}(t)) - \Phi_x(x(t),\dot{x}(t))) = \sum_{i=1}^{m}\lambda_i(t)\nabla\Psi_i(x(t))$$

for $t \in [t_a,t_b]$.

$$(2.5.25)$$

(If linear mappings or operators are identified with matrices, then they operate on column vectors. In the transposed Jacobian matrix, cf. (A.3), gradients of the components of a mapping are columns.)

In view of (A.24), the coefficients $\lambda_i(t)$ are unique, and by continuity of the left-hand side, $\lambda = (\lambda,\ldots,\lambda_m) \in (C[t_a,t_b])^m$.

Coming back to (2.5.9) and (2.5.10), we have the decompositions

$$h(t) = P(x(t))h(t) + Q(x(t))h(t) = a(t) + b(t),$$
$$\text{where} \quad a(t) \in T_{x(t)}M \text{ and } b(t) \in N_{x(t)}M.$$

$$(2.5.26)$$

Then finally:

$$\int_{t_a}^{t_b}(\frac{d}{dt}\Phi_{\dot{x}}(x,\dot{x}) - \Phi_x(x,\dot{x}) - \sum_{i=1}^{m}\lambda_i\nabla\Psi_i(x),h)dt$$

$$= \int_{t_a}^{t_b}(\frac{d}{dt}\Phi_{\dot{x}}(x,\dot{x}) - \Phi_x(x,\dot{x}) - \sum_{i=1}^{m}\lambda_i\nabla\Psi_i(x),a)dt$$

$$+ \int_{t_a}^{t_b}(\frac{d}{dt}\Phi_{\dot{x}}(x,\dot{x}) - \Phi_x(x,\dot{x}) - \sum_{i=1}^{m}\lambda_i\nabla\Psi_i(x),b)dt = 0.$$

$$(2.5.27)$$

The first integral $(2.5.27)_2$ vanishes by virtue of (2.5.24) and $(\sum_{i=1}^{m}\lambda_i\nabla\Psi_i(x),a) = 0$ pointwise for all $t \in [t_a,t_b]$. Observe that the vectors in the scalar product are orthogonal. The second integral $(2.5.27)_3$ vanishes by $(2.5.25)_3$: Using the properties (A.14) of the projection $Q(x(t))$, the integrand vanishes pointwise for all $t \in [t_a,t_b]$. Observe that $b = Q(x)b$.

Since $h \in (C_0^2[t_a,t_b])^n$ is arbitrary, Lemma 1.3.1 and (2.5.27) imply the claim (2.5.8).   □

Lagrangian Mechanics:

N point masses $m_1,\ldots,m_N$ in $\mathbb{R}^3$ have the coordinates $x=(x_1,y_1,z_1,\ldots,x_N,y_N,z_N) \in \mathbb{R}^{3N}$, where (x_k,y_k,z_k) are the coordinates of m_k. Their motion is governed by the kinetic energy

$$T(\dot{x}) = \sum_{k=1}^{N}\frac{1}{2}m_k(\dot{x}_k^2 + \dot{y}_k^2 + \dot{z}_k^2),$$

$$(2.5.28)$$

and by the potential energy

$$V(x) = V(x_1,y_1,z_1,\ldots,x_N,y_N,z_N)$$

$$(2.5.29)$$

under $m < 3N$ holonomic constraints

$$\Psi_i(x) = \Psi_i(x_1, y_1, z_1, \ldots, x_N, y_N, z_N) = 0, \quad i = 1, \ldots, m, \tag{2.5.30}$$

which describe exterior constraints as well as possibly interior links among the point masses.

The action along a trajectory of all N point masses is

$$J(x) = \int_{t_a}^{t_b} T(\dot{x}) - V(x) dt, \tag{2.5.31}$$

where $L = T - V$ is called the Lagrangian. According to Proposition 2.5.1, a trajectory that minimizes the action has to satisfy

$$m_k \ddot{x}_k = -V_{x_k}(x) + \sum_{i=1}^{m} \lambda_i \Psi_{i,x_k}(x),$$

$$m_k \ddot{y}_k = -V_{y_k}(x) + \sum_{i=1}^{m} \lambda_i \Psi_{i,y_k}(x), \tag{2.5.32}$$

$$m_k \ddot{z}_k = -V_{z_k}(x) + \sum_{i=1}^{m} \lambda_i \Psi_{i,z_k}(x) \quad \text{for} \quad k = 1, \ldots, N,$$

provided the hypotheses of Proposition 2.5.1 are fulfilled.

The number $3N - m$ is the dimension of the manifold M in $\mathbb{R}^{3N}$ defined by the constraints (2.5.30). It is the number of the degrees of freedom that the point masses have in moving freely on the manifold M. The $3N - m$ free coordinates on the manifold are called generalized coordinates. The additional terms in (2.5.32) compared to the free system (1.10.27) have the physical dimension of forces and are called constraining forces. They act orthogonally to the manifold M, cf. (2.5.25).

The simple **gravity pendulum**: The point mass m moves in the vertical (x_1, x_2)-plane subject to gravity mg in direction of the negative x_2-axis, and it is forced to move on a circle with radius ℓ.

The Lagrangian $L = T - V$ is given by

$$L(x, \dot{x}) = \frac{1}{2} m (\dot{x}_1^2 + \dot{x}_2^2) - mgx_2, \tag{2.5.33}$$

with the holonomic constraint

$$\Psi(x) = x_1^2 + x_2^2 - \ell^2 = 0. \tag{2.5.34}$$

We then obtain the system (cf. (2.5.32))

$$m\ddot{x}_1 = 2\lambda x_1,$$
$$m\ddot{x}_2 = -mg + 2\lambda x_2. \tag{2.5.35}$$

The constraining force acts in direction (x_1, x_2) orthogonal to the circle (2.5.34).

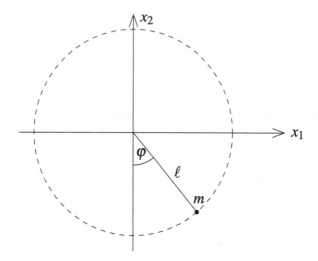

Fig. 2.7 The Gravity Pendulum

The equation of motion of a pendulum is commonly given in terms of the generalized coordinate φ, which is the angle sketched in Figure 2.7. Introducing polar coordinates

$$x_1 = r\sin\varphi,$$
$$x_2 = -r\cos\varphi, \tag{2.5.36}$$

the Lagrangian (2.5.33) becomes

$$\frac{1}{2}m(\dot{r}^2 + r^2\dot{\varphi}^2) + mgr\cos\varphi. \tag{2.5.37}$$

On the circle (2.5.34), we have $r = \ell$, $\dot{r} = 0$, and in terms of the generalized coordinate φ, the Lagrangian reads

$$L(\varphi, \dot{\varphi}) = \frac{1}{2}m\ell^2\dot{\varphi}^2 + mg\ell\cos\varphi. \tag{2.5.38}$$

Hamilton's principle says that solutions of the Euler-Lagrange equation with the Lagrangian (2.5.38) give solutions of the constrained system (2.5.35) and vice versa. We confirm Hamilton's principle for this example.

The Euler-Lagrange equation for (2.5.38) is the so-called pendulum equation

$$\frac{d}{dt}L_{\dot{\varphi}} = L_{\varphi}, \quad \text{or explicitly}$$
$$\ell\ddot{\varphi} + g\sin\varphi = 0. \tag{2.5.39}$$

Let φ be a solution of (2.5.39). According to Exercise 1.10.4, the conservation law

$$\frac{1}{2}\ell\dot{\varphi}^2 - g\cos\varphi = E = \text{const.} \tag{2.5.40}$$

is valid along any solution. Defining

$$x_1 = \ell\sin\varphi, \quad \dot{x}_1 = \ell\cos\varphi\dot{\varphi}, \quad \ddot{x}_1 = -\ell\sin\varphi\dot{\varphi}^2 + \ell\cos\varphi\ddot{\varphi},$$
$$x_2 = -\ell\cos\varphi, \quad \dot{x}_2 = \ell\sin\varphi\dot{\varphi}, \quad \ddot{x}_2 = \ell\cos\varphi\dot{\varphi}^2 + \ell\sin\varphi\ddot{\varphi}, \tag{2.5.41}$$

and using (2.5.39) and (2.5.40), the functions (2.5.41) solve the system (2.5.35) for

$$\lambda = -\frac{m}{2\ell}(3g\cos\varphi + 2E) = -\frac{m}{2\ell}(-\frac{3g}{\ell}x_2 + 2E). \tag{2.5.42}$$

On the other hand, let (x_1, x_2) be a solution of (2.5.35) satisfying $x_1^2 + x_2^2 = \ell^2$. We define

$$\varphi = \arctan\left(-\frac{x_1}{x_2}\right) \quad \text{or} \quad \begin{array}{l} x_1 = \ell\sin\varphi, \\ x_2 = -\ell\cos\varphi, \end{array} \tag{2.5.43}$$

and by (2.5.41), one obtains

$$m\ddot{x}_1 - 2\lambda x_1 = -m\ell\sin\varphi\dot{\varphi}^2 + m\ell\cos\varphi\ddot{\varphi} - 2\lambda\ell\sin\varphi = 0,$$
$$m\ddot{x}_2 + mg - 2\lambda x_2 = m\ell\cos\varphi\dot{\varphi}^2 + m\ell\sin\varphi\ddot{\varphi} + mg + 2\lambda\ell\cos\varphi = 0. \tag{2.5.44}$$

Multiplication of the first equation by $\cos\varphi$ and of the second equation by $\sin\varphi$ and addition of the two equations yield the differential equation (2.5.39) after division by m. Relation (2.5.42) gives the constraining force $2\lambda(x_1, x_2)$.

Another example is the **cycloidal pendulum**: The point mass m moves in a vertical (x_1, x_2)-plane subject to gravity mg, and it is forced on a cycloid. In order to use the formulas (1.8.17) for the cycloid, we orient the x_2-axis downward. The generalized coordinate φ on the cycloid is the angle of the generating wheel sketched in Figure 2.8. By (1.8.17), the point mass m has the coordinates

$$\begin{array}{l} x_1 = r(\varphi - \sin\varphi), \\ x_2 = r(1 - \cos\varphi), \end{array} \varphi \in [0, 2\pi], \tag{2.5.45}$$

which allows the Lagrangian to be expressed in terms of the generalized coordinate φ as follows:

$$\dot{x}_1 = r(1 - \cos\varphi)\dot{\varphi},$$
$$\dot{x}_2 = r\sin\varphi\dot{\varphi},$$
$$T = \frac{1}{2}m(\dot{x}_1^2 + \dot{x}_2^2) = mr^2(1 - \cos\varphi)\dot{\varphi}^2, \tag{2.5.46}$$
$$V = -mgx_2 = mgr(\cos\varphi - 1),$$
$$L = mr^2(1 - \cos\varphi)\dot{\varphi}^2 - mgr(\cos\varphi - 1).$$

This gives finally the Euler-Lagrange equation

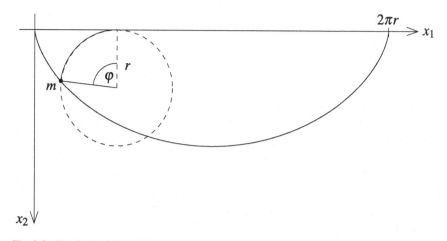

Fig. 2.8 The Cycloidal Pendulum

$$\frac{d}{dt}L_{\dot\varphi} = L_\varphi, \quad \text{or explicitly}$$

$$(1 - \cos\varphi)\ddot\varphi + \frac{1}{2}\sin\varphi\,\dot\varphi^2 = \frac{g}{2r}\sin\varphi. \tag{2.5.47}$$

Using the trigonometric formulas $1 - \cos\varphi = 2\sin^2\frac{\varphi}{2}$ and $\sin\varphi = 2\sin\frac{\varphi}{2}\cos\frac{\varphi}{2}$, one obtains after division by $2\sin\frac{\varphi}{2}$

$$\sin\frac{\varphi}{2}\ddot\varphi + \frac{1}{2}\cos\frac{\varphi}{2}\dot\varphi^2 = \frac{g}{2r}\cos\frac{\varphi}{2},$$

$$\text{or} \quad -2\frac{d^2}{dt^2}\cos\frac{\varphi}{2} = \frac{g}{2r}\cos\frac{\varphi}{2}, \tag{2.5.48}$$

$$\text{or} \quad \ddot q + \omega q = 0, \quad \text{where} \quad q = \cos\frac{\varphi}{2} \quad \text{and} \quad \omega = \frac{g}{4r}.$$

The last equation has the solutions $q(t) = c_1\sin\sqrt{\omega}t + c_2\cos\sqrt{\omega}t$ with the period

$$P = 2\pi\sqrt{\frac{4r}{g}}. \tag{2.5.49}$$

In contrast to the simple gravity pendulum with period

$$P \approx 2\pi\sqrt{\frac{\ell}{g}}, \tag{2.5.50}$$

valid only for small amplitudes, the period (2.5.49) of the cycloidal pendulum does not depend on the amplitude. Therefore, the cycloidal pendulum is isochronal.

Remark. *If N point masses are forced by a holonomic constraint on a* $(3N - m)$-*dimensional manifold, they have by definition* $3N - m = n$ *degrees of freedom. That n-dimensional manifold M can be described by n independent and free coordinates* $q_1, \ldots, q_n$. *(In (A.11), this is done locally.).*

If the Lagrangian $L = T - V$ *is expressed in terms of the so-called generalized coordinates* $q = (q_1, \ldots, q_n)$, *then by Hamilton's principle the motion of the N point masses is governed by the Euler-Lagrange equation of the action:*

$$q = (q_1, \ldots, q_n) \quad \text{are the generalized coordinates,}$$
$$L(q, \dot{q}) \quad \text{is the Lagrangian,} \tag{2.5.51}$$
$$\frac{d}{dt} L_{\dot{q}}(q, \dot{q}) = L_q(q, \dot{q}) \quad \text{is the Euler-Lagrange equation,}$$

whose solutions describe the motion of the N point masses. It is a system of n ordinary differential equations of second order.

By Proposition 1.4.2, the strong and weak versions of the Euler-Lagrange equation are equivalent, and the latter means that the first variation of the action functional vanishes. This is also shortly denoted as a principle of stationary or least action.

By the so-called Legendre transformation, new coordinates and a new function H are introduced:

$$p = L_{\dot{q}}(q, \dot{q}), \tag{2.5.52}$$

where the solution for $\dot{q}$ *is required to be unique, i.e.,*

$$p = L_{\dot{q}}(q, \dot{q}) \Leftrightarrow \dot{q} = h(p, q), \tag{2.5.53}$$

for a continuously totally differentiable function $h : \mathbb{R}^n \times \mathbb{R}^n \to \mathbb{R}^n$. *One defines*

$$H : \mathbb{R}^n \times \mathbb{R}^n \to \mathbb{R},$$
$$H(p, q) = L(q, h(p, q)) - (p, h(p, q)), \quad \text{or more concisely,} \tag{2.5.54}$$
$$H = L - \dot{q} L_{\dot{q}} = L - p\dot{q},$$

where the notation for the scalar product in $\mathbb{R}^n$, *used in (2.5.54)$_3$, is commonly used in physics.*

The coordinates p are called generalized momenta, and the function H is called the Hamiltonian. For a solution q of the Euler-Lagrange system (2.5.51)$_3$, one obtains by definitions (2.5.53) and (2.5.54)$_2$,

$$H_q = L_q + L_{\dot{q}} h_q - p h_q = L_q = \frac{d}{dt} L_{\dot{q}} = \dot{p},$$
$$H_p = L_{\dot{q}} h_p - h - p h_p = -h = -\dot{q}. \tag{2.5.55}$$

The so-called Hamiltonian system

$$\dot{p} = H_q(p,q),$$
$$\dot{q} = -H_p(p,q),$$

(2.5.56)

is equivalent to the Euler-Lagrange system $(2.5.51)_3$ if one redefines $p = L_{\dot{q}}$, $\dot{q} = h(p,q)$, and $L = H + p\dot{q}$. The Hamiltonian system is a system of 2n ordinary differential equations of first order.

The price for the reduction of the order is the doubling of the dimension. What is the advantage?

The special structure of the Hamiltonian system implies some useful properties of its flow. We mention an important conservation law:

$$\frac{d}{dt}H(p,q) = H_p(p,q)\dot{p} + H_q(p,q)\dot{q} = 0, \quad whence$$

(2.5.57)

$$H(p,q) = const. \quad along \; solutions \; of \; (2.5.56).$$

For $L = \frac{1}{2}m\dot{q}^2 - V(q)$, the Hamiltonian $H = L - \dot{q}L_{\dot{q}} = -\frac{1}{2}m\dot{q}^2 - V(q) = -\frac{1}{2m}p^2 - V(q) = -E$ is the negative total energy which is conserved along solutions of the Hamiltonian system. (We see in Exercise 1.10.4 that this is true also for solutions of the Euler-Lagrange equation.) In this case, the Hamiltonian system reads $\dot{p} = -V_q(q)$, $\dot{q} = \frac{1}{m}p$.

More conservation laws can be found as follows: A function $E = E(p,q)$ is constant along solutions of (2.5.56) if

$$\frac{d}{dt}E = E_p\dot{p} + E_q\dot{q} = E_pH_q - E_qH_p = 0 \quad or \; if$$

(2.5.58)

$$[E,H] = E_pH_q - E_qH_p = 0.$$

The expression $[E,H]$ is called Poisson bracket. Equation $(2.5.58)_2$ is a partial differential equation of first order.

Finally, we mention that the flow of the Hamiltonian system (2.5.56) preserves the volume: The unique solutions of the initial value problem

$$\dot{x} = f(x), \quad f : \mathbb{R}^n \rightarrow \mathbb{R}^n,$$
$$x(0) = z,$$

(2.5.59)

with a continuously totally differentiable vector field f are denoted $x(t) = \varphi(t,z)$. Uniqueness implies the relation $\varphi(s+t,z) = \varphi(s,\varphi(t,z))$ for all s,t and $s+t$ for which the solutions exist. Moreover $\varphi(0,z) = z$ for all $z \in \mathbb{R}^n$. The mapping $\varphi(t,\cdot)$: $\mathbb{R}^n \rightarrow \mathbb{R}^n$ is called the flow of the system (2.5.59). Let $\mu(\Omega)$ be the (Lebesgue-) measure of a measurable set $\Omega \subset \mathbb{R}^n$. The flow of the system (2.5.59) is volume preserving if

$$\mu(\varphi(t,\Omega)) = \mu(\varphi(0,\Omega)) = \mu(\Omega) \quad for \; all \; measurable \quad \Omega \in \mathbb{R}^n \quad (2.5.60)$$

and for all t for which the flow exists. Liouville's theorem reads as follows:

$$\text{If}\quad \mathrm{div} f(x) = 0 \quad \text{for all } x \in \mathbb{R}^n,$$

then the flow of (2.5.59) is volume preserving.

(2.5.61)

Here, $\mathrm{div} f(x) = \sum_{i=1}^{n} \frac{\partial f_i}{\partial x_i}(x)$ *is the divergence of the vector field* f. *We prove Liouville's theorem in the Appendix, (A.25)–(A.31).*

The 2n-dimensional vector field of the Hamiltonian system (2.5.56) satisfies

$$f = (H_q, -H_p) \quad \text{and} \quad \mathrm{div} f = \sum_{i=1}^{n}(H_{q_i p_i} - H_{p_i q_i}) = 0,$$

(2.5.62)

provided the Hamiltonian is twice continuously partially differentiable. Hence, the Hamiltonian flow is volume preserving. This has interesting consequences such as the exclusion of asymptotically stable equilibria (sinks) and the application of ergodic theory to Hamiltonian mechanics.

Let the parametric functional (2.5.1) be **invariant** in the sense of (1.10.20). Any admitted reparametrization of a curve $x \in D \subset (C^1[t_a, t_b])^n$ leaves, by Proposition 1.10.1, the functional as well as the holonomic constraint (2.5.2) invariant. Hence, Proposition 2.5.1 holds for a reparametrized local constrained minimizer with possibly different Lagrange multipliers. This is stated in the next proposition.

Proposition 2.5.2. *Let the functional (2.5.1) be invariant in the sense of (1.10.20) and let* $\tilde{x}(\tau) = x(\varphi(\tau))$ *be a reparametrization of* $x \in (C^1[t_a, t_b])^n$ *according to Definition 1.10.3. Assume that the functions* Φ *and* Ψ *are continuously totally differentiable. Then,*

$$\frac{d}{d\tau}\Phi_{\dot{x}}(\tilde{x}, \frac{d}{d\tau}\tilde{x}) = \Phi_x(\tilde{x}, \frac{d}{d\tau}\tilde{x}) + \sum_{i=1}^{m} \tilde{\lambda}_i \nabla \Psi_i(\tilde{x})$$

(2.5.63)

holds on $[\tau_a, \tau_b]$ *if and only if*

$$\frac{d}{dt}\Phi_{\dot{x}}(x, \dot{x}) = \Phi_x(x, \dot{x}) + \sum_{i=1}^{m} \lambda_i \nabla \Psi_i(x)$$

(2.5.64)

holds on $[t_a, t_b]$. *The Lagrange multipliers change as follows:*

$$\tilde{\lambda}_i(\tau) = \lambda_i(\varphi(\tau))\frac{d\varphi}{d\tau}(\tau) \quad \text{for} \quad \tau \in [\tau_a, \tau_b].$$

(2.5.65)

Proof. Relations like (1.10.14), following from (1.10.20) by differentiation, give identities, which are analogous to (1.10.15) and (1.10.16). By $\frac{d}{d\tau}\tilde{x}(\tau) = \dot{x}(\varphi(\tau))\frac{d\varphi}{d\tau}(\tau)$ with $\frac{d\varphi}{d\tau}(\tau) > 0$, one obtains

$$\frac{d}{d\tau}\Phi_{\dot{x}}(\tilde{x}(\tau),\frac{d}{d\tau}\tilde{x}(\tau)) - \Phi_x(\tilde{x}(\tau),\frac{d}{d\tau}\tilde{x}(\tau))$$

$$= \frac{d}{d\tau}\Phi_{\dot{x}}(x(\varphi(\tau)),\dot{x}(\varphi(\tau))) - \Phi_x(x(\varphi(\tau)),\dot{x}(\varphi(\tau)))\frac{d\varphi}{d\tau}(\tau) \qquad (2.5.66)$$

$$= \left(\frac{d}{dt}\Phi_{\dot{x}}(x(\varphi(\tau)),\dot{x}(\varphi(\tau))) - \Phi_x(x(\varphi(\tau)),\dot{x}(\varphi(\tau)))\right)\frac{d\varphi}{d\tau}(\tau),$$

proving the claimed equivalence. $\qquad\qquad\square$

A holonomic constraint requires $x(t_a), x(t_b) \in M$. If no further boundary conditions are prescribed, local constrained minimizers fulfill natural boundary conditions.

Proposition 2.5.3. *Let the curve $x \in (C^2[t_a,t_b])^n$ be a local minimizer of the functional (2.5.1) subject to the holonomic constraint (2.5.2). Assume the differentiability conditions on Φ and Ψ of Proposition 2.5.1. Then, x fulfills the system (2.5.8), and without further restrictions on the boundaries of x at $t = t_a$ and/or $t = t_b$, the natural boundary conditions hold:*

$$\Phi_{\dot{x}}(x(t_a),\dot{x}(t_a)) \in N_{x(t_a)}M \quad \text{and/or} \quad \Phi_{\dot{x}}(x(t_b),\dot{x}(t_b)) \in N_{x(t_b)}M, \qquad (2.5.67)$$

respectively.

Proof. Proposition 2.5.1 holds regardless of any boundary conditions, meaning that x fulfills the system (2.5.8). Assume now no further restriction on $x(t_a) \in M$. Choosing $h \in (C^2[t_a,t_b])^n$ with arbitrary $h(t_a)$ and $h(t_b) = 0$, we obtain in (2.5.10) a curve $a(t)$ with an arbitrary vector $a(t_a) \in T_{x(t_a)}M$ and $a(t_b) = 0$. Constructing a perturbation like in (2.5.12)–(2.5.21), we end up with $h(s,t)$ having the properties (2.5.21) except that in place of $h(s,t_a) = 0$, we have $x(t_a) + h(s,t_a) \in M$. Using this perturbation, one obtains from (2.5.23) after intergration by parts,

$$\int_{t_a}^{t_b}(\Phi_x(x,\dot{x}) - \frac{d}{dt}\Phi_{\dot{x}}(x,\dot{x}),a)dt - (\Phi_{\dot{x}}(x(t_a),\dot{x}(t_a)),a(t_a)) = 0. \qquad (2.5.68)$$

By $\Phi_x(x(t),\dot{x}(t)) - \frac{d}{dt}\Phi_{\dot{x}}(x(t),\dot{x}(t)) = -\sum_{i=1}^{m}\lambda_i(t)\nabla\Psi_i(x(t)) \in N_{x(t)}M$, the integral in (2.5.68) vanishes due to $a(t) \in T_{x(t)}M$. Hence,

$$(\Phi_{\dot{x}}(x(t_a),\dot{x}(t_a)),a(t_a)) = 0 \quad \text{for arbitrary} \quad a(t_a) \in T_{x(t_a)}M, \qquad (2.5.69)$$

which proves the natural boundary condition (2.5.67) at $t = t_a$. $\qquad\square$

Differentiating the holonomic constraint $\Psi(x(t)) = 0$ for $t \in [t_a,t_b]$ gives

$$D\Psi(x(t))\dot{x}(t) = 0,$$
$$\text{or} \quad \dot{x}(t) \in \text{Kern } D\Psi(x(t)) = T_{x(t_a)}M \quad \text{for} \quad t \in [t_a,t_b]. \qquad (2.5.70)$$

We study that restriction for the mechanical model (2.5.28)–(2.5.32): The natural boundary conditions together with (2.5.70) yield

$$L_{\dot{x}}(x(t_a), \dot{x}(t_a)) \in N_{x(t_a)}M \quad \text{and}$$
$$\dot{x}(t_a) \in T_{x(t_a)}M. \tag{2.5.71}$$

By orthogonality, (2.5.71) implies for $L(x, \dot{x}) = T(\dot{x}) - V(x)$

$$(L_{\dot{x}}(x(t_a), \dot{x}(t_a)), \dot{x}(t_a)) = \sum_{k=1}^{N} m_k(\dot{x}_k^2 + \dot{y}_k^2 + \dot{z}_k^2) = 0, \quad \text{and hence,}$$
$$\dot{x}(t_a) = 0. \tag{2.5.72}$$

With or without holonomic constraints, the velocity vanishes at a free boundary.

We generalize (2.5.1) to an explicit dependence on the independent variable, i.e.,

$$J(x) = \int_{t_a}^{t_b} \Phi(t, x, \dot{x}) dt, \tag{2.5.73}$$

defined on $D \subset (C^1[t_a, t_b])^n$. We further assume that the holonomic constraints now depend explicitly on t:

$$\Psi(t, x) = 0, \tag{2.5.74}$$

where $\Psi : [t_a, t_b] \times \mathbb{R}^n \to \mathbb{R}^m$. Since reparametrizations are not an issue for these variational problems, we treat them in a different notation, namely

$$J(y) = \int_a^b F(x, y, y') dx, \tag{2.5.75}$$

and

$$G(x, y) = 0, \tag{2.5.76}$$

where $G : [a, b] \times \mathbb{R}^n \to \mathbb{R}^m$ with $n > m$. For a function $y : [a, b] \to \mathbb{R}^n$, the constraints (2.5.76) read $G(x, y(x)) = 0$ for $x \in [a, b]$.

We assume that F is twice and that G is three times continuously partially differentiable with respect to all variables, and that

$$D_y G(x, y) = \left(\frac{\partial G_i}{\partial y_j}(x, y) \right)_{\substack{i=1,\dots,m \\ j=1,\dots,n}} \in \mathbb{R}^{m \times n} \tag{2.5.77}$$

has the maximal rank m for all $(x, y) \in [a, b] \times \mathbb{R}^n$ where $G(x, y) = 0$. As shown in the Appendix, for each $x \in [a, b]$,

$M_x = \{y \in \mathbb{R}^n | G(x,y) = 0\}$ is a $(n-m)$ -dimensional continuously differentiable manifold, and

$$M = \bigcup_{x \in [a,b]} (\{x\} \times M_x) \quad \text{is a} \quad (n+1-m) \text{ -dimensional} \qquad (2.5.78)$$

continuously differentiable manifold

with boundary $(\{a\} \times M_a) \cup (\{b\} \times M_b)$.

Observe that the Jacobian matrix $DG(x,y) \in \mathbb{R}^{m \times (n+1)}$ has rank m, which is maximal.

By the holonomic constraints $G(x,y) = 0$, the graphs of functions $y \in D \subset (C^1[a,b])^n$ belong to M and $y(x) \in M_x$ for $x \in [a,b]$. Additional boundary conditions can be imposed on $\{a\} \times M_a$ and $\{b\} \times M_b$, respectively.

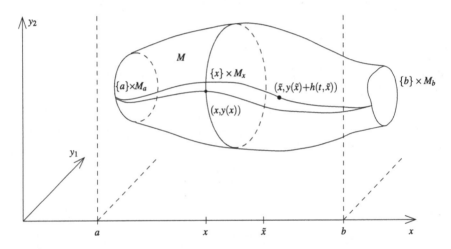

Fig. 2.9 An Admitted Perturbation on a Manifold

Proposition 2.5.4. *Let $y \in D \subset (C^2[a,b])^n$ be a local minimizer of the functional (2.5.75), subject to the holonomic constraints (2.5.76). In particular,*

$$J(y) \leq J(\tilde{y}) \quad \text{for all} \quad \tilde{y} \in D \cap \{G(x,y) = 0\}$$
$$\text{satisfying} \quad \max_{k=1,\dots,n} \|y_k - \tilde{y}_k\|_1 < d. \qquad (2.5.79)$$

Then, there is a continuous function

$$\lambda = (\lambda_1,\dots,\lambda_m) : [a,b] \to \mathbb{R}^m, \qquad (2.5.80)$$

such that y is a solution of the following system of differential equations:

$$\frac{d}{dx}F_{y'}(\cdot,y,y') = F_y(\cdot,y,y') + \sum_{i=1}^{m}\lambda_i\nabla_y G_i(\cdot,y) \quad on \quad [a,b]. \qquad (2.5.81)$$

Here, we employ the notation $F_{y'} = (F_{y'_1},\dots,F_{y'_n})$, $F_y = (F_{y_1},\dots,F_{y_n})$, $y = (y_1,\dots,y_n)$, $y' = (y'_1,\dots,y'_n)$, *and* $\nabla_y G_i = (G_{i,y_1},\dots,G_{i,y_n})$.

Proof. Since the proof is analogous to that of Proposition 2.5.1, we merely give a sketch. As a first step, we construct an admitted perturbation of the minimizing function y in $D \cap \{G(x,y) = 0\}$ such that $y(x) + h(t,x) \in M_x$. For that purpose, we use the orthogonal projections

$$P_x(y) : \mathbb{R}^n \to T_y M_x \quad \text{and} \quad Q_x(y) : \mathbb{R}^n \to N_y M_x$$
on the tangential and normal spaces to M_x $\qquad (2.5.82)$
in $y \in M_x$, $x \in [a,b]$, respectively.

For any $h \in (C_0^2[a,b])^n$,

$$f(x) = P_x(y(x))h(x) \in T_{y(x)}M_x \quad \text{for all} \quad x \in [a,b],$$
$$f \in (C^2[a,b])^n \quad \text{and} \quad f(a) = 0, \ f(b) = 0. \qquad (2.5.83)$$

A construction analogous to (2.5.12)–(2.5.21) (with a function $H(t,x,z)=G(x,y(x)+tf(x)+Q_x(y(x))z)$ for $(t,x,z) \in \mathbb{R} \times [a,b] \times N_{y(x_0)}M_{x_0}$) yields:

There is a twice continuously differentiable perturbation
$h : (-\varepsilon_0,\varepsilon_0) \times [a,b] \to \mathbb{R}^n$,   satisfying
$y(x) + h(t,x) \in M_x$ for $(-\varepsilon_0,\varepsilon_0) \times [a,b]$.
The perturbation is of the form $h(t,x) = tf(x) + g(t,x)$,   where
$f(x) \in T_{y(x)}M_x$, $g(t,x) \in N_{y(x)}M_x$, $\qquad (2.5.84)$
$h(t,a) = 0$, $h(t,b) = 0$ for $t \in (-\varepsilon_0,\varepsilon_0)$,
$h(0,x) = 0$, $\dfrac{\partial}{\partial t}h(0,x) = f(x)$,   and $\dfrac{\partial^2}{\partial t \partial x}h(0,x) = f'(x)$
for $x \in [a,b]$.

Since $J(y + h(t,\cdot))$ is locally minimal at $t = 0$,

$$\frac{d}{dt}J(y+h(t,\cdot))|_{t=0} = 0 \quad \text{or explicitly}$$

$$\int_a^b \sum_{k=1}^n (F_{y_k}(\cdot,y,y')\frac{\partial}{\partial t}h_k(0,\cdot) + F_{y_k'}(\cdot,y,y')\frac{\partial^2}{\partial t\partial x}h_k(0,\cdot))dx \tag{2.5.85}$$

$$= \int_a^b (F_y(\cdot,y,y'),f) + (F_{y'}(\cdot,y,y'),f')dx = 0.$$

Integration by parts gives

$$\int_a^b (F_y(\cdot,y,y') - \frac{d}{dx}F_{y'}(\cdot,y,y'),f)dx = 0 \tag{2.5.86}$$

$$\text{for all} \quad f \in (C_0^2[a,b])^n \quad \text{satisfying} \quad f(x) \in T_{y(x)}M_x.$$

As in (2.5.25), the properties of the tangent and normal space and their projections entail

$$Q_x(y(x))(\frac{d}{dx}F_{y'}(x,y(x),y'(x)) - F_y(x,y(x),y'(x))) = \sum_{i=1}^m \lambda_i(x)\nabla_y G_i(x,y(x)), \tag{2.5.87}$$

where the functions λ_i are uniquely determined and continuous on $[a,b]$. Let $h \in (C_0^2[a,b])^n$ be arbitrary. Then, the decomposition

$$h(x) = P_x(y(x))h(x) + Q_x(y(x))h(x) = f(x) + g(x), \quad \text{where} \tag{2.5.88}$$
$$f(x) \in T_{y(x)}M_x \quad \text{and} \quad g(x) \in N_{y(x)}M_x,$$

shows by the same arguments given after (2.5.27) that

$$\int_a^b (\frac{d}{dx}F_{y'}(\cdot,y,y') - F_{y'}(\cdot,y,y') - \sum_{i=1}^m \lambda_i\nabla_y G_i(\cdot,y),h)dx = 0 \tag{2.5.89}$$

holds. Since $h \in (C_0^2[a,b])^n$ is arbitrary, this proves the claim (2.5.81). □

If for admitted functions $y \in D \cap \{G(x,y) = 0\}$ only $y(a) \in M_a$ and $y(b) \in M_b$ are required, then local constrained minimizers fulfill natural boundary conditions.

Proposition 2.5.5. *Let $y \in (C^2[a,b])^n$ be a local minimizer of the functional (2.5.75) under the holonomic constraints (2.5.76). Assume the same differentiability properties of F and G as given in Proposition 2.5.4. If no further boundary conditions are imposed on y at $x = a$ and/or $x = b$, then y fulfills the natural boundary conditions*

$$F_{y'}(a,y(a),y'(a)) \in N_{y(a)}M_a \quad \text{and/or}$$
$$F_{y'}(b,y(b),y'(b)) \in N_{y(b)}M_b, \quad \text{respectively.} \tag{2.5.90}$$

Proof. Suppose that no further boundary condition is imposed at $x = a$. We construct perturbations $h(t,x)$ as in (2.5.84), but now $f(a) \in T_{y(a)}M_a$ is arbitrary and $f(b) = 0$.

Then, the perturbation is only constrained by $y(a) + h(t,a) \in M_a$. Integration by parts of (2.5.85) gives

$$\int_a^b \left(F_y(\cdot,y,y') - \frac{d}{dx}F_{y'}(\cdot,y,y'), f\right)dx - \left(F_{y'}(a,y(a),y'(a)), f(a)\right) = 0, \quad (2.5.91)$$

by $f(b) = 0$. By (2.5.81), the integral vanishes, and since $f(a) \in T_{y(a)}M_a$ is arbitrary, the vector $F_{y'}(a,y(a),y'(a))$ is in the orthogonal complement of $T_{y(a)}M_a$, which is $N_{y(a)}M_a$.   □

The holonomic constraints $G(x,y(x)) = 0$ imply

$$G_x(x,y(x)) + D_y G(x,y(x))y'(x) = 0 \quad \text{for} \quad x \in [a,b], \quad (2.5.92)$$

which gives additional constraints at the boundaries $x = a$ and $x = b$.

Exercises

2.5.1. Adopt the notation employed in (2.5.28)–(2.5.32), and show that the total energy $E(x,\dot{x}) = T(\dot{x}) + V(x)$ is constant along a trajectory that fulfills both the holonomic constraints and the Euler-Lagrange equations of the action.

2.5.2. Compute the trajectory $x = x(t)$ of a point mass m running on the inclined plane $x_1 + x_3 - 1 = 0$ from $x(0) = (0,0,1)$ with initial speed $\dot{x}(0) = (0,v_2,0)$ in the plane $x_3 = 0$. The only force acting on the point mass is the gravity g in direction of the negative x_3-axis. Compute the running time and compare it to the time of a free fall from $(0,0,1)$ to $(0,0,0)$. Does the running time depend on the initial speed $(0,v_2,0)$?

2.5.3. Compute the Euler-Lagrange equations for the spherical pendulum moving on the sphere $x_1^2 + x_2^2 + x_3^2 - \ell^2 = 0$ under gravity that acts in direction of the negative x_3-axis.
Compute also the Euler-Lagrange equations with respect to the generalized coordinates $\varphi \in [0,2\pi]$ and $\theta \in [-\frac{\pi}{2}, \frac{\pi}{2}]$, where

$$x_1 = r\sin\theta\cos\varphi,$$
$$x_2 = r\sin\theta\sin\varphi,$$
$$x_3 = r\cos\theta,$$

are the spherical coordinates.

2.5.4. Show that any solution $x = x(t)$ of the Euler-Lagrange equations for the spherical pendulum fulfills the conservation law

$$\frac{d}{dt}(x_1\dot{x}_2 - \dot{x}_1 x_2) = 0, \quad \text{or} \quad x_1\dot{x}_2 - \dot{x}_1 x_2 = c,$$

and gives a geometric interpretation.

Hint: Use formula (2.2.4) along x and describe the area that it bounds.

2.6 Geodesics

We show in the Appendix that for a continuously totally differentiable mapping $\Psi : \mathbb{R}^n \to \mathbb{R}^m$, where $n > m$, the nonempty zero set $M = \{x \in \mathbb{R}^n \,|\, \Psi(x) = 0\}$ is a continuously differentiable manifold of dimension $n - m$.

Definition 2.6.1. *A shortest path connection between two points on a manifold M is called a geodesic.*

A geodesic is a curve $x \in D = (C^1[t_a, t_b])^n \cap \{x(t_a) = A, x(t_b) = B\}$ that minimizes the functional

$$J(x) = \int_{t_a}^{t_b} \|\dot{x}\| dt, \tag{2.6.1}$$

measuring its length, under the holonomic constraint

$$\Psi(x) = 0. \tag{2.6.2}$$

The hypotheses of Proposition 2.5.1 are that Ψ is three times continuously partially differentiable and that admitted curves are twice continuously differentiable. Furthermore, since the Lagrange function $\Phi(\dot{x}) = \|\dot{x}\|$ is not differentiable at $\dot{x} = 0$, only curves x with nonvanishing tangent vectors are admitted.

By Proposition 2.5.1, a geodesic has to fulfill the system of Euler-Lagrange equations

$$\frac{d}{dt} \frac{\dot{x}}{\|\dot{x}\|} = \sum_{i=1}^{m} \lambda_i \nabla \Psi_i(x) \quad \text{on} \quad [t_a, t_b], \tag{2.6.3}$$

where the continuous Lagrange multipliers $\lambda = (\lambda_1, \ldots, \lambda_m) \in (C[t_a, t_b])^m$ depend on t.

Since the functional (2.6.1) is invariant, we can apply Proposition 2.5.2. A suitable reparametrization simplifies the system (2.6.3) as follows: The arc length

$$s(t) = \int_{t_a}^{t} \|\dot{x}(\sigma)\| d\sigma, \quad \dot{s}(t) = \|\dot{x}(t)\| > 0, \tag{2.6.4}$$

has an inverse function $t = \varphi(s)$ which is a continuously differentiable mapping $\varphi : [0, L] \to [t_a, t_b]$. Here, $L = L(x)$ is the length of the admitted curve $x \in (C^2[t_a, t_b])^n$.

A reparametrization of x by the arc length

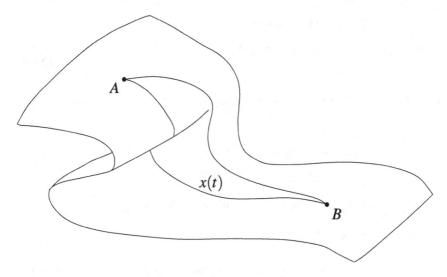

Fig. 2.10 Path Connections on a Manifold

$$x(\varphi(s)) = \tilde{x}(s), \quad \text{where}$$
$$\frac{d\varphi}{ds}(s) = \frac{1}{\dot{s}(\varphi(s))} > 0, \tag{2.6.5}$$

is admitted according to Definition 1.10.3, and relations (2.6.4), (2.6.5) imply

$$\left\| \frac{d\tilde{x}}{ds}(s) \right\| = \|\dot{x}(\varphi(s))\| \frac{d\varphi}{ds}(s) = 1. \tag{2.6.6}$$

Observe that the interval of the new parameter is given by the length of the curve. Nonetheless Proposition 2.5.2 is applicable, since the equivalence of (2.5.63) and of (2.5.64) holds for any fixed curve.

Due to (2.6.6), a geodesic parametrized by its arc length has to fulfill the following system, where we replace s by t and omit the tilde:

$$\ddot{x} = \sum_{i=1}^{m} \lambda_i \nabla \Psi_i(x) \quad \text{on} \quad [0, L], \tag{2.6.7}$$
$$\|\dot{x}(t)\| = 1.$$

Examples:

We compute the geodesics on a sphere S_R in $\mathbb{R}^3$ given by

$$\Psi(x) = x_1^2 + x_2^2 + x_3^2 - R^2 = 0. \tag{2.6.8}$$

According to (2.6.7), we have to solve

$$\ddot{x} = 2\lambda x,$$
$$\dot{x}_1^2 + \dot{x}_2^2 + \dot{x}_3^2 = 1. \tag{2.6.9}$$

We differentiate (2.6.8) twice, and we obtain by (2.6.9)

$$2(x_1\dot{x}_1 + x_2\dot{x}_2 + x_3\dot{x}_3) = 0,$$
$$\dot{x}_1^2 + \dot{x}_2^2 + \dot{x}_3^2 + x_1\ddot{x}_1 + x_2\ddot{x}_2 + x_3\ddot{x}_3 = 0,$$
$$\text{or} \quad 1 + 2\lambda(x_1^2 + x_2^2 + x_3^2) = 1 + 2\lambda R^2 = 0, \tag{2.6.10}$$
$$\text{or} \quad 2\lambda = -\frac{1}{R^2}, \quad \text{and (2.6.9)}_1 \text{ becomes finally} \quad \ddot{x} + \frac{1}{R^2}x = 0.$$

The general solution of $(2.6.10)_4$ is given by

$$x(s) = a\cos\frac{1}{R}s + b\sin\frac{1}{R}s \quad \text{where} \quad a,b \in \mathbb{R}^3. \tag{2.6.11}$$

The constraints imply

$$\|x(0)\| = \|a\| = R, \quad \|\dot{x}(0)\| = \frac{1}{R}\|b\| = 1,$$
$$(x(0),\dot{x}(0)) = (a,\frac{1}{R}b) = 0 \quad \text{by } (2.6.10)_1. \tag{2.6.12}$$

The vectors a and b have length R, and they are orthogonal. Let $c \in \mathbb{R}^3$ be a third vector (of length 1) that is orthogonal to a and to b. Then,

$$(x(s),c) = 0 \quad \text{for all} \quad s \in [0,L]. \tag{2.6.13}$$

On the other hand, the set

$$E = \{x \in \mathbb{R}^3 | (x,c) = 0\}, \tag{2.6.14}$$

describes a plane in $\mathbb{R}^3$ containing the center 0 of the sphere S_R. This gives the result that any geodesic fulfills

$$x(s) \in S_R \cap E \quad \text{for all} \quad s \in [0,L]. \tag{2.6.15}$$

The intersection of the sphere S_R and of the plane E through the center 0 is called a **great circle**.

Given two points A and B on the sphere S_R, the three points 0, A, and B span a plane E, and the geodesic between A and B runs in $S_R \cap E$. The points A and B are connected by two arcs on a great circle, one of which is shorter than the other, or both have the same length if A and B are antipodes.

Shortest connections between two points on a sphere are important for the route planning of flights. If, for instance, two cities in Europe and North America are on the same circle of latitude, then the aircraft does not follow that circle of latitude, but rather it takes a seeming detour over the North Atlantic.

If the manifold M is described by generalized coordinates, the geodesics are minimizers of a variational problem without holonomic constraints. We discuss an example of a surface of revolution in $\mathbb{R}^3$:

$$M = \{(r\cos\varphi,\ r\sin\varphi, f(r)) | 0 \leq r_1 < r < r_2 \leq \infty, \varphi \in [0, 2\pi]\}. \qquad (2.6.16)$$

We assume that f is twice continuously differentiable, and for $r_1 = 0$, that $f'(0) \neq 0$. Thus, the surface M has a cusp at $(0, 0, f(0))$, cf. Figure 2.11.

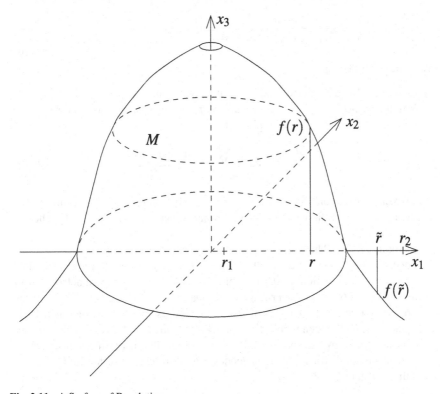

Fig. 2.11 A Surface of Revolution

Given two points $A=(r_a\cos\varphi_a,\ r_a\sin\varphi_a, f(r_a))$ and $B=(r_b\cos\varphi_b,\ r_b\sin\varphi_b, f(r_b))$ on M, we determine the geodesic between A and B. For that purpose, we parametrize connecting curves on M as follows:

$$\{x(t) = (r(t)\cos\varphi(t),\ r(t)\sin\varphi(t), f(r(t))) | t \in [t_a, t_b]\} \qquad (2.6.17)$$

where $r(t_a) = r_a$, $\varphi(t_a) = \varphi_a$, $r(t_b) = r_b$, $\varphi(t_b) = \varphi_b$. The length of the curves is given by the functional

$$J(x) = \int_{t_a}^{t_b} \sqrt{\dot{x}_1^2 + \dot{x}_2^2 + \dot{x}_3^2}\, dt$$

$$= \int_{t_a}^{t_b} \sqrt{\dot{r}^2(1 + (f'(r))^2) + r^2\dot{\varphi}^2}\, dt = J(r, \varphi), \tag{2.6.18}$$

with Lagrange function $\Phi(r, \varphi, \dot{r}, \dot{\varphi}) = \sqrt{\dot{r}^2(1 + (f'(r))^2) + r^2\dot{\varphi}^2}$. Assuming that the tangent $\dot{x}(t)$ does not vanish, the Euler-Lagrange equations for a minimizer read as follows:

$$\frac{d}{dt}\Phi_{\dot{r}} = \frac{d}{dt}\frac{\dot{r}(1 + (f'(r))^2)}{\sqrt{\dot{r}^2(1 + (f'(r))^2) + r^2\dot{\varphi}^2}} = \frac{\dot{r}^2 f'(r)f''(r) + r\dot{\varphi}^2}{\sqrt{\dot{r}^2(1 + (f'(r))^2) + r^2\dot{\varphi}^2}} = \Phi_r,$$

$$\frac{d}{dt}\Phi_{\dot{\varphi}} = \frac{d}{dt}\frac{r^2\dot{\varphi}}{\sqrt{\dot{r}^2(1 + (f'(r))^2) + r^2\dot{\varphi}^2}} = 0 = \Phi_\varphi, \quad t \in [t_a, t_b].$$

$$\tag{2.6.19}$$

We distinguish two cases:

1. The points A and B have the same angular coordinates $\varphi_a = \varphi_b$.
    With the ansatz $\varphi(t) = \varphi_a = \varphi_b$ for all $t \in [t_a, t_b]$, the system (2.6.19) for $r = r(t)$ with $\dot{r}(t) \neq 0$ and $\dot{\varphi}(t) = 0$ simplifies to

    $$\frac{d}{dt}\sqrt{1 + (f'(r))^2} = \frac{\dot{r} f'(r)f''(r)}{\sqrt{1 + (f'(r))^2}}. \tag{2.6.20}$$

    Any parametrization (2.6.17) with a continuously differentiable function $r(t)$ fulfills (2.6.20), and the geodesic is a meridian $\{(r\cos\varphi_a, r\sin\varphi_a, f(r))\}$ where r runs from r_a to r_b.
2. The points A and B have different angular coordinates $\varphi_a \neq \varphi_b$.
    Since the Lagrange function Φ in (2.6.18) is invariant according to Definition 1.10.2, Proposition 1.10.4 is applicable, allowing any reparametrization in the system (2.6.19) of Euler-Lagrange equations. Choosing the arc length as a parameter (cf. (2.6.4)–(2.6.6)), the denominators in (2.6.19) are equal to 1. Equation (2.6.19)$_2$ then yields $r^2\dot{\varphi} = c \neq 0$, since the angular coordinate φ is not constant in this case. Consequently, φ is monotonically increasing or decreasing. This means geometrically that the geodesic winds round the surface of revolution with a constant angular velocity.

Due to $\dot{\varphi}(t) \neq 0$ for all $t \in [t_a, t_b]$, there exists the inverse function $t = \tau(\varphi)$, and $r(t) = r(\tau(\varphi)) = \tilde{r}(\varphi)$ is now parametrized by $\varphi \in [\varphi_a, \varphi_b]$, assuming that $\varphi_a < \varphi_b$.

For a curve (2.6.17) parametrized by φ, the Lagrange function of the functional (2.6.18) transforms as follows: Replace r by $\tilde{r}$, $\dot{r}$ by $\frac{d\tilde{r}}{d\varphi}$, and $\dot{\varphi}$ by 1, and integrate over $[\varphi_a, \varphi_b]$. By Proposition 1.10.1, the functional describes the length as before. Omitting the tilde and setting $\frac{dr}{d\varphi} = r'$, we obtain the new Lagrange function $\Phi(r, r') = \sqrt{(r')^2(1 + (f'(r))^2) + r^2}$. Since it does not depend explicitly on the parameter φ, we can proceed as in case 7 of Paragraph 1.5: Any solution of $\Phi(r, r') - r'\Phi_{r'} = c_1$ is also a solution of the Euler-Lagrange equation, or it is constant. In this case, we obtain

$$\frac{r^2}{\sqrt{(r')^2(1+(f'(r))^2)+r^2}} = c_1 \quad \text{for} \quad r = r(\varphi), \quad \varphi \in [\varphi_a, \varphi_b]. \quad (2.6.21)$$

We establish a geometric property of a curve $x(\varphi) = (x_1(\varphi), x_2(\varphi), x_3(\varphi)) = (r(\varphi)\cos\varphi, r(\varphi)\sin\varphi, f(r(\varphi)))$ satisfying (2.6.21). For that purpose, we use the tangent

$$x'(\varphi) = (r'(\varphi)\cos\varphi - r(\varphi)\sin\varphi, \ r'(\varphi)\sin\varphi + r(\varphi)\cos\varphi, \ f'(r(\varphi))r'(\varphi)).$$
$$(2.6.22)$$

The circle of latitude on M intersecting the curve in $x(\varphi)$ is

$$\{z(\psi) = (r(\varphi)\cos\psi, r(\varphi)\sin\psi, \ f(r(\varphi)))|\psi \in [0, 2\pi]\}, \quad (2.6.23)$$

with tangent in $x(\varphi)$ given by

$$z'(\varphi) = (-r(\varphi)\sin\varphi, r(\varphi)\cos\varphi, 0), \quad (\ )' = \frac{d}{d\psi}. \quad (2.6.24)$$

The scalar product of the two tangents (2.6.22) and (2.6.24) in $\mathbb{R}^3$ yields

$$(x'(\varphi), z'(\varphi)) = r(\varphi)^2 = \|x'(\varphi)\| \ \|z'(\varphi)\| \cos\beta(\varphi), \quad (2.6.25)$$

where $\beta(\varphi)$ is the angle between the vectors $x'(\varphi)$ and $z'(\varphi)$. This is the equivalent geometric definition of the scalar product. Now, $\|x'(\varphi)\| = \Phi(r(\varphi), r'(\varphi)) = \sqrt{(r'(\varphi))^2(1+(f'(r(\varphi)))^2)+(r(\varphi))^2}$ and $\|z'(\varphi)\| = r(\varphi)$. Combining this with (2.6.21) and (2.6.25), we finally obtain

$$r(\varphi)\cos\beta(\varphi) = c_1 \quad \text{for all} \quad \varphi \in [\varphi_a, \varphi_b]. \quad (2.6.26)$$

Formula (2.6.26) is called **Clairaut's relation** (Clairaut, 1713–1765). It relates the radius of a circle of latitude on M and the angle of an intersecting geodesic between the circle of latitude. If, for instance, $r(\varphi)$ decreases, then the angle $\beta(\varphi)$ decreases as well, and for $r(\varphi) = c_1$, Clairaut's relation gives $\beta(\varphi) = 0$. Apparently $r(\varphi) \geq c_1$ for all $\varphi \in [\varphi_a, \varphi_b]$, and $r(\varphi) = c_1$ for a single value of φ only. (The Euler-Lagrange system (2.6.19) has no locally constant solution $r = c_1$ for φ in an interval; observe that (2.6.21) is not equivalent to the Euler-Lagrange equation for the Lagrange function $\Phi(r, r')$).

The geodesic connecting the points A and B, sketched in Figure 2.12, is continued as a solution of (2.6.21) beyond the interval $[\varphi_a, \varphi_b]$. It satisfies $r \geq c_1$ globally, and it cannot form a loop since in this case it is tangent to a second circle of latitude with a radius $r > c_1$. By $\cos\beta = 1$ in the point of contact, Clairaut's relation (2.6.26) would be violated. The global geodesic winds around the surface of revolution and disappears finally at "infinity." This is to be shown in Exercise 2.6.2 for a cone: The geodesic comes from infinity, winds around the cone, touches a circle of latitude, and winds around to infinity again.

There is another type of a surface of revolution other than (2.6.16), namely

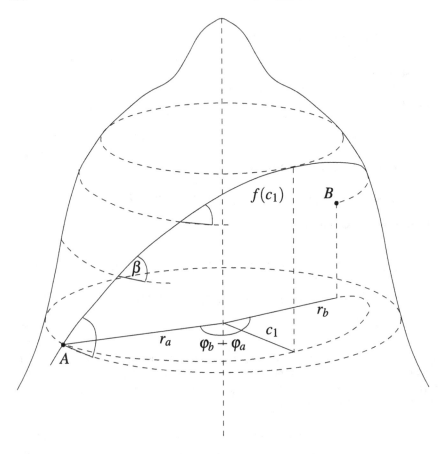

Fig. 2.12 A Geodesic on a Surface of Revolution

$$M = \{(g(x_3)\cos\varphi, g(x_3)\sin\varphi, x_3)|-\infty \le c < x_3 < d \le \infty\}. \qquad (2.6.27)$$

The function $g : (c,d) \to \mathbb{R}$ is assumed to be positive and twice continuously differentiable. In contrast to the surface depicted in Figure 2.12 that surface of revolution might also narrow downward, and in this case the condition $r \ge c_1$ means also a bound of the geodesic from below. Since the surface between two circles of latitude is compact, the global geodesic exists for $\varphi \in (-\infty, \infty)$ and it winds infinitely often around the surface, and it approaches each point of it arbitrarily close. A detailed discussion can be found in [13], p. 138.

Exercises

2.6.1. Compute the geodesic from $A = (1,0,0)$ to $B = (-1,0,1)$ on the cylinder $Z = \{x = (x_1,x_2,x_3)|x_1^2 + x_2^2 = 1\}$. Is it unique?

2.6.2. Consider the cone $K = \{x = (x_1,x_2,x_3)|x_1^2 + x_2^2 = x_3^2, x_3 \geq 0\} = \{(r\cos\varphi, r\sin\varphi, r)|0 \leq r < \infty, \varphi \in [0,2\pi]\}$. Assuming that geodesics exist globally on K, show that all global geodesics except meridians are unbounded in both direction. To be more precise: Fix a point $x_0 = x(t_0)$ on a geodesic $x = x(t)$ and parametrize the geodesic in direction $t \geq t_0$ by the arc length and in direction $t \leq t_0$ by the negative arc length. Then, $x = \tilde{x}(s)$ for $s \in (s_-, s_+)$, $\tilde{x}(0) = x_0$. Since segments of the geodesic of finite length can be continued, we can assume that $s_- = -\infty$ and $s_+ = +\infty$. The global geodesic $\{x(s)|s \in (-\infty,\infty)\}$ satisfies the Euler-Lagrange equation in Cartesian coordinates with a holonomic constraint or in generalized coordinates without constraint. Show that $\lim_{s\to\infty} \|x(s)\| = \lim_{s\to\infty} \sqrt{2}r(s) = \infty$ and $\lim_{s\to-\infty} \|x(s)\| = \lim_{s\to-\infty} \sqrt{2}r(s) = \infty$. Make use of the fact that the Euler-Lagrange equations simplify for curves parametrized by the (positive or negative) arc length.

2.7 Nonholonomic Constraints

In physics, in particular in mechanics, a constraint is called nonholonomic if it restricts not only the coordinates of position but also of velocity.

Definition 2.7.1. *For a parametric functional*

$$J(x) = \int_{t_a}^{t_b} \Phi(t,x,\dot{x})dt \tag{2.7.1}$$

or, in a different notation, for a nonparametric functional

$$J(y) = \int_a^b F(x,y,y')dx, \tag{2.7.2}$$

defined on $D \subset (C^1[t_a,t_b])^n$ or on $D \subset (C^1[a,b])^n$, respectively, the constraints

$$\begin{aligned} \Psi(t,x,\dot{x}) &= 0 \quad \text{for } t \in [t_a,t_b] \text{ or} \\ G(x,y,y') &= 0 \quad \text{for } x \in [a,b], \text{ respectively,} \end{aligned} \tag{2.7.3}$$

are called nonholonomic constraints. The functions

$$\begin{aligned} \Psi &: [t_a,t_b] \times \mathbb{R}^n \times \mathbb{R}^n \to \mathbb{R}^m \\ \text{and} \quad G &: [a,b] \times \mathbb{R}^n \times \mathbb{R}^n \to \mathbb{R}^m \quad \text{for} \quad n > m, \end{aligned} \tag{2.7.4}$$

are three times continuously partially differentiable with respect to all variables, and the matrices

$$D_p\Psi(t,x,p) = \left(\frac{\partial \Psi_i}{\partial p_j}(t,x,p)\right)_{\substack{i=1,\ldots,m \\ j=1,\ldots,n}}$$

$$and \quad D_pG(x,y,p) = \left(\frac{\partial G_i}{\partial p_j}(x,y,p)\right)_{\substack{i=1,\ldots,m \\ j=1,\ldots,n}}$$

(2.7.5)

have maximal rank m at all points $(t,x,p) \in [t_a,t_b] \times \mathbb{R}^n \times \mathbb{R}^n$ and all $(x,y,p) \in [a,b] \times \mathbb{R}^n \times \mathbb{R}^n$ that solve (2.7.3) with $\dot{x} = p$ and $y' = p$, respectively.

Apparently, the parametric and nonparametric versions are equivalent, and they differ only by notation. We give the general results only for (2.7.2) with (2.7.3)$_2$.

The nonholonomic constraints (2.7.3)$_2$ are no longer geometric constraints on the coordinates of admitted functions as are holonomic constraints, but they link the coordinates of the functions to their derivatives. Such equations are differential equations, but the form (2.7.3)$_2$ differs from those that are commonly taught in courses on ordinary differential equations: They are implicit differential equations. Nonetheless condition (2.7.5) implies by the implicit function theorem that near a solution, equation (2.7.3) is solved for m components of the derivative. In the resulting system of m explicit differential equations, the remaining $n - m$ components of the derivative play the role of "free parameters."

The theory of nonholonomic constraints does not give the existence of solutions of (2.7.3) on the entire interval $[t_a,t_b]$ or $[a,b]$, respectively, but it assumes the existence of a minimizing (or maximizing) function of the functional (2.7.1) or (2.7.2), respectively, that fulfills (2.7.3). The theory investigates mainly the class of admitted perturbations in order to derive a necessary system of Euler-Lagrange equations. The following example shows that admitted perturbations do not necessarily exist.

For $n = 2$ and $m = 1$, let

$$G(x,y,y') = y_2' - \sqrt{1+(y_1')^2} = 0 \quad \text{for } x \in [0,1].$$

(2.7.6)

Then,

$$D_pG(x,y,p) = \left(-\frac{p_1}{\sqrt{1+p_1^2}}, 1\right)$$

(2.7.7)

has rank $m = 1$. An admitted extremal must fulfill the boundary conditions

$$y(0) = (y_1(0),y_2(0)) = (0,0),$$
$$y(1) = (y_1(1),y_2(1)) = (0,1),$$

(2.7.8)

and also the nonholonomic constraint (2.7.6). By

$$y_2'(x) = \sqrt{1 + (y_1'(x))^2} \geq 1,$$

$$y_2(1) = \int_0^1 y_2'(x)dx = 1 \quad \text{is fulfilled only by} \tag{2.7.9}$$

$$y_1'(x) = 0, \quad \text{or } y_1(x) = y_1(0) = 0. \quad \text{Hence,}$$

$$y_2(x) = x.$$

Accordingly, the only admitted function is $y(x) = (y_1(x), y_2(x)) = (0, x)$ for $x \in [0, 1]$, and there is no admitted perturbation. Hence, a variational problem with nonholonomic constraint (2.7.6) and boundary conditions (2.7.8) is ill-posed.

In order to allow perturbations, an extremal satisfying a nonholonomic constraint must be free in the following sense:

Definition 2.7.2. *A solution* $y \in (C^1[a,b])^n$ *of* $G(x, y, y') = 0$ *satisfying* $y(a) = A$ *and* $y(b) = B$ *is free if the boundary value problem*

$$\begin{aligned} G(x, y, y') &= 0 \quad \text{for} \quad x \in [a, b] \\ y(a) &= A, \quad y(b) = \tilde{B}, \end{aligned} \tag{2.7.10}$$

has a solution for all $\tilde{B} \in \mathbb{R}^n$ *in a ball* $\|B - \tilde{B}\| < d$ *for some positive d. If y is not free, it is bound.*

Here are two examples in parametric form:

1. For $m = 1$, let

$$\Psi(x, \dot{x}) = \dot{x}_1^2 + \cdots + \dot{x}_n^2 - 1 = 0 \quad \text{for } t \in [t_a, t_b] \tag{2.7.11}$$

and assume $n \geq 2$. Then,

$$D_p\Psi(x, p) = 2(p_1, \ldots, p_n) \tag{2.7.12}$$

has rank $m = 1$ for all $\|p\|^2 = 1$, for which (2.7.11) is solvable with $\dot{x} = p$. All solutions of

$$\begin{aligned} \Psi(x, \dot{x}) &= 0, \\ x(t_a) &= A, \quad x(t_b) = B, \end{aligned} \tag{2.7.13}$$

have length $L = \int_{t_a}^{t_b} \|\dot{x}\| dt = t_b - t_a$, and therefore

> the straight line segment satisfying (2.7.13) with
> $\|B - A\| = t_b - t_a$ is bound,
> and any continuously differentiable curve satisfying (2.7.13) with
> $\|B - A\| < t_b - t_a$ is free,   (2.7.14)

cf. Figure 2.13.

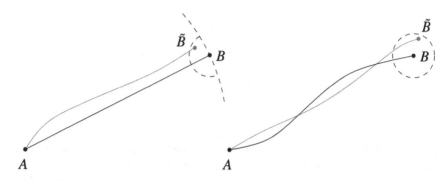

Fig. 2.13 A Bound Curve and a Free Curve

2. Consider the nonholonomic constraint

$$\tilde{\Psi}(x,\dot{x}) = D\Psi(x)\dot{x} = 0 \quad \text{for} \quad t \in [t_a,t_b], \tag{2.7.15}$$

where the mapping $\Psi : \mathbb{R}^n \to \mathbb{R}^m$, with $m < n$, is continuously totally differentiable and whose Jacobian matrix $D\Psi(x) \in \mathbb{R}^{m \times n}$ has maximal rank m for all $x \in \mathbb{R}^n$ satisfying $\Psi(x) = \Psi(A)$ for some $A \in \mathbb{R}^n$. In view of $D_p\tilde{\Psi}(x,p) = D\Psi(x)$, the nonholonomic constraint (2.7.15) fulfills the maximal rank condition (2.7.5) for the same points $x \in \mathbb{R}^n$. All solutions of

$$\tilde{\Psi}(x,\dot{x}) = D\Psi(x)\dot{x} = 0 \quad \text{for} \quad t \in [t_a,t_b],$$
$$x(t_a) = A, \quad x(t_b) = B, \tag{2.7.16}$$

are on the $(n-m)$-dimensional manifold $M = \{x \in \mathbb{R}^n | \Psi(x) = \Psi(A)\}$, since (2.7.16)$_1$ is equivalent to $\frac{d}{dt}\Psi(x) = 0$. Solutions exist only if the endpoints B and $\tilde{B}$ are on M, and therefore any curve satisfying the nonholonomic constraint (2.7.16) is bound.

This example shows that the holonomic constraint $\Psi(x) = \Psi(A)$ can be expressed as a nonholonomic constraint $\frac{d}{dt}\Psi(x) = D\Psi(x)\dot{x} = 0$ that binds solution curves.

Isoperimetric constraints can be expressed as nonholonomic constraints as well. For

$$K_i(y) = \int_a^b G_i(x,y,y')dx = c_i, \quad i = 1,\ldots,m, \tag{2.7.17}$$

with $G = (G_1,\ldots,G_m)$ and $c = (c_1,\ldots,c_m)$ for $(y,z) \in (C^1[a,b])^{n+m}$, define the nonholonomic constraint

$$\tilde{G}(x,y,z,y',z') = z'(x) - G(x,y,y') = 0, \, x \in [a,b],$$
$$z(a) = 0, \quad z(b) = c. \tag{2.7.18}$$

This is equivalent to the isoperimetric constraint in that sense that a function y satisfies (2.7.17) if and only if the function (y, z) satisfies (2.7.18). In view of

$$D_{(p,q)}\tilde{G}(x, y, z, p, q) = (-D_p G(x, y, p) \quad E) \in \mathbb{R}^{m \times (n+m)}, \tag{2.7.19}$$

where $E \in \mathbb{R}^{m \times m}$ denotes the unit matrix, the Jacobian matrix (2.7.19) has maximal rank m throughout. We see in Exercise 2.7.4 that a function $y \in (C^1[a,b])^n$ is not critical for the isoperimetric constraint (2.7.17), in the sense of Proposition 2.1.3, if and only if the function $(y, z) \in (C^1[a,b])^{n+m}$ satisfying (2.7.18) is normal according to Definition 2.7.3 below. Finally, a function is normal for a nonholonomic constraint if and only if it is free for it. That equivalence is proven in [11], IV, 4.

We quote the main proposition on nonholonomic constraints.

Proposition 2.7.1. *Let $y \in (C^2[a,b])^n$ be a local minimizer of the functional (2.7.2) under the nonholonomic constraint $(2.7.3)_2$. The functions F and G are three times continuously partially differentiable with respect to all variables, and the matrix $(2.7.5)_2$ has maximal rank m for all (x, y, p) in a neighborhood of the graph of (y, y') in $[a,b] \times \mathbb{R}^n \times \mathbb{R}^n$.*
Then, there is a continuously differentiable function

$$\lambda = (\lambda_1, \dots, \lambda_m) : [a,b] \to \mathbb{R}^m \quad \text{and}$$
$$\lambda_0 \in \{0, 1\}, \tag{2.7.20}$$

such that y solves the system of differential equations

$$\frac{d}{dx}(\lambda_0 F + \sum_{i=1}^{m} \lambda_i G_i)_{y'} = (\lambda_0 F + \sum_{i=1}^{m} \lambda_i G_i)_y \quad \text{on} \quad [a,b]. \tag{2.7.21}$$

Here, $F_{y'} = (F_{y'_1}, \dots, F_{y'_n})$, $F_y = (F_{y_1}, \dots, F_{y_n})$, and $G_{i,y'}$, $G_{i,y}$ are defined analogously. The argument of all functions in (2.7.21) is $(x, y(x), y'(x))$.
Finally

$$(\lambda_0, \lambda) \neq (0, 0) \in \mathbb{R} \times (C^1[a,b])^m. \tag{2.7.22}$$

The case $\lambda_0 = 1$ is of special interest, since for $\lambda_0 = 0$ the Lagrange function F does not appear in the system (2.7.21). In view of (2.7.22) in case $\lambda_0 = 0$, some of the multipliers $\lambda_1, \dots, \lambda_m$ do not vanish. That motivates the following definition:

Definition 2.7.3. *A solution $y \in (C^2[a,b])^n$ of $G(x, y, y') = 0$ on $[a,b]$ is called normal if*

$$\frac{d}{dx}(\sum_{i=1}^{m} \lambda_i G_i(\cdot, y, y'))_{y'} = (\sum_{i=1}^{m} \lambda_i G_i(\cdot, y, y'))_y \quad \text{on} \quad [a,b] \tag{2.7.23}$$

for $\lambda = (\lambda_1, \ldots, \lambda_m) \in (C^1[a,b])^m$ *has the only solution* $\lambda = 0 \in \mathbb{R}^m$. *If* (2.7.23) *is solvable for some nontrivial* λ, *then y is not normal.*

As remarked before, a function y satisfying a nonholonomic constraint is normal if and only if it is free. Without using this terminology, this is also proved in [3], Paragraph 69, and in [13], Chaps. 2, 3.

In view of these definitions, we complete Proposition 2.7.1 as follows:

Corollary 2.7.1. *If the local minimizer in Proposition 2.7.1 is free or normal with respect to the nonholonomic constraint, then it fulfills the system* (2.7.21) *with* $\lambda_0 = 1$.

The proof of Proposition 2.7.1 and of Corollary 2.7.1 is elaborate, and we refer to the literature [3], [11], [13], for example.

Proposition 2.7.1 is the most general in covering all constraints considered so far:

For isoperimetric constraints, the Lagrange multipliers $\lambda_1, \ldots, \lambda_m$ are constant and $\lambda_0 = 1$. Then, Proposition 2.7.1 yields Proposition 2.1.3, cf. Exercise 2.7.4 for details.

Curves that satisfy holonomic constraints in the nonholonomic form (2.7.15) are not free, cf. example 2 above and Exercise 2.7.2. Nonetheless system (2.7.21) is equivalent to system (2.5.8) of Proposition 2.5.1, cf. Exercise 2.7.3. This shows that for the choice $\lambda_0 = 1$ in (2.7.21), the local minimizer is not necessarily free or normal. That condition is sufficient but not necessary.

We apply Proposition 2.7.1 to the hanging chain, which we have considered already in Paragraph 2.3:

We parametrize the graph $\{(x,y(x))|x \in [a,b]\}$ representing the hanging chain by its arc length s, i.e.,

$$(x,y(x)) = (\tilde{x}(s),\tilde{y}(s)) \quad \text{for } s \in [0,L] \text{ where}$$
$$(\tilde{x}(0),\tilde{y}(0)) = (a,A), \quad (\tilde{x}(L),\tilde{y}(L)) = (b,B), \tag{2.7.24}$$

cf. Figure 2.3. Then, its potential energy is given by (up to the factor $g\rho$)

$$J(\tilde{x},\tilde{y}) = \int_0^L \tilde{y}(s)ds \tag{2.7.25}$$

which has to be minimized under the boundary conditions (2.7.24)$_2$ and the nonholonomic constraint

$$\dot{\tilde{x}}^2 + \dot{\tilde{y}}^2 = 1, \quad \cdot = \frac{d}{ds}, \tag{2.7.26}$$

cf. (2.6.6).

Assuming that $L > \sqrt{(b-a)^2 + (B-A)^2}$, then any admitted curve from (a,A) to (b,B) is not a straight line segment and therefore it is free according to (2.7.14). However, we do not use that information, but we apply Proposition 2.7.1 formally excluding $\lambda_0 = 0$. For the parametric versions (2.7.25) and (2.7.26), the system (2.7.21) reads

$$2\frac{d}{ds}\lambda_1\dot{\tilde{x}} = 0,$$

$$2\frac{d}{ds}\lambda_1\dot{\tilde{y}} = \lambda_0, \quad s \in [0,L]. \tag{2.7.27}$$

This yields $\lambda_1\dot{\tilde{x}} = c_1$, $\lambda_1\dot{\tilde{y}} = \frac{1}{2}\lambda_0 s + c_2$, and using (2.7.26), we obtain

$$\lambda_1^2 = \lambda_1^2(\dot{\tilde{x}}^2 + \dot{\tilde{y}}^2) = c_1^2 + (\frac{1}{2}\lambda_0 s + c_2)^2. \tag{2.7.28}$$

For $\lambda_0 = 0$, this gives $\lambda_1 = \sqrt{c_1^2 + c_2^2} \neq 0$ (by (2.7.22)) and

$$\tilde{x}(s) = \frac{c_1}{\lambda_1}s + c_3, \quad \tilde{y}(s) = \frac{c_2}{\lambda_1}s + c_4, \tag{2.7.29}$$

which describes a straight line. However, in view of $L > \sqrt{(b-a)^2 + (B-A)^2}$, it is not admitted.

Therefore, $\lambda_0 = 1$, and from the above calculations,

$$\lambda_1(s) = \sqrt{c_1^2 + (\frac{1}{2}s + c_2)^2},$$

$$\dot{\tilde{y}}(s) = (\frac{1}{2}s + c_2)/\sqrt{c_1^2 + (\frac{1}{2}s + c_2)^2},$$

$$\tilde{y}(s) = 2\sqrt{c_1^2 + (\frac{1}{2}s + c_2)^2} + c_4,$$

$$\dot{\tilde{x}}(s) = c_1/\sqrt{c_1^2 + (\frac{1}{2}s + c_2)^2}, \tag{2.7.30}$$

$$\tilde{x}(s) = 2c_1\,\mathrm{Arsinh}\left(\frac{1}{c_1}(\frac{1}{2}s + c_2)\right) + c_3,$$

$$c_1\sinh\left(\frac{\tilde{x}(s) - c_3}{2c_1}\right) = \frac{1}{2}s + c_2 \quad \text{and}$$

$$\tilde{y}(s) = 2c_1\cosh\left(\frac{\tilde{x}(s) - c_3}{2c_1}\right) + c_4.$$

We obtain a catenary with three constants as in (2.3.7).

Remark. *The general Proposition 2.7.1 with $\lambda_0 = 1$ goes back to Lagrange. We sketch roughly his arguments for $n = 2$ and $m = 1$: For a minimizer $y \in (C^1[a,b])^2$ of (2.7.2) under the constraint (2.7.3)$_2$, let a function $y + th$ with any $h \in (C_0^1[a,b])^2$ be an admitted perturbation, i.e., $G(x, y + th, y' + th') = 0$ for $t \in (-\varepsilon, \varepsilon)$. Then, $\frac{d}{dt}G(x, y + th, y' + th')|_{t=0} = 0$ or $(G_y, h) + (G_{y'}, h') = 0$ where $(\ ,\ )$ denotes the scalar product in $\mathbb{R}^2$. By assumption, $J(y + th)$ is minimal at $t = 0$, and hence, $\delta J(y)h = 0$. Therefore,*

$$\int_a^b (F_y + \lambda G_y, h) + (F_{y'} + \lambda G_{y'}, h') dx = 0 \qquad (2.7.31)$$

for any arbitrary continuously differentiable function λ, and integration by parts yields

$$\int_a^b (F_y + \lambda G_y - \frac{d}{dx}(F_{y'} + \lambda G_{y'}), h) dx = 0. \qquad (2.7.32)$$

The function λ is then determined by the scalar differential equation

$$F_{y_1} + \lambda G_{y_1} - \frac{d}{dx}(F_{y'_1} + \lambda G_{y'_1}) = 0 \quad \text{on } [a,b], \qquad (2.7.33)$$

and since the component h_2 is arbitrary, the fundamental lemma of the calculus of variations finally yields

$$F_{y_2} + \lambda G_{y_2} - \frac{d}{dx}(F_{y'_2} + \lambda G_{y'_2}) = 0 \quad \text{on } [a,b]. \qquad (2.7.34)$$

It is not appropriate to criticize the weak points of that argument, but rather we judge Lagrange's merit by the pioneering and useful result.

Exercises

2.7.1. For $n \geq 2$, let

$$\Psi(t,x,\dot{x}) = \dot{x}_1^2 + \cdots + \dot{x}_n^2 - 1 = 0 \quad \text{on } [t_a, t_b].$$

Show that any curve satisfying that nonholonomic constraint is normal, provided it is not a straight line.

2.7.2. Let

$$\tilde{\Psi}(x,\dot{x}) = D\Psi(x)\dot{x},$$

where $\Psi : \mathbb{R}^n \to \mathbb{R}^m$ with $m < n$, be a twice continuously partially differentiable function whose Jacobian matrix $D\Psi(x)$ has a maximal rank for all $x \in M = \{x \in \mathbb{R}^n | \Psi(x) = \Psi(A)\}$.

Show that a curve $x \in (C^1[t_a, t_b])^n$ satisfying $x(t_a) = A$ and $\tilde{\Psi}(x, \dot{x}) = 0$ is not normal.

2.7.3. Let $\tilde{\Psi}(x,\dot{x}) = D\Psi(x)\dot{x}$, where Ψ satisfies the hypotheses required in Exercise 2.7.2. Let $\Phi \colon \mathbb{R}^n \times \mathbb{R}^n \to \mathbb{R}$ be a twice continuously partially differentiable Lagrange function that fulfills system (2.5.8), viz.,

$$\frac{d}{dt}\Phi_{\dot{x}} = \Phi_x + \sum_{i=1}^m \lambda_i \nabla \Psi_i \quad \text{on } [t_a, t_b],$$

where $\lambda = (\lambda_1,\ldots,\lambda_m) : [t_a,t_b] \to \mathbb{R}^m$ is continuous.
Show that system (2.5.8) is equivalent to system (2.7.21), given by

$$\frac{d}{dt}\left(\Phi + \sum_{i=1}^m \tilde{\lambda}_i \tilde{\Psi}_i\right)_{\dot{x}} = \left(\Phi + \sum_{i=1}^m \tilde{\lambda}_i \tilde{\Psi}_i\right)_x,$$

where $\lambda_0 = 1$ and $\tilde{\lambda} = (\tilde{\lambda}_1,\ldots,\tilde{\lambda}_m) : [t_a,t_b] \to \mathbb{R}^m$ is continuously differentiable.

2.7.4. Formulate the isoperimetric constraints

$$K_i(y) = \int_a^b G_i(x,y,y')dx = c_i, \quad i = 1,\ldots,m,$$

with functions $G_i : [a,b] \times \mathbb{R}^n \times \mathbb{R}^n \to \mathbb{R}$, which are continuous and continuously partially differentiable with respect to the last $2n$ variables, as equivalent nonholonomic constraints

$$\tilde{G}(x,y,z,y',z') = z'(x) - G(x,y,y') = 0 \quad \text{for } x \in [a,b],$$
$$z(a) = 0, \ z(b) = c,$$

with $G = (G_1,\ldots,G_m)$, $c = (c_1,\ldots,c_m)$, and $(y,z) \in (C^1[a,b])^{n+m}$. Then, $D_{(p,q)}\tilde{G}(x,y,z,p,q)$ has rank m, cf. (2.7.19).

a) Show that a function $y \in (C^1[a,b])^n$ is not critical for the isoperimetric constraints in the sense of Proposition 2.1.3 or of Exercise 2.1.1, if and only if the function $(y,z) \in (C^1[a,b])^{n+m}$, satisfying the nonholonomic constraints, is normal.

b) Show that for a local minimizer $y \in (C^2[a,b])^n$, of the functional (2.7.2) or (2.1.15), under the nonholonomic constraints (2.7.18), the Lagrange multipliers in system (2.7.21) are constant and that for $\lambda_0 = 1$, system (2.7.21) is converted to system (2.1.20).

2.7.5. A functional of second order for $y \in C^2[a,b]$,

$$J(y) = \int_a^b F(x,y,y',y'')dx,$$

where the Lagrange function $F : [a,b] \times \mathbb{R} \times \mathbb{R} \times \mathbb{R} \to \mathbb{R}$ is three times continuously differentiable with respect to all variables, is formulated as a functional of first order for $(y,z) \in (C^1[a,b])^2$ via

$$J(y,z) = \int_a^b F(x,y,y',z')dx,$$

under the nonholonomic constraint

$$G(x,y,y',z') = y' - z = 0.$$

Prove the following statements:

a) The nonholonomic constraint fulfills the rank condition (2.7.5).
b) Any function fulfilling the nonholonomic constraint is free.
c) Any function fulfilling the nonholonomic constraint is normal.
d) Give the Euler-Lagrange equation for a local minimizer $y \in C^2[a,b]$ of $J(y)$ by
   application of Proposition 2.7.1.

2.7.6. Consider the nonholonomic constraint

$$G(y,y') = g_1(y)y_1' + g_2(y)y_2' = 0 \quad \text{on } [a,b] \,,$$

for $n = 2$ and $m = 1$. Assume that the vector field $g = (g_1,g_2) : \mathbb{R}^2 \to \mathbb{R}^2$ is continuously totally differentiable and that $g(y) \neq 0$ for all $y \in \mathbb{R}^2$. Then,

$$D_p G(y,p) = \nabla_p G(y,p) = g(y) \neq 0$$

has rank $m = 1$ for all $y \in \mathbb{R}^2$. Prove the statements:

a) Any solution $y \in (C^1[a,b])^2$ of $G(y,y') = 0$ is not normal.
b) Any solution $y \in (C^2[a,b])^2$ of $G(y,y') = 0$ satisfying $y(a) = A$ and $y(b) = B$ is
   bound.

2.8 Transversality

For the problem of finding a shortest connection between two disjoint surfaces in
space, all rectifiable curves having free boundaries on the surfaces are admitted. We
generalize this scenario as follows:

Definition 2.8.1. *For a parametric functional*

$$J(x) = \int_{t_a}^{t_b} \Phi(t,x,\dot{x})dt \tag{2.8.1}$$

curves $x \in (C^{1,pw}[t_a,t_b])^n$ for $n \geq 2$ are admitted, whose boundaries satisfy $x(t_a) \in M_a$ and/or $x(t_b) \in M_b$. Here, M_a and M_b are disjoint manifolds in $\mathbb{R}^n$ given by

$$M = \{x \in \mathbb{R}^n | \Psi(x) = 0\}, \tag{2.8.2}$$

where the function $\Psi : \mathbb{R}^n \to \mathbb{R}^m$ is twice continuously totally differentiable and whose Jacobian matrix $D\Psi(x)$ has maximal rank m. The dimension $n - m$ of the manifolds $M = M_a$ and $M = M_b$ may be different as long as $n - m > 0$.

If $x \in (C^{1,pw}[t_a,t_b])^n$ is admitted, then $x + sh$ for any $h \in (C_0^{1,pw}[t_a,t_b])^n$ is admitted
as well. Therefore, we can state by the same arguments employed before leading to
(1.10.22):

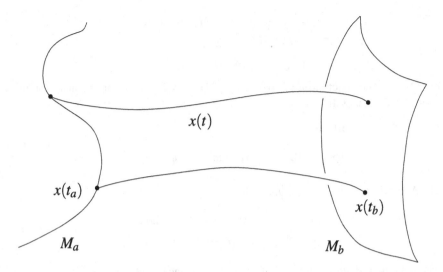

Fig. 2.14 Curves Connecting Two Manifolds

Proposition 2.8.1. *Let $x \in (C^{1,pw}[t_a,t_b])^n \cap \{x(t_a) \in M_a$ and/or $x(t_b) \in M_b\}$ be a local minimizer of the functional (2.8.1) whose Lagrange function $\Phi : [t_a,t_b] \times \mathbb{R}^n \times \mathbb{R}^n \to \mathbb{R}$ is continuous and continuously partially differentiable with respect to the last $2n$ variables. Then, x fulfills the system of Euler-Lagrange equations*

$$\Phi_{\dot{x}}(\cdot,x,\dot{x}) \in (C^{1,pw}[t_a,t_b])^n \quad and$$
$$\frac{d}{dt}\Phi_{\dot{x}}(\cdot,x,\dot{x}) = \Phi_x(\cdot,x,\dot{x}) \quad piecewise\ on\ \ [t_a,t_b]. \tag{2.8.3}$$

It is of interest which "natural boundary condition" a local minimizer has to fulfill at a free boundary on a manifold. We use the terminology of the Appendix; in particular, we refer to the definitions of the tangent and normal spaces at a point x of a manifold, cf. (A.7).

Proposition 2.8.2. *Under the hypotheses of Proposition 2.8.1, a local minimizer $x \in (C^{1,pw}[t_a,t_b])^n \cap \{x(t_a) \in M_a$ and/or $x(t_b) \in M_b\}$ of the functional (2.8.1) satisfies*

$$\Phi_{\dot{x}}(t_a,x(t_a),\dot{x}(t_a)) \in N_{x(t_a)}M_a \quad and/or$$
$$\Phi_{\dot{x}}(t_b,x(t_b),\dot{x}(t_b)) \in N_{x(t_b)}M_b, \quad respectively. \tag{2.8.4}$$

Proof. Let $M_a = M$ be given by (2.8.2). Then, for $x(t_a) \in M_a$, there exists a perturbation

$$x(t_a) + sy + \varphi(sy) \in M_a, \quad \text{where}$$
$$y \in T_{x(t_a)}M_a, \; \|y\| \leq 1, \tag{2.8.5}$$
$$\varphi(sy) \in N_{x(t_a)}M_a \, , \; s \in (-r, r),$$

cf. (A.11). The function $\varphi : B_r(0) \subset T_{x(t_a)}M_a \to N_{x(t_a)}M_a$ is continuously totally differentiable and satisfies

$$\varphi(0) = 0 \quad \text{and}$$
$$D\varphi(0) = 0; \quad \text{in particular,} \quad \frac{d}{ds}\varphi(sy)|_{s=0} = 0, \tag{2.8.6}$$

cf. (A.13). For $s \in (-r, r)$, $t \in [t_a, t_b]$, let

$$h(s,t) = \eta(t)(sy + \varphi(sy)), \quad \text{where}$$
$$\eta \in C^1[t_a, t_b] \, , \; \eta(t_a) = 1, \eta(t_b) = 0. \tag{2.8.7}$$

Then, the perturbation $x + h(s, \cdot)$ of x is admitted since it fulfills

$$h(s,\cdot) \in (C^1[t_a, t_b])^n,$$
$$x(t_a) + h(s, t_a) = x(t_a) + sy + \varphi(sy) \in M_a, \quad h(s, t_b) = 0,$$
$$h(0, t) = 0, \tag{2.8.8}$$
$$\frac{\partial}{\partial s}h(0, t) = \eta(t)y, \; \frac{\partial}{\partial s}\dot{h}(0, t) = \dot{\eta}(t)y.$$

Since the functional $J(x + h(s, \cdot))$ is locally minimal at $s = 0$, we derive

$$\frac{d}{ds}J(x + h(s, \cdot))|_{s=0} = 0$$
$$= \int_{t_a}^{t_b} (\Phi_x, \eta y) + (\Phi_{\dot{x}}, \dot{\eta} y)dt \tag{2.8.9}$$
$$= \int_{t_a}^{t_b} (\Phi_x - \frac{d}{dt}\Phi_{\dot{x}}, \eta y)dt + (\Phi_{\dot{x}}, \eta y)\big|_{t_a}^{t_b}$$
$$= -(\Phi_{\dot{x}}(t_a, x(t_a), \dot{x}(t_a)), y)$$

where we use also (2.8.3) and (2.8.7)$_2$. Since $y \in T_{x(t_a)}M_a$ fulfills only $\|y\| \leq 1$ but is otherwise arbitrary, the claim (2.8.4)$_1$ follows from the fact that $N_{x(t_a)}M_a$ is the orthogonal complement of $T_{x(t_a)}M_a$. The claim (2.8.4)$_2$ is proved in an analogous way. $\square$

The requirements (2.8.4) are **transversality** conditions.
For an example, consider

$$J(x) = \int_{t_a}^{t_b} \varphi(x)\|\dot{x}\|dt, \tag{2.8.10}$$

where $\varphi : \mathbb{R}^n \to \mathbb{R}$ is a continuously differentiable function. If $\varphi(x) > 0$, it is called a "density" of a "weighted" length of a curve x. The Euler-Lagrange equation reads

$$\frac{d}{dt}\left(\varphi(x)\frac{\dot{x}}{\|\dot{x}\|}\right) = \nabla\varphi(x)\|\dot{x}\|, \qquad (2.8.11)$$

and due to the invariance (1.10.20) of the Lagrange function, the Euler-Lagrange equation (2.8.11) is invariant under reparametrizations. A parametrization by the arc length gives $\|\dot{x}\| = 1$, cf. (2.6.6), such that (2.8.11) and (2.8.4) become

$$\frac{d}{dt}(\varphi(x)\dot{x}) = \nabla\varphi(x), \quad \|\dot{x}\| = 1,$$
$$\varphi(x(t_a))\dot{x}(t_a) \in N_{x(t_a)}M_a, \qquad (2.8.12)$$
$$\varphi(x(t_b))\dot{x}(t_b) \in N_{x(t_b)}M_b.$$

If $\varphi(x(t_a)) \neq 0$ or $\varphi(x(t_b)) \neq 0$, the conditions $(2.8.12)_{2,3}$ mean that the curve is orthogonal to M_a or to M_b. If $\varphi(x) \equiv 1$, the minimizing curve is a straight line.

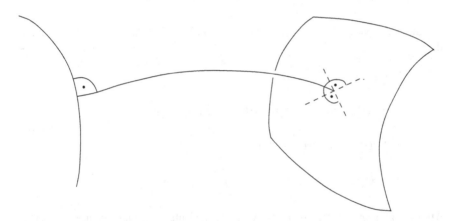

Fig. 2.15 A Transversal Curve

Next we derive a modified transversality if the Lagrange function depends explicitly on the starting point and/or on the endpoint. This condition is used for the generalized brachistochrone problem studied below.

Proposition 2.8.3. *Let* $x \in (C^{1,pw}[t_a,t_b])^n \cap \{x(t_a) \in M_a \text{ and/or } x(t_b) \in M_b\}$ *be a local minimizer of*

$$J(x) = \int_{t_a}^{t_b} \Phi(x,\dot{x},x(t_a))dt, \qquad (2.8.13)$$

where $\Phi : \mathbb{R}^n \times \mathbb{R}^n \times \mathbb{R}^n \to \mathbb{R}$ *is continuously totally differentiable. Then,* x *satisfies the Euler-Lagrange equation (2.8.15), and at* $t = t_a$ *a modified transversality condition, viz.,*

$$\Phi_{\dot{x}}(x(t_a),\dot{x}(t_a),x(t_a)) - \int_{t_a}^{t_b} \Phi_{x(t_a)}(x,\dot{x},x(t_a))dt \in N_{x(t_a)}M_a. \tag{2.8.14}$$

Here, $\Phi_{x(t_a)}$ *denotes the vector of the derivatives of* Φ *with respect to the last* n
variables. If $x(t_b) \in M_b$, *then* x *satisfies the transversality condition* (2.8.4)$_2$.

Proof. For all $h \in (C_0^{1,pw}[t_a,t_b])^n$, the perturbation $x + sh$ is admitted as well, and by
assumption,

$$J(x + sh) = \int_{t_a}^{t_b} \Phi(x + sh, \dot{x} + sh, x(t_a))dt$$

is minimal at $s = 0$. Therefore, x satisfies

$$\Phi_{\dot{x}}(x,\dot{x},x(t_a)) \in (C^{1,pw}[t_a,t_b])^n \quad \text{and}$$
$$\frac{d}{dt}\Phi_{\dot{x}}(x,\dot{x},x(t_a)) = \Phi_x(x,\dot{x},x(t_a)) \quad \text{piecewise on } [t_a,t_b]. \tag{2.8.15}$$

For the perturbation $h(s,t)$ given in (2.8.7) and (2.8.8), we have $\frac{\partial}{\partial s}h(0,t_a) = y$, and
thus,

$$\frac{d}{ds}J(x + h(s,\cdot))|_{s=0} = 0$$
$$= \int_{t_a}^{t_b} (\Phi_x(x,\dot{x},x(t_a)),\eta y) + (\Phi_{\dot{x}}(x,\dot{x},x(t_a)),\dot{\eta}y)dt$$
$$+ \int_{t_a}^{t_b} (\Phi_{x(t_a)}(x,\dot{x},x(t_a)),y)dt \tag{2.8.16}$$
$$= -(\Phi_{\dot{x}}(x(t_a),\dot{x}(t_a),x(t_a)),y) + \left(\int_{t_a}^{t_b} \Phi_{x(t_a)}(x,\dot{x},x(t_a))dt,y\right).$$

After integration (2.8.16)$_2$ by parts, we apply the Euler-Lagrange equation (2.8.15),
cf. (2.8.9). The fact that $y \in T_{x(t_a)}M_a$ satisfies only $\|y\| \leq 1$ but is otherwise arbitrary
implies the claim (2.8.14).

The last claim is proved with a perturbation h that is analogous to (2.8.7) where
t_a and M_a are replaced by t_b and M_b, respectively. Then, $h(s,t_a) = x(t_a)$ for all $s \in$
$(-r,r)$, whence $\frac{\partial}{\partial s}h(0,t_a) = 0$. Therefore, the integral (2.8.16)$_3$ vanishes.   □

We study the following **brachistochrone problem**: Given two disjoint curves
M_a and M_b in a vertical plane, determine the curve on which a point mass acted on
only by gravity runs from M_a to M_b in shortest time.

The curve is certainly a cycloid, and the problem is where the minimizing cycloid
starts on M_a and where it ends on M_b.

We establish the running time of a point mass m on a curve (x,y) parametrized
by the time t in a vertical (x,y)-plane, where the y-axis points downward. We can
use now our knowledge on holonomic constraints, namely

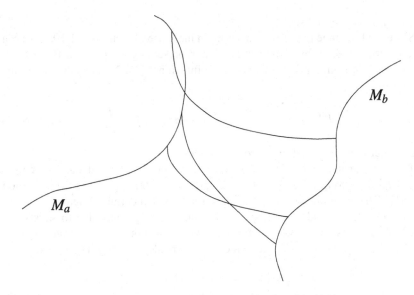

Fig. 2.16   Admitted Curves for a Brachistochrone connecting Two Manifolds

$$T = \frac{1}{2}m(\dot{x}^2 + \dot{y}^2) \quad \text{is the kinetic energy,}$$

$$V = -mgy \qquad \text{is the potential energy, and} \qquad (2.8.17)$$

$$\Psi(x,y) = 0 \quad \text{is the holonomic constraint}$$

that determines the minimizing curve. According to (2.5.28)–(2.5.32), the equations of motion derived from the Euler-Lagrange equations are

$$m\ddot{x} = \lambda\Psi_x(x,y),$$
$$m\ddot{y} = mg + \lambda\Psi_y(x,y). \qquad (2.8.18)$$

Due to $\frac{d}{dt}\Psi(x,y) = \Psi_x(x,y)\dot{x} + \Psi_y(x,y)\dot{y} = 0$ and by $\ddot{x}\dot{x} + \ddot{y}\dot{y} = \frac{1}{2}\frac{d}{dt}(\dot{x}^2 + \dot{y}^2) = \frac{1}{2}\frac{d}{dt}v^2$, (2.8.18) implies

$$\frac{d}{dt}(\frac{1}{2}v^2 - gy) = 0. \quad \text{Hence}$$

$$\frac{1}{2}v(t)^2 - gy(t) = \frac{1}{2}v(t_a)^2 - gy(t_a), \quad \text{and} \qquad (2.8.19)$$

$$\sqrt{\dot{x}(t)^2 + \dot{y}(t)^2} = \sqrt{2g(y(t) - y(t_a)) + v(t_a)^2}.$$

This gives the running time on the curve $\{(x(t),y(t))|t \in [t_a,t_b]\}$:

$$T = t_b - t_a = \int_{t_a}^{t_b} 1\,dt = \int_{t_a}^{t_b} \sqrt{\frac{\dot{x}^2 + \dot{y}^2}{2g(y - y(t_a)) + v(t_a)^2}}\,dt, \qquad (2.8.20)$$

where $v(t_a)$ is the initial speed.

Since the Lagrange function is invariant in the sense of Definition 1.10.2, we can reparametrize (2.8.20) by τ used in (1.8.17) representing the cycloid. By (1.8.19) $t = \alpha\tau$, $\alpha > 0$. Defining $k = v(t_a)^2/2g$, the functional (2.8.20) becomes (up to the factor $\frac{1}{\sqrt{2g}}$)

$$J(x,y) = \int_{\tau_a}^{\tau_b} \sqrt{\frac{\dot{x}^2 + \dot{y}^2}{y - y(\tau_a) + k}}\, d\tau, \quad (\dot{\ }) = \frac{d}{d\tau}, \tag{2.8.21}$$

where, after reparametrization, we omit the tilde.

The functional (2.8.21) is to be minimized among admitted curves $(x,y) \in (C^1[\tau_a, \tau_b])^2 \cap \{(x(\tau_a),y(\tau_a)) \in M_a, (x(\tau_b),y(\tau_b)) \in M_b\} \cap \{y - y(\tau_a) + k > 0$ on $(\tau_a, \tau_b]\}$. The minimizing curve satisfies for $k > 0$ the Euler-Lagrange system (2.8.15) on $[\tau_a, \tau_b]$, and according to Proposition 2.8.3, it has also to satisfy the modified transversality (2.8.14) at $\tau = \tau_a$. We know that the Euler-Lagrange system is solved by a cycloid (and we recommend to the reader to verify it). We obtain

$$
\begin{aligned}
x(\tau) &= x(\tau_a) - c + r(\tau - \sin\tau), \\
y(\tau) &= y(\tau_a) - k + r(1 - \cos\tau), \\
c &= r(\tau_a - \sin\tau_a), \quad k = r(1 - \cos\tau_a).
\end{aligned}
\tag{2.8.22}
$$

The modified transversality is evaluated using the derivatives of the Lagrange function of (2.8.21) in the family (2.8.22), which gives at $\tau = \tau_a$,

$$
\begin{aligned}
&\frac{1}{\sqrt{2r}}\left((1, \cot\frac{\tau_a}{2}) - \int_{\tau_a}^{\tau_b}\left(0, \frac{1}{1 - \cos\tau}\right)d\tau\right) \\
&= \frac{1}{\sqrt{2r}}(1, \cot\frac{\tau_b}{2}) \in N_{(x(\tau_a),y(\tau_a))}M_a.
\end{aligned}
\tag{2.8.23}
$$

At $\tau = \tau_b$, the transversality $(2.8.4)_2$ must be satisfied, meaning

$$\frac{1}{\sqrt{2r}}(1, \cot\frac{\tau_b}{2}) \in N_{(x(\tau_b),y(\tau_b))}M_b. \tag{2.8.24}$$

Since

$$\cot\frac{\tau_b}{2} = \frac{\sin\tau_b}{1 - \cos\tau_b} = \frac{\dot{y}(\tau_b)}{\dot{x}(\tau_b)}, \tag{2.8.25}$$

the vector in $(2.8.23)_2$ and in (2.8.24) is tangent to the cycloid at its endpoint $(x(\tau_b),y(\tau_b))$ on M_b. Geometrically, (2.8.24) means that the cycloid is orthogonal to M_b.

The remarkable property is that by $(2.8.23)_2$, the tangent vector in the endpoint is also orthogonal to M_a in its starting point $(x(\tau_a),y(\tau_a))$, see Figure 2.17.

The limit $v(t_a) \searrow 0$ yields $k = 0$, $\tau_a = 0$, and finally $c = 0$ in (2.8.22). The cycloid has then a vertical tangent in its starting point on M_a, as expected.

We determine for special curves M_a and M_b the starting and endpoints of the minimizing cycloid. We choose

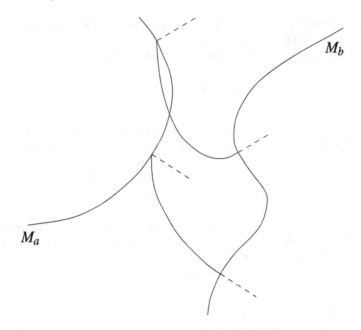

Fig. 2.17 Conditions on a Brachistochrone Connecting Two Manifolds

$$M_a = \{(x,y)\,|\,x^2 + y^2 = 1\}, \quad M_b = \{(x,y)\,|\,(x-x_0)^2 + (y-1)^2 = 1\}, \quad (2.8.26)$$

which are two circles with centers $(0,0)$ and $(x_0,1)$ and radii 1. At the points $(x(0),y(0)) \in M_a$ and $(x(\tau_b),y(\tau_b)) \in M_b$, the manifolds M_a and M_b have the same normal vectors, which are radial vectors in this case, cf. (2.8.23)$_2$, (2.8.24):

$$
\begin{aligned}
(x(0),y(0)) &= (\cos\alpha, \sin\alpha), \\
(x(\tau_b),y(\tau_b)) &= (x_0,1) - (\cos\alpha, \sin\alpha).
\end{aligned}
\qquad (2.8.27)
$$

Substitution into (2.8.22) (where $c = k = 0$) gives

$$
\begin{aligned}
r(\tau_b - \sin\tau_b) &= x_0 - 2\cos\alpha, \\
r(1 - \cos\tau_b) &= 1 - 2\sin\alpha \quad \text{and} \\
f(\tau_b) = \frac{\tau_b - \sin\tau_b}{1 - \cos\tau_b} &= \frac{x_0 - 2\cos\alpha}{1 - 2\sin\alpha}, \quad \text{see Fig. 1.14.}
\end{aligned}
\qquad (2.8.28)
$$

From (2.8.24) and (2.8.25), the tangent at the end is normal to the circle, which means

$$\cot\frac{\tau_b}{2} = \tan\alpha. \qquad (2.8.29)$$

Finally, the equation

$$r = \frac{1 - 2\sin\alpha}{1 - \cos\tau_b} \qquad (2.8.30)$$

gives the parameter r, where τ_b and α are determined by $(2.8.28)_3$ and $(2.8.29)$. In Figure 2.18, we sketch three typical cases.

Proposition 2.8.1 applied to nonparametric functionals

$$J(y) = \int_a^b F(x, y, y')dx, \tag{2.8.31}$$

for functions $y \in (C^{1,pw}[a,b])^n$ with boundary points on manifolds M_a and/or M_b in $\mathbb{R}^n$, $n \geq 2$, shows that local minimizers satisfy the following transversalities:

$$\begin{aligned} F_{y'}(a, y(a), y'(a)) &\in N_{y(a)}M_a \quad \text{and/or} \\ F_{y'}(b, y(b), y'(b)) &\in N_{y(b)}M_b, \end{aligned} \tag{2.8.32}$$

where $F_{y'} = (F_{y'_1}, \dots, F_{y'_n})$. We mention that the natural boundary conditions on the boundaries of a manifold M in $\mathbb{R}^n$ given in Proposition 2.5.5 are precisely the transversalities conditions $(2.8.32)$.

Another problem is the following (we confine ourselves to the case $n = 1$).

Definition 2.8.2. *For the functional*

$$J(y) = \int_{x_a}^{x_b} F(x, y, y')dx, \tag{2.8.33}$$

functions $y \in C^{1,pw}[x_a, x_b]$ *are admitted, whose boundaries* $(x_a, y(x_a))$ *and/or* $(x_b, y(x_b))$ *are on one-dimensional manifolds (curves)* M_a *and/or* M_b, *respectively.* M_a *and* M_b *are disjoint curves in* $\mathbb{R}^2$ *and represented by*

$$\begin{aligned} M &= \{(x,y) \in \mathbb{R}^2 | \Psi(x,y) = 0\}, \quad \text{or in particular,} \\ M_\psi &= \{(x,y) \in \mathbb{R}^2 | y - \psi(x) = 0\}. \end{aligned} \tag{2.8.34}$$

The functions $\Psi : \mathbb{R}^2 \to \mathbb{R}$ *and* $\psi : \mathbb{R} \to \mathbb{R}$ *are twice continuously (partially) differentiable, and the Jacobian matrix* $D\Psi(x)$ *has maximal rank 1 for all* $x \in M$. *In the case of* $(2.8.34)_2$, *the set* M_ψ *is the graph of* ψ, *and* M *and* M_ψ *are continuously differentiable manifolds of dimension 1, i.e., curves, cf. the Appendix. It is also possible that one boundary is prescribed by* $x_a = a$ *and* $y(a) = A$ *and/or* $x_b = b$ *and* $y(b) = B$.

Since the intervals $[x_a, x_b]$, where admitted functions are defined, are variable, the distance between two admitted functions is not defined as usual. We reparametrize the graph of a function $y \in C^{1,pw}[x_a, x_b]$ by a parameter on a fixed interval $[\tau_a, \tau_b]$, cf. $(1.11.1)$, $(1.11.2)$:

$$\{(x, y(x)) | x \in [x_a, x_b]\} = \{(\tilde{x}(\tau), \tilde{y}(\tau)) | \tau \in [\tau_a, \tau_b]\}, \tag{2.8.35}$$

and we obtain a parametric functional

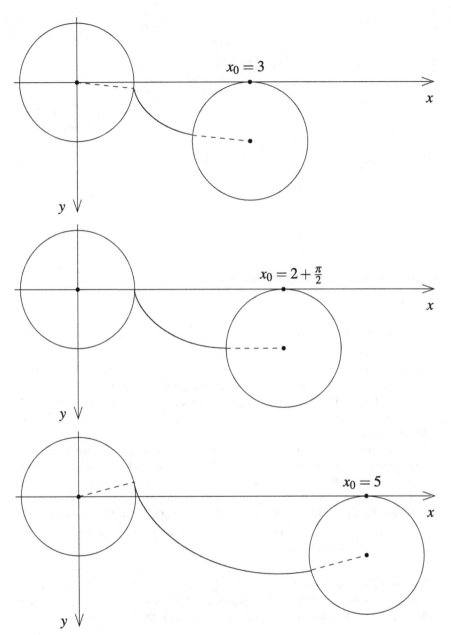

Fig. 2.18 Brachistochrones Connecting Two Circles

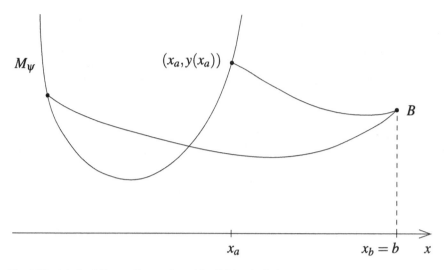

Fig. 2.19 Admitted Curves Connecting a Manifold and a Point

$$J(y) = \int_{x_a}^{x_b} F(x,y,y')dx = \int_{\tau_a}^{\tau_b} F\left(\tilde{x}, \tilde{y}, \frac{\dot{\tilde{y}}}{\dot{\tilde{x}}}\right)\dot{\tilde{x}}d\tau = J(\tilde{x}, \tilde{y}), \qquad (2.8.36)$$

cf. (1.11.4). For the parametric functional, curves $(\tilde{x}, \tilde{y}) \in (C^1[\tau_a, \tau_b] \times C^{1,pw}[\tau_a, \tau_b]) \cap$ $\{\dot{\tilde{x}}(\tau) > 0 \text{ for } \tau \in [\tau_a, \tau_b]\}$ are admitted, which are graphs of functions $y \in C^{1,pw}[x_a, x_b]$, cf. Exercise 1.10.1. The boundaries of a function y are on manifolds M_a and/or M_b according to Definition 2.8.2 if and only if the endpoints of the curve (2.8.35) are on M_a and/or M_b.

As in Proposition 1.11.1, it follows that an admitted global minimizer y of (2.8.33) yields, via a parametrization (2.8.35), an admitted local minimizer of the parametric functional (2.8.36). Like the graph of y, the admitted perturbations (1.11.8) of the curve must have endpoints on M_a and/or M_b.

Applying the Propositions 2.8.1 and 2.8.2, the invariance of the Lagrange function $\Phi(\tilde{x}, \tilde{y}, \dot{\tilde{x}}, \dot{\tilde{y}}) = F(\tilde{x}, \tilde{y}, \frac{\dot{\tilde{y}}}{\dot{\tilde{x}}})\dot{\tilde{x}}$ in the sense of Definition 1.10.2 gives necessary conditions on the global minimizer.

Proposition 2.8.4. *Let* $y \in C^{1,pw}[x_a, x_b] \cap \{(x_a, y(x_a)) \in M_a$ *and/or* $(x_b, y(x_b)) \in M_b\}$ *be a global minimizer of the functional (2.8.33) whose Lagrange function* $F : \mathbb{R}^3 \to \mathbb{R}$ *is continuously totally differentiable. Then, the following hold:*

$$F_{y'}(\cdot,y,y') \in C^{1,pw}[x_a,x_b] \subset C[x_a,x_b],$$

$$\frac{d}{dx}F_{y'}(\cdot,y,y') = F_y(\cdot,y,y') \quad \text{piecewise on } [x_a,x_b],$$

$$F(\cdot,y,y') - y'F_{y'}(\cdot,y,y') \in C[x_a,x_b],$$

$$((F - y'F_{y'})(x_a,y(x_a),y'(x_a)), F_{y'}(x_a,y(x_a),y'(x_a))) \in N_{(x_a,y(x_a))}M_a, \tag{2.8.37}$$

and in case $(2.8.34)_2$,

$$F(x_a,y(x_a),y'(x_a)) + (\psi'(x_a) - y'(x_a))F_{y'}(x_a,y(x_a),y'(x_a)) = 0$$

and/or an analog transversality at $(x_b,y(x_b))$.

Proof. As mentioned before, we apply Propositions 2.8.1 and 2.8.2 to the parametric functional (2.8.36). The derivatives of the Lagrange function read

$$\Phi_{\dot{\tilde{x}}}(\tilde{x},\tilde{y},\dot{\tilde{x}},\dot{\tilde{y}}) = F(\tilde{x},\tilde{y},\frac{\dot{\tilde{x}}}{\dot{\tilde{y}}}) - F_{y'}(\tilde{x},\tilde{y},\frac{\dot{\tilde{x}}}{\dot{\tilde{y}}})\frac{\dot{\tilde{x}}}{\dot{\tilde{y}}},$$

$$\Phi_{\dot{\tilde{y}}}(\tilde{x},\tilde{y},\dot{\tilde{x}},\dot{\tilde{y}}) = F_{y'}(\tilde{x},\tilde{y},\frac{\dot{\tilde{x}}}{\dot{\tilde{y}}}), \tag{2.8.38}$$

$$\Phi_{\tilde{x}}(\tilde{x},\tilde{y},\dot{\tilde{x}},\dot{\tilde{y}}) = F_x(\tilde{x},\tilde{y},\frac{\dot{\tilde{x}}}{\dot{\tilde{y}}})\dot{\tilde{x}},$$

$$\Phi_{\tilde{y}}(\tilde{x},\tilde{y},\dot{\tilde{x}},\dot{\tilde{y}}) = F_y(\tilde{x},\tilde{y},\frac{\dot{\tilde{x}}}{\dot{\tilde{y}}})\dot{\tilde{x}}.$$

Due to the invariance of Φ, the regularity and the Euler-Lagrange equation (2.8.3) hold for any parametrization. In particular, we have

$$\tilde{x} = x, \ \dot{\tilde{x}} = 1, \ \tilde{y} = y, \ \dot{\tilde{y}} = y', \ \tau_a = x_a, \ \tau_b = x_b, \tag{2.8.39}$$

whence $(2.8.37)_1$–$(2.8.37)_3$, cf. also Proposition 1.10.4. The transversality $(2.8.4)_1$ implies $(2.8.37)_4$, and the last statement follows from the fact that a vector tangent to M_ψ in $(x_a,y(x_a)) = (x_a,\psi(x_a))$ is given by $(1,\psi'(x_a))$. $\qquad\square$

In $(2.8.37)_1$ and $(2.8.37)_3$, we recognize the Weierstraß-Erdmann corner conditions; the transversalities $(2.8.37)_4$ and $(2.8.37)_7$ are called **free transversalities**.

Exercises

2.8.1. Let $x \in (C^{1,pw}[t_a,t_b])^n$ be a local minimizer of

$$J(x) = \int_{t_a}^{t_b} \Phi(x,\dot{x})dt,$$

subject to isoperimetric constraints

$$K_i(x) = \int_{t_a}^{t_b} \Psi_i(x,\dot{x})dt = c_i \,, \; i = 1,\dots,m,$$

whose boundary fulfills $x(t_a) \in M_a = \{x \in \mathbb{R}^n | \Psi(x) = 0\}$. Find and prove its transversality at its boundary $x(t_a)$.

We assume that $\Psi : \mathbb{R}^n \to \mathbb{R}^{m_a}$, where $0 < m_a < n$, is twice continuously totally differentiable and that the Lagrange functions $\Phi, \Psi_i : \mathbb{R}^n \times \mathbb{R}^n \to \mathbb{R}$ are continuously totally differentiable. We assume also that the Jacobian matrix $D\Psi(x)$ has maximal rank m_a for all $x \in M_a$ and that x is not critical for the constraints in the sense that $\delta K(x) : (C_0^{1,pw}[t_a,t_b])^n \to \mathbb{R}^m$ is surjective.

2.8.2. Let $x \in (C^2[t_a,t_b])^n$ be a local minimizer of

$$J(x) = \int_{t_a}^{t_b} \Phi(x,\dot{x})dt,$$

subject to the holonomic constraint

$$\Psi(x) = 0 \quad \text{or}$$
$$x \in M = \{x \in \mathbb{R}^n | \Psi(x) = 0\},$$

whose boundary fulfills $x(t_a) \in M_a = \{x \in M | \Psi_a(x) = 0\} \subset M$. Find and prove its transversality in its boundary $x(t_a)$.

We assume that $\Psi : \mathbb{R}^n \to \mathbb{R}^m$ and $\Psi_a : \mathbb{R}^n \to \mathbb{R}^{m_a}$, where $0 < m + m_a < n$, are three times and the Lagrange function $\Phi : \mathbb{R}^n \times \mathbb{R}^n \to \mathbb{R}$ is twice continuously totally differentiable. We assume also that the Jacobian matrices of $\Psi : \mathbb{R}^n \to \mathbb{R}^m$ and $(\Psi, \Psi_a) : \mathbb{R}^n \to \mathbb{R}^{m+m_a}$ have maximal rank m and $m + m_a$ in all $x \in M$ and in all $x \in M_a$, respectively. Then, $M_a \subset M$ is a $(n - (m + m_a))$-dimensional submanifold of the $(n - m)$-dimensional manifold M.

2.8.3. Determine $(x_b, y(x_b))$, $x_b > 1$, on the graph of

$$\psi(x) = \frac{2}{x^2} - 3\,,$$

such that

$$J(y) = \int_1^{x_b} x^3(y')^2 dx, \quad y(1) = 0,$$

is extremal. (Only the necessary conditions on extremals have to be verified.)

2.8.4. Determine an extremal of

$$J(y) = \int_0^{x_b} y^2 + (y')^2 dx, \quad y(0) = 1,$$

whose boundary fulfills $(x_b, y(x_b)) = (x_b, 2)$, $x_b > 0$.

2.8.5. Determine an extremal $y \in C^{1,pw}[0,1]$ of

$$J(y) = \int_0^1 \frac{1}{2}(y')^2 + yy' + y' + y \, dx,$$

whose boundaries are on $M_0 = \{(0,y)|y \in \mathbb{R}\}$ and $M_1 = \{(1,y)|y \in \mathbb{R}\}$.

2.8.6. Check whether the functions determined in Exercises 2.8.3–2.8.5 are local or global minimizers or maximizers among admitted functions of the respective functionals.

Hint: Observe that the boundary $(\tilde{x}_b, y(\tilde{x}_b) + h(\tilde{x}_b))$ of admitted perturbations $y + h$ is necessarily on the respective given manifold, where $\tilde{x}_b$ and x_b might be different.

2.9 Emmy Noether's Theorem

"Invariants" are important for mathematics as well as theoretical physics, meaning that certain quantities or physical laws are invariant under special transformations. These are, in particular, actions of the "orthogonal group," or a special symmetry group, under which a given physical law is invariant, and possibly translations. The invariance restricts functions or functionals and often simplifies their analysis. The invariance of differential equations, like the Euler-Lagrange equation, means that one solution defines a family of solutions, thus allowing, as we shall see, to establish conservation laws. These, in turn, help the mathematical analysis and provide physical insights.

Definition 2.9.1. *A continuously totally differentiable Lagrange function* $\Phi : \mathbb{R}^n \times \mathbb{R}^n \to \mathbb{R}$ *of a parametric functional*

$$J(x) = \int_{t_a}^{t_b} \Phi(x, \dot{x}) dt, \qquad (2.9.1)$$

is invariant under a family of local diffeomorphisms

$$h^s : \mathbb{R}^n \to \mathbb{R}^n, \quad s \in (-\delta, \delta), \qquad (2.9.2)$$

if

$$\Phi(h^s(x), Dh^s(x)\dot{x}) = \Phi(x, \dot{x}) \quad \text{for all} \quad (x, \dot{x}) \in \mathbb{R}^n \times \mathbb{R}^n \qquad (2.9.3)$$

and for all $s \in (-\delta, \delta)$. *The mappings* h^s *are twice continuously partially differentiable with respect to the parameters* $s \in (-\delta, \delta)$ *and* $x \in \mathbb{R}^n$. *The Jacobian matrices*

$$Dh^s(x) \in \mathbb{R}^{n \times n} \quad \text{are regular for all} \quad s \in (-\delta, \delta) \quad \text{and all} \quad x \in \mathbb{R}^n. \qquad (2.9.4)$$

Inserting a curve $x \in (C^1[t_a, t_b])^n$ into h^s, we then have $\frac{d}{dt} h^s(x) = Dh^s(x)\dot{x}$, and hence, the invariance (2.9.3) implies

$$\Phi(h^s(x), \frac{d}{dt}h^s_{\cdot}(x)) = \Phi(x, \dot{x}) \quad \text{for all} \quad t \in [t_a, t_b] \tag{2.9.5}$$

and for all $s \in (-\delta, \delta)$. Therefore, a minimizing curve x defines a family $h^s(x)$ of minimizing curves of the functional (2.9.1). Before profiting from that fact, we give the following formulas obtained by differentiation of (2.9.3):

$$\begin{aligned} \Phi_x(h^s(x), Dh^s(x)\dot{x})Dh^s(x) + \Phi_{\dot{x}}(h^s(x), Dh^s(x)\dot{x})D^2h^s(x)\dot{x} &= \Phi_x(x, \dot{x}), \\ \Phi_{\dot{x}}(h^s(x), Dh^s(x)\dot{x})Dh^s(x) &= \Phi_{\dot{x}}(x, \dot{x}). \end{aligned} \tag{2.9.6}$$

Here, we employ the notation

$$\Phi_x = (\Phi_{x_1}, \ldots, \Phi_{x_n}), \quad \Phi_{\dot{x}} = (\Phi_{\dot{x}_1}, \ldots, \Phi_{\dot{x}_n}),$$

$$Dh^s(x) = \left(\frac{\partial h^s_i}{\partial x_j}(x)\right)_{\substack{i=1,\ldots,n \\ j=1,\ldots,n}}, \quad D^2h^s(x)\dot{x} = \left(\sum_{k=1}^n \frac{\partial^2 h^s_i}{\partial x_j \partial x_k}(x)\dot{x}_k\right)_{\substack{i=1,\ldots,n \\ j=1,\ldots,n}}, \tag{2.9.7}$$

and that the product of a row vector and a matrix is again a row vector. For a curve $x \in (C^1[t_a, t_b])^n$, we obtain

$$\frac{d}{dt}h^s(x) = Dh^s(x)\dot{x} \quad \text{and} \quad \frac{d}{dt}Dh^s(x) = D^2h^s(x)\dot{x} \quad \text{for} \quad t \in [t_a, t_b]. \tag{2.9.8}$$

Proposition 2.9.1. *Assume that* $\Phi : \mathbb{R}^n \times \mathbb{R}^n \to \mathbb{R}$ *is continuously totally differentiable and invariant under a family of local diffeomorphisms* $h^s : \mathbb{R}^n \to \mathbb{R}^n$, $s \in (-\delta, \delta)$, *in the sense of Definition 2.9.1. If* $x \in (C^1[t_a, t_b])^n$ *is a solution of the Euler-Lagrange system, i.e.,*

$$\frac{d}{dt}\Phi_{\dot{x}}(x, \dot{x}) = \Phi_x(x, \dot{x}) \quad \text{on} \quad [t_a, t_b], \tag{2.9.9}$$

then $h^s(x) \in (C^1[t_a, t_b])^n$ *is a solution as well, i.e.,*

$$\frac{d}{dt}\Phi_{\dot{x}}(h^s(x), \frac{d}{dt}h^s(x)) = \Phi_x(h^s(x), \frac{d}{dt}h^s(x)) \quad \text{on} \quad [t_a, t_b]. \tag{2.9.10}$$

Proof. By (2.9.6) and (2.9.8), we obtain

$$\frac{d}{dt}\Phi_{\dot{x}}(x,\dot{x}) - \Phi_x(x,\dot{x}) = 0$$

$$= \frac{d}{dt}(\Phi_{\dot{x}}(h^s(x), Dh^s(x)\dot{x})Dh^s(x))$$

$$- \Phi_x(h^s(x), Dh^s(x)\dot{x})Dh^s(x) - \Phi_{\dot{x}}(h^s(x), Dh^s(x)\dot{x})D^2h^s(x)\dot{x}$$

$$= \frac{d}{dt}\Phi_{\dot{x}}(h^s(x), Dh^s(x)\dot{x})Dh^s(x) + \Phi_{\dot{x}}(h^s(x), Dh^s(x)\dot{x})D^2h^s(x)\dot{x} \qquad (2.9.11)$$

$$- \Phi_x(h^s(x), Dh^s(x)\dot{x})Dh^s(x) - \Phi_{\dot{x}}(h^s(x), Dh^s(x)\dot{x})D^2h^s(x)\dot{x}$$

$$= \left(\frac{d}{dt}\Phi_{\dot{x}}(h^s(x), Dh^s(x)\dot{x}) - \Phi_x(h^s(x), Dh^s(x)\dot{x})\right)Dh^s(x)$$

$$= \left(\frac{d}{dt}\Phi_{\dot{x}}(h^s(x), \frac{d}{dt}h^s(x)) - \Phi_x(h^s(x), \frac{d}{dt}h^s(x))\right)Dh^s(x),$$

which implies the claim (2.9.10), given that $Dh^s(x)$ is assumed to be regular for all $t \in [t_a, t_b]$, cf. (2.9.4). □

In 1918, E. Noether (1882–1935) proved the following conservation law. It is important in mathematics as well as theoretical physics.

Proposition 2.9.2. *Assume that* $\Phi : \mathbb{R}^n \times \mathbb{R}^n \to \mathbb{R}$ *is continuously totally differentiable and invariant under a family of local diffeomorphisms* $h^s : \mathbb{R}^n \to \mathbb{R}^n$, $s \in (-\delta, \delta)$, *in the sense of Definition 2.9.1. If* $x \in (C^1[t_a, t_b])^n$ *is a solution of the Euler-Lagrange system of the functional (2.9.1), i.e.,*

$$\frac{d}{dt}\Phi_{\dot{x}}(x,\dot{x}) = \Phi_x(x,\dot{x}) \quad on \quad [t_a, t_b], \qquad (2.9.12)$$

then for each $s \in (-\delta, \delta)$

$$\Phi_{\dot{x}}(h^s(x), Dh^s(x)\dot{x})\frac{\partial}{\partial s}h^s(x) = const. \quad for \quad t \in [t_a, t_b]. \qquad (2.9.13)$$

In the special case $h^0(x) = x$ *and* $Dh^0(x) = E$, *(2.9.13) yields*

$$\Phi_{\dot{x}}(x,\dot{x})\frac{\partial}{\partial s}h^s(x)|_{s=0} = const. \quad for \quad t \in [t_a, t_b]. \qquad (2.9.14)$$

Here, the product of the vectors $\Phi_{\dot{x}}$ *and* $\frac{\partial}{\partial s}h^s$ *is the Euclidean scalar product in* $\mathbb{R}^n$.

Different from our usual notation for the scalar product, we adopt here the notation of physicists.

Proof. By differentiation of (2.9.3) with respect to $s \in (-\delta, \delta)$, we obtain

$$\Phi_x(h^s(x), Dh^s(x)\dot{x})\frac{\partial}{\partial s}h^s(x) + \Phi_{\dot{x}}(h^s(x), Dh^s(x)\dot{x})\frac{\partial}{\partial s}Dh^s(x)\dot{x} = 0. \qquad (2.9.15)$$

Using

$$\frac{\partial}{\partial s}Dh^s(x)\dot{x} = \frac{\partial}{\partial s}\frac{d}{dt}h^s(x) = \frac{d}{dt}\frac{\partial}{\partial s}h^s(x) \tag{2.9.16}$$

and the Euler-Lagrange system (2.9.10), differentiation by the product rule yields for all $t \in [t_a, t_b]$:

$$\begin{aligned}
\frac{d}{dt}&\left(\Phi_{\dot{x}}(h^s(x), Dh^s(x)\dot{x})\frac{\partial}{\partial s}h^s(x)\right)\\
&= \frac{d}{dt}\Phi_{\dot{x}}(h^s(x), \frac{d}{dt}h^s(x))\frac{\partial}{\partial s}h^s(x) + \Phi_{\dot{x}}(h^s(x), Dh^s(x)\dot{x})\frac{d}{dt}\frac{\partial}{\partial s}h^s(x)\\
&= \Phi_x(h^s(x), \frac{d}{dt}h^s(x))\frac{\partial}{\partial s}h^s(x) + \Phi_{\dot{x}}(h^s(x), Dh^s(x)\dot{x})\frac{\partial}{\partial s}Dh^s(x)\dot{x}\\
&= \Phi_x(h^s(x), Dh^s(x)\dot{x})\frac{\partial}{\partial s}h^s(x) + \Phi_{\dot{x}}(h^s(x), Dh^s(x)\dot{x})\frac{\partial}{\partial s}Dh^s(x)\dot{x} = 0
\end{aligned} \tag{2.9.17}$$

by (2.9.15). $\square$

We give some **applications** to Lagrangian mechanics of Noether's Theorem.

1. Here and in the next two applications, we return to the mechanical model introduced in (2.5.28), (2.5.29), (2.5.31): N point masses $m_1, \ldots, m_N$ in $\mathbb{R}^3$ have the coordinates $x = (x_1, y_1, z_1, \ldots, x_N, y_N, z_N) \in \mathbb{R}^{3N}$. The Lagrangian is the free energy

$$L(x, \dot{x}) = T(\dot{x}) - V(x) \quad \text{where}$$

$$T(\dot{x}) = \sum_{k=1}^{N}\frac{1}{2}m_k(\dot{x}_k^2 + \dot{y}_k^2 + \dot{z}_k^2). \tag{2.9.18}$$

We assume invariance under a simultaneous shift of all point masses in the direction of one axis, i.e.,

$$h^s : \mathbb{R}^{3N} \to \mathbb{R}^{3N} \quad \text{is defined by}$$

$$h^s(x_1, y_1, z_1, \ldots, x_N, y_N, z_N) = (x_1 + s, y_1, z_1, \ldots, x_N + s, y_N, z_N). \tag{2.9.19}$$

The invariance of the kinetic energy is obvious, and we assume only the invariance of the potential energy, which means

$$V(h^s(x)) = V(x). \tag{2.9.20}$$

Then, (2.9.14) of Proposition 2.9.2 yields

$$L_{\dot{x}}(x, \dot{x})\frac{\partial}{\partial s}h^s(x)|_{s=0} = \sum_{k=1}^{N}m_k\dot{x}_k = const. \tag{2.9.21}$$

along each solution x of the Euler-Lagrange system of the action (2.5.31). These curves x are solutions of the equations of motion (the Euler-Lagrange system)

$$m_k \ddot{x}_k = -V_{x_k}(x),$$
$$m_k \ddot{y}_k = -V_{y_k}(x), \qquad\qquad\qquad (2.9.22)$$
$$m_k \ddot{z}_k = -V_{z_k}(x) \quad \text{on} \quad [t_a, t_b], \ k = 1, \dots, N.$$

If the potential energy is invariant under a simultaneous shift in the direction of one axis, the conservation law (2.9.21) means that the combined total momentum of all point masses in that direction is conserved.

2. Assume that the potential energy of the N point masses $m_1, \dots, m_N$ depends only on the differences $(x_i, y_i, z_i) - (x_k, y_k, z_k)$ of the coordinates of m_i and of m_k, $i, k = 1, \dots, N$. Then, V is invariant under simultaneous shifts in all directions of $\mathbb{R}^3$, and as seen in Example 1, the total momentum of all point masses is constant in all directions:

$$\sum_{k=1}^{N} m_k(\dot{x}_k, \dot{y}_k, \dot{z}_k) = a \quad \text{or}$$

$$\qquad\qquad\qquad\qquad\qquad\qquad (2.9.23)$$

$$\sum_{k=1}^{N} m_k(x_k, y_k, z_k) = at + b \quad \text{where} \quad a, b \in \mathbb{R}^3.$$

The coordinates of the barycenter of all N point masses are $\sum_{k=1}^{N} m_k(x_k, y_k, z_k) / \sum_{k=1}^{N} m_k$, and (2.9.23) means that the barycenter moves with constant speed in one direction of the three-dimensional space.

3. Assume that the potential energy of the N point masses depends only on the distances $\sqrt{(x_i - x_k)^2 + (y_i - y_k)^2 + (z_i - z_k)^2}$ of m_i and m_k, $i, k = 1, \dots, N$, then the total linear momentum is constant in all directions.

However, there is still another conservation law. The free energy $L(x, \dot{x}) = T(\dot{x}) - V(x)$ is also invariant under simultaneous rotations of all point masses. Take for instance the rotation about the z-axis by an angle s

$$R^s = \begin{pmatrix} \cos s & -\sin s & 0 \\ \sin s & \cos s & 0 \\ 0 & 0 & 1 \end{pmatrix},$$

$$\qquad\qquad\qquad\qquad\qquad\qquad (2.9.24)$$

$$h^s(x) = (R^s(x_1, y_1, z_1), \dots, R^s(x_N, y_N, z_N)).$$

Then, linearity and orthogonality of h^s yield

$$Dh^s(x)\dot{x} = (R^s(\dot{x}_1, \dot{y}_1, \dot{z}_1), \dots, R^s(\dot{x}_N, \dot{y}_N, \dot{z}_N)),$$
$$T(Dh^s(x)\dot{x}) = T(\dot{x}), \quad \text{since} \qquad\qquad\qquad (2.9.25)$$
$$\|R^s(\dot{x}_k, \dot{y}_k, \dot{z}_k)\|^2 = \|(\dot{x}_k, \dot{y}_k, \dot{z}_k)\|^2 = \dot{x}_k^2 + \dot{y}_k^2 + \dot{z}_k^2$$

for $k = 1, \dots, N$. Furthermore, the orthogonality of R^s implies that simultaneous rotations do not change the distances of the point masses. Hence, they leave the potential energy invariant:

$$V(h^s(x)) = V(x). \qquad\qquad\qquad (2.9.26)$$

Since $L(x,\dot{x}) = T(\dot{x}) - V(x)$ is invariant under simultaneous rotations (2.9.24), Noether's Theorem (Proposition 2.9.2) yields:

$$\frac{\partial}{\partial s}R^s|_{s=0} = \begin{pmatrix} 0 & -1 & 0 \\ 1 & 0 & 0 \\ 0 & 0 & 0 \end{pmatrix},$$

$$L_{\dot{x}}(x,\dot{x})\frac{\partial}{\partial s}h^s(x)|_{s=0} = \sum_{k=1}^{N} m_k(\dot{y}_k x_k - \dot{x}_k y_k) \qquad (2.9.27)$$

$$= \sum_{k=1}^{N} ((x_k,y_k,z_k) \times m_k(\dot{x}_k,\dot{y}_k,\dot{z}_k), e_3) = const.,$$

where "$\times$" denotes the vector product, $(\ ,\ )$ denotes the Euclidean scalar product in $\mathbb{R}^3$, and $e_3 = (0,0,1)$. The expression $(2.9.27)_3$ describes the total angular momentum of the point masses about the z-axis.

Since the free energy $L(x,\dot{x}) = T(\dot{x}) - V(x)$ is (by assumption) invariant under simultaneous rotations about all axes, the total angular momentum is constant about all axes, i.e.,

$$\sum_{k=1}^{N}(x_k,y_k,z_k) \times m_k(\dot{x}_k,\dot{y}_k,\dot{z}_k) = c \in \mathbb{R}^3. \qquad (2.9.28)$$

Finally, the conservation of the total energy of one point mass, cf. Exercise 1.10.4, holds for N point masses as well.

We summarize: If the potential energy V of the N point masses depends only on their distances, the following conservation laws hold for any motion governed by (2.9.22):

$$\sum_{k=1}^{N} m_k(\dot{x}_k,\dot{y}_k,\dot{z}_k) = a$$

(conservation of linear momentum),

$$\sum_{k=1}^{N}(x_k,y_k,z_k) \times m_k(\dot{x}_k,\dot{y}_k,\dot{z}_k) = c \qquad (2.9.29)$$

(conservation of angular momentum),

$$\sum_{k=1}^{N}\frac{1}{2}m_k(\dot{x}_k^2 + \dot{y}_k^2 + \dot{z}_k^2) + V(x_1,y_1,z_1,\ldots,x_N,y_N,z_N) = E$$

(conservation of total energy).

A prominent and interesting example is the solar system, consisting of the sun and its planets considered as point masses. By Newton's law of gravitation, the force between two point masses m_i and m_k depends only on the masses and their mutual distance. If N masses are involved, the system gives rise to a so-called N-body problem. For $N \geq 3$, the theoretical analysis of their motion is an extremely hard problem, even when exploiting the conservation laws (2.9.29). Up to the present time, there is no proof that the motion of the solar system is stable. This was the prize question asked by the Swedish king Oskar II in the year 1885. The prize was won by H. Poincaré although he did not give a definite answer. But he opened the door to a promising mathematical theory.

The earth with the moon, the International Space Station (ISS), and countless satellites form a similar system, where, however, the influence of the sun and the other planets cannot be neglected. The success and precision of all space flights rely on accurate numerical simulations.

In the next paragraph, we study the two-body problem.

Exercise

2.9.1. Assume that a continuously partially differentiable Lagrange function Φ : $\mathbb{R}^n \times \mathbb{R}^n \to \mathbb{R}$ is invariant under a family of local diffeomorphisms $h^s : \mathbb{R}^n \to \mathbb{R}^n$, $s \in (-\delta, \delta)$, in the sense of Definition 2.9.1. Prove the following conservation law:

Let $x \in (C^2[t_a, t_b])^n$ be a local minimizer of the functional

$$J(x) = \int_{t_a}^{t_b} \Phi(x, \dot{x}) dt,$$

subject to a holonomic constraint

$$\Psi(x) = 0, \quad \text{where} \quad \Psi : \mathbb{R}^n \to \mathbb{R}^m, \ n > m,$$

is three times continuously partially differentiable. Assume that the Jacobian matrix $D\Psi(x) \in \mathbb{R}^{m \times n}$ has a maximal rank m for all $x \in M = \{x \in \mathbb{R}^n | \Psi(x) = 0\}$ and that Ψ is invariant, i.e.,

$$\Psi(h^s(x)) = \Psi(x) \quad \text{for all} \quad x \in \mathbb{R}^n.$$

Then,

$$\Phi_{\dot{x}}(h^s(x), Dh^s(x)\dot{x}) \frac{\partial}{\partial s} h^s(x) = \text{const.} \quad \text{for} \quad t \in [t_a, t_b] \quad \text{and for all} \quad s \in (-\delta, \delta),$$

and in particular for $h^0(x) = x$ and $Dh^0(x) = E$,

$$\Phi_{\dot{x}}(x, \dot{x}) \frac{\partial}{\partial s} h^s(x)|_{s=0} = \text{const.} \quad \text{for} \quad t \in [t_a, t_b] \quad \text{and for all} \quad s \in (-\delta, \delta).$$

2.10   The Two-Body Problem

Two point masses m_1 and m_2 move in $\mathbb{R}^3$ acted on only by the gravitational force on each other. Isaac Newton (1643–1727) gave the formula for this force: If the coordinates of m_k are (x_k, y_k, z_k), then the distance between the two point masses m_1 and m_2 is $r = \sqrt{(x_1 - x_2)^2 + (y_1 - y_2)^2 + (z_1 - z_2)^2}$, and the magnitude of the force acting on m_1 as well as on m_2 amounts to

$$G = \gamma \frac{m_1 m_2}{r^2}, \quad \text{where } \gamma \text{ is the gravitational constant.} \tag{2.10.1}$$

In the equations of motion (2.9.22), the force is the negative gradient of the potential energy. According to (2.10.1), for $x = (x_1, y_1, z_1, x_2, y_2, z_2)$, this potential energy is given by

$$V(x) = -\frac{k}{r}, \quad k = \gamma m_1 m_2, \tag{2.10.2}$$

and the equations of motion for m_1 and m_2 read

$$m_1 \ddot{x}_1 = -\frac{k}{r^3}(x_1 - x_2), \quad m_2 \ddot{x}_2 = \frac{k}{r^3}(x_1 - x_2),$$

$$m_1 \ddot{y}_1 = -\frac{k}{r^3}(y_1 - y_2), \quad m_2 \ddot{y}_2 = \frac{k}{r^3}(y_1 - y_2), \tag{2.10.3}$$

$$m_1 \ddot{z}_1 = -\frac{k}{r^3}(z_1 - z_2), \quad m_2 \ddot{z}_2 = \frac{k}{r^3}(z_1 - z_2).$$

The force on m_1 acts in direction of m_2, and the force on m_2 acts in direction of m_1, with the same magnitude given by (2.10.1). Since the potential energy depends only on the distance between m_1 and m_2, the conservation laws (2.9.29) are valid, and in particular (2.9.23)$_2$ shows that the barycenter of m_1 and m_2 moves with a constant speed in a fixed direction of $\mathbb{R}^3$. This motion, however, is not of interest. Only the motion of the relative position vector

$$(x, y, z) = (x_1 - x_2, y_1 - y_2, z_1 - z_2), \tag{2.10.4}$$

describing the relative position of the two point masses is to be analyzed. By (2.10.3), the components of the relative position vector satisfy the following system:

$$m_1 m_2 \ddot{x} = -(m_1 + m_2)\frac{k}{r^3}x,$$

$$m_1 m_2 \ddot{y} = -(m_1 + m_2)\frac{k}{r^3}y, \tag{2.10.5}$$

$$m_1 m_2 \ddot{z} = -(m_1 + m_2)\frac{k}{r^3}z.$$

The system (2.10.5) corresponds to the Euler-Lagrange equations of the Lagrangian

$$L = \frac{1}{2}m(\dot{x}^2 + \dot{y}^2 + \dot{z}^2) - V(x, y, z), \quad \text{where}$$

$$V = -\frac{k}{r}, \quad r = \sqrt{x^2 + y^2 + z^2}, \quad m = \frac{m_1 m_2}{m_1 + m_2}. \tag{2.10.6}$$

Again, this potential energy $(2.10.6)_2$ depends only on the length r of the vector (x, y, z). Hence for solutions of (2.10.5), the conservation laws (2.9.29) are valid; in particular, conservation of the angular momentum holds:

$$(x, y, z) \times m(\dot{x}, \dot{y}, \dot{z}) = c \in \mathbb{R}^3. \tag{2.10.7}$$

Therefore, any solution (x, y, z) of (2.10.5) fulfills

$$((x, y, z), (x, y, z) \times (\dot{x}, \dot{y}, \dot{z})) = ((x, y, z), \frac{c}{m}) = 0, \tag{2.10.8}$$

because the vector product is orthogonal to the relative position vector. For $c = 0$, the solution fulfills $(\dot{x}, \dot{y}, \dot{z}) = \alpha(x, y, z)$ for all $t \in [t_a, t_b]$, describing a straight line. (Equations (2.10.5) do not admit an equilibrium.) For $c \neq 0$, the solution curve is in a plane orthogonal to the vector c. In both cases, we see that

$$\text{any solution of (2.10.5) is planar.} \tag{2.10.9}$$

Without loss of generality, we assume that the motion takes place in the (x, y)-plane, i.e., $z = z(t) = 0$ for all t. The conservation of angular momentum (2.10.7) yields in this case

$$(x, y, 0) \times (\dot{x}, \dot{y}, 0) = (x\dot{y} - \dot{x}y)e_3 = \frac{c}{m}, \quad e_3 = (0, 0, 1),$$

$$x\dot{y} - \dot{x}y = \beta \in \mathbb{R}, \quad \text{where} \quad \beta e_3 = \frac{c}{m} \text{ for all } t \in [t_a, t_b]. \tag{2.10.10}$$

If $\beta = 0$, the solution runs in a straight line through $(0, 0)$. For $\beta \neq 0$, we consider the domain $F = F(t)$ sketched in Figure 2.20.

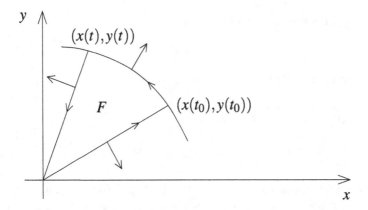

Fig. 2.20 On Green's Formula

By Green's formula (2.2.3), the area F is given by the line integral

$$F = \frac{1}{2} \int_{s_a}^{s_b} \tilde{x}\dot{\tilde{y}} - \dot{\tilde{x}}\tilde{y}\, ds \tag{2.10.11}$$

where $\{(\tilde{x}(s), \tilde{y}(s)) | s \in [s_a, s_b]\}$ is a parametrization of the boundary of F, cf. (2.2.4). Along the lines from $(0,0)$ to $(x(t_0), y(t_0))$ and from $(x(t), y(t))$ to $(0,0)$, the integrand vanishes; the vectors $(\tilde{x}, \tilde{y})$ and $(\dot{\tilde{y}}, -\dot{\tilde{x}})$ are orthogonal. From $(x(t_0), y(t_0))$ to $(x(t), y(t))$, we parametrize the curve by $(\tilde{x}, \tilde{y}) = (x, y)$, and $(2.10.10)_2$ implies

$$F = \frac{1}{2} \int_{t_0}^{t} x\dot{y} - \dot{x}y\, dt = \frac{1}{2}\beta(t - t_0) \quad \text{and}$$

$$\frac{dF}{dt} = \frac{1}{2}\beta. \tag{2.10.12}$$

The last expression is called "areal velocity," and formulas (2.10.12) mean the following:

> The areal velocity of any solution is constant.
> The line segment from $(0,0)$ to $(x(t), y(t))$ sweeps out   (2.10.13)
> equal areas during equal time intervals.

We introduce polar coordinates in the (x, y)-plane:

$$\begin{aligned} x &= r\cos\varphi, & r &\geq 0, \\ y &= r\sin\varphi, & \varphi &\in \mathbb{R}. \end{aligned} \tag{2.10.14}$$

The conserved total energy $E = T + V$, cf. (2.9.29), of a solution $(r, \varphi) = (r(t), \varphi(t))$ is given by

$$\frac{1}{2}m(\dot{r}^2 + r^2\dot{\varphi}^2) + V(r) = E, \tag{2.10.15}$$

and the angular momentum, cf. (2.10.10), reads

$$r^2\dot{\varphi} = \beta. \tag{2.10.16}$$

Replacing $\dot{\varphi}$ in (2.10.15) by $\dot{\varphi}$ from (2.10.16) yields

$$\frac{1}{2}m(\dot{r}^2 + \frac{\beta^2}{r^2}) - \frac{k}{r} = E, \quad \text{or}$$

$$\dot{r} = \sqrt{\frac{2}{m}(E - U(r))} = F(r), \quad \text{where} \quad U(r) = \frac{\beta^2 m}{2r^2} - \frac{k}{r}. \tag{2.10.17}$$

The function $U(r)$ is called "effective potential energy." Obviously, it is necessary that $U(r) \leq E$ for a solution $r = r(t)$. For $\beta \neq 0$, solutions with $-\frac{k^2}{2\beta^2 m} \leq E < 0$ are bounded, there is no solution for $E < -\frac{k^2}{2\beta^2 m}$, and for $E \geq 0$ all solutions are unbounded. For $\beta = 0$, solutions for $E < 0$ are bounded, and for $E \geq 0$, they might be unbounded.

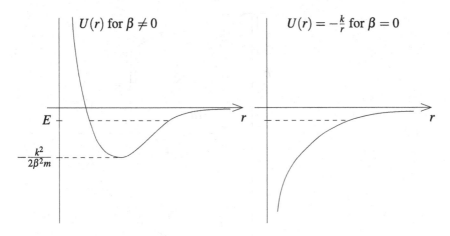

Fig. 2.21 The Effective Potential Energy

Now, let $-\frac{k^2}{2\beta^2 m} < E < 0$. Then,

$$\varphi = G(r), \qquad \text{where} \qquad \frac{d}{dr}G(r) = \frac{\beta}{r^2 F(r)}, \tag{2.10.18}$$

which follows from the fact that the expressions (2.10.16) and (2.10.17)$_2$ yield $\dot{\varphi} = \frac{d}{dr}G(r)\dot{r} = \frac{d}{dt}G(r)$. The function

$$\frac{\beta\sqrt{m}/r^2}{\sqrt{2(E-U(r))}} \qquad \text{has a primitive,}$$

$$G(r) = \arccos\frac{(\beta\sqrt{m}/r) - (k/\beta\sqrt{m})}{\sqrt{2E + (k^2/\beta^2 m)}}, \qquad \text{and} \quad \varphi = G(r) \text{ yields} \tag{2.10.19}$$

$$r = \frac{p}{1 + e\cos\varphi}, \qquad \text{where} \quad p = \beta^2 m/k, \quad e = \sqrt{1 + E\frac{2\beta^2 m}{k^2}}.$$

The assumption on E gives $0 < e < 1$, and (2.10.19)$_3$ corresponds to an ellipse, where

$$a = \frac{1}{2}\left(\frac{p}{1+e} + \frac{p}{1-e}\right) = \frac{p}{1-e^2} \quad \text{is the semimajor axis,}$$

$$b = a\sqrt{1-e^2} \quad \text{is the semi-minor axis,} \tag{2.10.20}$$

and the origin is a focus, cf. Figure 2.22.

A transition from φ to $\varphi + c$ (an integration constant in (2.10.19)) corresponds to a rotation of the ellipse about the origin.

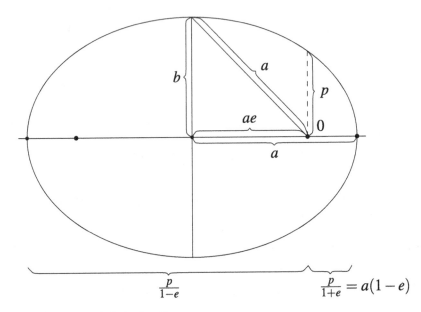

Fig. 2.22 An Ellipse in Polar Coordinates

All solutions with nonvanishing angular momentum and with energy allowing only bounded solutions run on ellipses. According to (2.10.4), the vector $(x,y,0)$ is the vector from m_2 to m_1 with length r, which means:

$$\text{The point mass } m_1 \text{ runs on an ellipse} \\ \text{where the point mass } m_2 \text{ is at one of its foci.} \tag{2.10.21}$$

The period T of m_1 on the ellipse is related to its area F, cf. (2.10.12):

$$F = \frac{1}{2}\beta T = \pi ab = \pi a^2 \sqrt{1 - e^2}. \tag{2.10.22}$$

By $(2.10.19)_3$ and (2.10.20), we obtain

$$\beta^2 = \frac{k}{m}p = \frac{k}{m}a(1 - e^2) \quad \text{and}$$
$$\frac{T^2}{a^3} = 4\pi^2\frac{m}{k}. \tag{2.10.23}$$

By the definitions (2.10.2) and $(2.10.6)_2$, we can rewrite $(2.10.23)_2$ as follows:

$$\frac{T^2}{a^3} = \frac{4\pi^2}{\gamma(m_1 + m_2)}. \tag{2.10.24}$$

In 1609, Johannes Kepler (1571–1630) published in his treatise "Astronomia Nova" two laws, which are known as the first and second Kepler's laws:

1. The orbit of a planet is an ellipse with the sun at one of the two foci.
2. A line segment joining the sun and a planet sweeps out equal areas during equal intervals of time.

   We find both laws in (2.10.21) and in (2.10.13).

   In 1619, Kepler published in an article "Harmonices Mundi" his so-called third law:
3. The square of the orbital period of a planet is proportional to the cube of the semimajor axis of its ellipse.

To be more precise, the proportionality constant is the same for all planets. To Kepler, this demonstrated the harmony of the world.

We find the third law in (2.10.24) - almost. But due to the big mass m_2 of the sun, the sum $m_1 + m_2$ is almost equal for all planets.

We have to admit an inaccuracy too: The above proof of the three Kepler's laws is valid only if one planet orbits the sun. But there are eight big planets, Mercury, Venus, Earth, Mars, Jupiter, Saturn, Uranus, Neptune, an asteroid belt, and a trans-Neptunian region with Pluto and some more. Since all celestial bodies attract each other according to Newton's law, the solar system has to be treated as an N-body problem with $N \geq 9$. However, an accurate analysis like that for the two-body problem is not (yet) possible. As mentioned before, we do not even know if the present constellation of the solar system is stable.

To Kepler and his contemporaries, the harmony of the solar system was undoubted. The order of the cosmos was a proof of God for Newton.

Some comments on other solutions of (2.10.17): For $\beta \neq 0$ and $E \geq 0$, there is a minimal $r > 0$ with $\dot{r} = 0$, cf. Figure 2.21. If m_1 starts with $\dot{r} < 0$, $\dot{r}$ changes sign for the minimal r and $\dot{r} > 0$ is conserved for the future time. The point mass m_1 passes m_2 and disappears at infinity.

For $\beta = 0$, the orbits are straight lines through the origin $(0,0)$, due to $\dot{\varphi} = 0$, by (2.10.16). For $E < 0$, there is a maximal $r > 0$ where $\dot{r} = 0$ can vanish. If m_1 starts with $\dot{r} > 0$ below that bound, then $\dot{r}$ changes sign and m_1 falls with $\dot{r} < 0$ on a straight line on m_2. If $E \geq 0$, the sign of $\dot{r}$ does not change: The point mass m_1 falls on m_2 if $\dot{r} < 0$, or it disappears at infinity if $\dot{r} > 0$.

Exercises

2.10.1. Assume that at time $t = 0$, the distance of the point mass m_1 from m_2 is R and that its velocity at time $t = 0$ is zero. Give a formula for the time T that the point mass takes to fall on a straight line on m_2, and estimate the time T that m_1 takes from $r = R$ to $r = 0$.

2.10.2. Assume that a point mass m_1 falls on a straight line from infinity on m_2 with vanishing energy $E = 0$.

a) Give its velocity at the distance R from m_2.
b) Give the time that it takes to fall from height R on m_2.

Chapter 3
Direct Methods
in the Calculus of Variations

3.1 The Method

The Euler-Lagrange calculus was created to determine extremals of functionals. If the solution of the Euler-Lagrange equation is unique among all admitted functions, then physical or geometric insights into the problem might lead to the conclusion that it is indeed the desired extremal. In addition, the second variation provides necessary and also sufficient conditions on extremals.

History shows that this approach is quite effective, but it hits its limit if the solvability of the Euler-Lagrange equation is open. In particular, if the admitted functions depend on more than one variable, the Euler-Lagrange equation is a partial differential equation, for which it is not generally possible to give an explicit solution. But also for boundary value problems governed by ordinary differential equations, the existence of a solution is generally not as obvious, as shown for some examples given in this book. In these cases, the calculus of variations provides a converse argument: If the variational problem has an extremal then the Euler-Lagrange equation has a solution.

This argument was used by Dirichlet to "prove" in an elegant way the existence of harmonic functions having prescribed boundary values. The minimizer among all functions, having the same boundary values, of the so-called Dirichlet integral solves Laplace's equation since it is the corresponding Euler-Lagrange equation. Apart from its regularity, which is not trivial in this case, the crucial problem is its existence. For Dirichlet the answer was "evident" since the Dirichlet integral is convex and bounded from below. As we mention in the Introduction, Weierstraß' example, discussed in Paragraph 1.5, shows that convex functionals, which are bounded from below, do not have necessarily admitted minimizers.

The key question of the direct methods in the calculus of variations is the following.

Under which general conditions do functionals have minimizers?

We discuss this questions for an abstract functional

© Springer International Publishing AG 2018
H. Kielhöfer, *Calculus of Variations*, Texts in Applied Mathematics 67,
https://doi.org/10.1007/978-3-319-71123-2_3

$$J : D \subset X \to \mathbb{R} \cup \{+\infty\} \qquad (3.1.1)$$

where X is a normed linear space over $\mathbb{R}$. For global minimizers, which are of main interest, a first necessary condition is that the functional J is bounded from below on D:

(H1) $\qquad\qquad\qquad J(y) \geq c \in \mathbb{R} \quad$ for all $\quad y \in D$.

Then the infimum exists in $\mathbb{R}$ (if J is not identically $+\infty$ on D):

$$\inf\{J(y)|y \in D\} = m > -\infty. \qquad (3.1.2)$$

According to the definition of the infimum, there exists a so-called minimizing sequence $(y_n)_{n \in \mathbb{N}} \subset D$ satisfying

$$\lim_{n \to \infty} J(y_n) = m \quad \text{in} \quad \mathbb{R}. \qquad (3.1.3)$$

Here are two more hypotheses:

(H2) The minimizing sequence $(y_n)_{n \in \mathbb{N}}$ contains a subsequence $(y_{n_k})_{k \in \mathbb{N}}$, which converges to an element $y_0 \in D$ with respect to a suitable definition of convergence:

$$\lim_{k \to \infty} y_{n_k} = y_0 \in D. \qquad (3.1.4)$$

(H3) The functional J is lower semi-continuous with respect to the convergence (3.1.4), i.e.,

$$\lim_{n \to \infty} \tilde{y}_n = \tilde{y}_0 \quad \Rightarrow \quad \liminf_{n \to \infty} J(\tilde{y}_n) \geq J(\tilde{y}_0). \qquad (3.1.5)$$

We then have:

Proposition 3.1.1. *Under the hypotheses (H1), (H2), and (H3), the functional J possesses a global minimizer $y_0 \in D$.*

Proof. For the subsequence of the minimizing sequence, we have

$$\lim_{k \to \infty} J(y_{n_k}) = m = \liminf_{k \to \infty} J(y_{n_k}) \geq J(y_0) \geq m. \quad \text{Hence}$$
$$J(y_0) = m = \inf\{J(y)|y \in D\}. \qquad\qquad (3.1.6)$$

$\qquad\qquad\qquad\qquad\qquad\qquad\qquad\qquad\qquad\qquad\qquad\qquad\qquad\qquad\qquad\qquad\qquad\square$

The hypotheses (H2) and (H3) compete: The more general is the definition of the convergence in (H2), i.e., the more sequences converge, the more restrictive is the lower semi-continuity for the functional. One has to find a balance in the following sense: Under which natural definition of convergence do minimizing sequences contain convergent subsequences such that a sufficiently large class of funtionals are lower semi-continuous with respect to that convergence.

We give the definition of a suitable convergence and show next that the class of functionals satisfying (H3) is sufficiently large.

In functional analysis two definitions of convergence are introduced in a normed linear space X: convergence with respect to the norm, cf. Definition 1.1.4, called "strong convergence," and the so-called weak convergence defined as follows:

$X' = \{\ell : X \to \mathbb{R} | \ell$ is linear and continuous$\}$ is the dual space of X.

A sequence $(y_n)_{n \in \mathbb{N}} \subset X$ converges weakly to y_0 $\qquad$ (3.1.7)

if $\lim\limits_{n \to \infty} \ell(y_n) = \ell(y_0)$ for all $\ell \in X'$.

The weak convergence is denoted $w\text{-}\lim_{n \to \infty} y_n = y_0$. Obviously strong convergence implies weak convergence.

Remark. *In $X = \mathbb{R}^n$ endowed with the Euclidean norm the weak convergence means a convergence componentwise, which, in turn, implies the strong convergence. Therefore in a finite-dimensional normed linear space, strong and weak convergence are equivalent.*
In infinite-dimensional normed linear spaces strong convergence implies weak convergence but not vice versa. We provide a counterexample in the next remark.

A (real) Hilbert space is defined as follows:

> In a (real) Hilbert space X, the norm $\| \quad \|$
>
> is defined by a scalar product$(\quad , \quad)$
>
> in the following way: $\|y\| = \sqrt{(y,y)}$.
>
> A Hilbert space is complete in the following sense:
>
> Any Cauchy sequence converges
>
> to an element in X (strongly with respect to the norm). $\qquad$ (3.1.8)

As in the case of a finite-dimensional Euclidean space, a norm defined by a scalar product fulfills the axioms of Definition 1.1.3. The proof relies on the Cauchy-Schwarz inequality.

The Riesz representation theorem (F. Riesz, 1880-1956) reads as follows:

> For any $\ell \in X'$ there exists a unique $z \in X$
>
> such that $\ell(y) = (y,z)$ for all $y \in X$. $\qquad$ (3.1.9)

We prove this theorem in Proposition 3.1.2.

In view of (3.1.9), weak convergence in a real Hilbert space can be defined as follows:

$$w\text{-}\lim_{n \to \infty} y_n = y_0 \quad \Leftrightarrow \quad \lim_{n \to \infty} (y_n, z) = (y_0, z) \quad \text{for all} \quad z \in X. \qquad (3.1.10)$$

Remark. *As stated by Hilbert, the prototype of a Hilbert space is the space ℓ^2, which results from a Fourier analysis with a complete and countable orthonormal system:*

$$\ell^2 = \{y = (\eta_k)_{k \in \mathbb{N}} | \sum_{k=1}^{\infty} \eta_k^2 < \infty\} \quad \text{with the scalar product}$$

$$(y,z) = \sum_{k=1}^{\infty} \eta_k \zeta_k \quad \text{and norm} \quad \|y\| = \sqrt{\sum_{k=1}^{\infty} \eta_k^2}. \tag{3.1.11}$$

In this space we give an example of a weakly convergent sequence, which does not converge strongly:

$$e_n = (\delta_{kn})_{k \in \mathbb{N}} = (0,0,\ldots,1,0,\ldots) \quad \text{with the number 1 at the nth place,}$$

$$(e_n,z) = \zeta_n, \lim_{n \to \infty} \zeta_n = 0, \quad \text{since} \quad \sum_{k=1}^{\infty} \zeta_k^2 < \infty. \quad \text{On the other hand,} \tag{3.1.12}$$

$$\|e_n - e_m\| = \sqrt{2} \quad \text{for all } n,m \in \mathbb{N} \text{ with } n \neq m.$$

$$\text{Therefore} \quad \text{w-}\lim_{n \to \infty} e_n = 0 \quad \text{but} \quad \lim_{n \to \infty} e_n \quad \text{does not exist.}$$

The sequence $(e_n)_{n \in \mathbb{N}} \subset \ell^2$ is bounded since $\|e_n\| = 1$. This example shows that in infinite-dimensional normed spaces, the theorem of Bolzano-Weierstraß, saying that a bounded sequence contains a convergent subsequence, is not necessarily true.

We quote from functional analysis the following theorem, which goes back to Hilbert. It is crucial for the direct methods in the calculus of variations:

In a "reflexive" space, in a Hilbert space, e.g.,

any bounded sequence contains a weakly convergent subsequence. $\tag{3.1.13}$

Remark. *The definition of a reflexive normed linear space goes beyond the context of this introduction. All notions and spaces of functional analysis used in this and in the next paragraph are found in introductory textbooks on analysis and functional analysis, such as [18], [19], [20], [21] and [25]. We prove the sequential weak compactness in the Appendix.*

Our excursion into functional analysis allows us to specify hypothesis (H2) more precisely and, at the same time, to give a sufficient condition under which it is fulfilled:

(H2)′ The minimizing sequence $(y_n)_{n \in \mathbb{N}}$ is bounded in a reflexive space X, in a Hilbert space, e.g., which means $\|y_n\| \leq C$ for all $n \in \mathbb{N}$.

Then $(y_n)_{n \in \mathbb{N}}$ contains a subsequence $(y_{n_k})_{k \in \mathbb{N}}$ that converges weakly to an element $y_0 \in X$:

$$\text{w-}\lim_{k \to \infty} y_{n_k} = y_0 \in X. \tag{3.1.14}$$

If $D \subset X$ is closed and convex, for instance if $D = z_0 + X_0$, where X_0 is a closed subspace of X, then the weak limit y_0 of $(y_{n_k})_{k \in \mathbb{N}} \subset D$ is in D as well.

A suitable hypothesis on the functional J should entail the boundedness of a minimizing sequence. The common hypothesis is "coercivity," which in its general
form reads as follows:

(H2)'' For all unbounded sequences $(\tilde{y}_n)_{n\in\mathbb{N}} \subset D \subset X$ the sequence $(J(\tilde{y}_n))_{n\in\mathbb{N}}$ is
unbounded from above. This is the case if

$$J(y) \geq f(\|y\|) \quad \text{for all} \quad y \in D \subset X$$
$$\text{where} \quad \lim_{r\to\infty} f(r) = \infty. \tag{3.1.15}$$

What does hypothesis (H3) mean under (H2)' and (H2)''?

(H3)' The functional J is *weakly* lower semi-continuous, meaning

$$w\text{-}\lim_{n\to\infty} \tilde{y}_n = \tilde{y}_0 \quad \Rightarrow \quad \liminf_{n\to\infty} J(\tilde{y}_n) \geq J(\tilde{y}_0). \tag{3.1.16}$$

Which conditions on J imply (H3)'?

A sufficient condition is a partial convexity of J, which in the concrete case
(1.1.1) is given by the partial convexity of the Lagrange function F with respect
to the variable y', cf. Proposition 3.2.5. We recall Exercise 1.4.3, where for a local
minimizer the necessary Legendre condition, which is a local partial convexity with
respect to y', is to be proved. Therefore this sufficient condition to guarantee (H3)'
is natural.

Remark. *If the Lagrange function F is not partially convex with respect to the
variable y', the functional J in (1.1.1) does not necessarily have a global minimizer,
even if it is bounded from below and if minimizing sequences are bounded. A typical
example is given by $J(y) = \int_a^b y^2 + ((y')^2 - 1)^2 dx$. Its infimum zero is not a minimum:
Obviously the conditions $y = 0$ and $y' = \pm 1$ are exclusive. In view of the relevance
of energy functionals with W-potentials depending on y' of this type, which we
mention in Paragraph 2.4, a way out was created by extending the class of admitted
functions to measures, the so-called Young measures. For the example above, a minimizer has the derivatives ± 1 in each point with probability $1/2$, respectively. For
details we refer for instance to J. M. Ball "A version of the fundamental theorem
for Young measures" in "Partial Differential Equations and Continuum Models of
Phase Transitions," Lecture Notes in Physics 359, 207–215, Springer-Verlag Berlin,
Heidelberg (1989).*

So far the functional $J : X \to \mathbb{R} \cup \{+\infty\}$ is as general as possible. Next we restrict
ourselves to quadratic functionals for two reasons: The direct methods give good
results, and the Euler-Lagrange equations are linear. However, in Paragraph 3.3 we
demonstrate the strength of the direct method by solving nonlinear problems.

Definition 3.1.1. *Let X be a (real) Hilbert space. A bilinear form $B : X \times X \to \mathbb{R}$ is equivalent to the scalar product $(\, ,)$ on X if the following conditions hold for all $y, z \in X$:*

$$B(y,z) = B(z,y), \quad \textit{i.e., } B \textit{ is symmetric,}$$
$$|B(y,z)| \le C_1 \|y\| \|z\| \quad \textit{where } C_1 > 0, \textit{ i.e., } B \textit{ is continuous,} \qquad (3.1.17)$$
$$B(y,y) \ge C_2 \|y\|^2 \quad \textit{where } C_2 > 0, \textit{ i.e., } B \textit{ is positive definite.}$$

If a bilinear form fulfills all conditions (3.1.17), then B is a scalar product on X, and in view of $C_2 \|y\|^2 \le B(y,y) \le C_1 \|y\|^2$, the expression $\sqrt{B(y,y)}$ defines a norm that is equivalent to $\sqrt{(y,y)}$, cf. (3.1.8). The following proposition contains Riesz' representation theorem (3.1.9) as a special case.

Proposition 3.1.2. *For a bilinear form $B : X \times X \to \mathbb{R}$ on a real Hilbert X space, assume the hypotheses (3.1.17). Then the following holds:*

$$\textit{For any } \ell \in X' \textit{there is a unique } y_0 \in X \textit{such that}$$
$$B(y_0, h) = \ell(h) \quad \textit{for all } h \in X. \qquad (3.1.18)$$

This element $y_0 \in X$ is the global minimizer of the functional

$$J(y) = \tfrac{1}{2} B(y,y) - \ell(y), \quad \textit{i.e.,}$$
$$J(y_0) = \inf\{J(y) | y \in X\} = m. \qquad (3.1.19)$$

All minimizing sequences $(y_n)_{n \in \mathbb{N}}$ of the functional $(3.1.19)_1$ converge strongly in X to the global minimizer y_0:

$$\lim_{n \to \infty} J(y_n) = m = J(y_0) \quad \textit{implies}$$
$$\lim_{n \to \infty} y_n = y_0 \quad \textit{in} \quad X. \qquad (3.1.20)$$

Proof. The continuity of $\ell : X \to \mathbb{R}$ is equivalent to $|\ell(y)| \le C_3 \|y\|$ for all $y \in X$, and due to the positive definiteness of B, the functional is coercive and bounded from below:

$$J(y) \ge \tfrac{1}{2} C_2 \|y\|^2 - C_3 \|y\| \ge c \quad \text{for all } y \in X. \qquad (3.1.21)$$

Using the symmetry $(3.1.17)_1$ and the positive definiteness $(3.1.17)_2$ of B, we obtain for any minimizing sequence $(y_n)_{n \in \mathbb{N}} \subset X$:

$$C_2\|y_n - y_m\|^2 \leq B(y_n - y_m, y_n - y_m)$$
$$= 2B(y_n, y_n) + 2B(y_m, y_m) - B(y_n + y_m, y_n + y_m)$$
$$= 4\left(\tfrac{1}{2}B(y_n, y_n) - \ell(y_n) + \tfrac{1}{2}B(y_m, y_m) - \ell(y_m)\right.$$
$$\left. - B(\tfrac{1}{2}(y_n + y_m), \tfrac{1}{2}(y_n + y_m)) + 2\ell(\tfrac{1}{2}(y_n + y_m)))\right) \qquad (3.1.22)$$
$$= 4(J(y_n) + J(y_m) - 2J(\tfrac{1}{2}(y_n + y_m)))$$
$$= 4(J(y_n) - m + J(y_m) - m - 2(J(\tfrac{1}{2}(y_n + y_m)) - m))$$
$$\leq 4(\varepsilon + \varepsilon) \quad \text{for} \quad n, m \geq n_o(\varepsilon),$$

where we use $J(\tfrac{1}{2}(y_n + y_m)) - m \geq 0$. Thus any minimizing sequence is a Cauchy sequence in X, and due to completeness of the Hilbert space, it possesses a limit $y_0 \in X$. By $J(y) - J(\tilde{y}) = \tfrac{1}{2}B(y - \tilde{y}, y + \tilde{y}) - \ell(y - \tilde{y})$, the functional $J : X \to \mathbb{R}$ is continuous, which proves (3.1.20).

Apparently,

$$J(y_0 + th) = \tfrac{1}{2}B(y_0 + th, y_0 + th) - \ell(y_0 + th)$$
$$= J(y_0) + t(B(y_0, h) - \ell(h)) + \tfrac{1}{2}t^2 B(h, h) \qquad (3.1.23)$$
$$\geq J(y_0) \quad \text{for all } h \in X \text{ and for all } t \in \mathbb{R}.$$

Therefore

$$\frac{d}{dt}J(y_0 + th)|_{t=0} = \delta J(y_0)h = B(y_0, h) - \ell(h) = 0 \quad \text{for all } h \in X, \qquad (3.1.24)$$

which proves (3.1.18).

To prove uniqueness, we assume

$$B(y_0, h) = \ell(h) = B(\tilde{y}_0, h) \quad \text{or}$$
$$B(y_0 - \tilde{y}_0, h) = 0 \quad \text{for all } h \in X, \text{ and in particular}, \qquad (3.1.25)$$
$$B(y_0 - \tilde{y}_0, y_0 - \tilde{y}_0) = 0,$$

which implies $y_0 = \tilde{y}_0$ by positive definiteness. $\qquad \square$

Remark. *The representation theorem* (3.1.18) *is also valid for bilinear forms satisfying* (3.1.17) *without the symmetry* (3.1.17)$_1$. *In this case,* (3.1.18) *is called the* **Lax-Milgram Theorem**, *which is not proved with methods of the calculus of variations, but which follows from the Riesz representation theorem as shown below: For any $u \in X$, there exist unique v and v^* in X, such that for the scalar product, we have*

$$B(u, y) = (v, y), \quad \text{and}$$
$$B(y, u) = (y, v^*) \quad \text{for all} \quad y \in X,$$

which follows from the fact that $B(u, \cdot)$, $B(\cdot, u) \in X'$. Setting $v = Lu$ and $v^* = L^*u$, we define linear operators $L, L^* : X \to X$, which fulfill

$$(Ly, u) = (y, L^*u) \quad \text{for all} \quad y, u \in X.$$

Both operators L and L^* are continuous:

$$\|Lu\|^2 = (Lu, Lu) = B(u, Lu) \leq C_1 \|u\| \|Lu\|, \quad or$$
$$\|Lu\| \leq C_1 \|u\|, \quad \text{and analogously,} \quad \|L^*u\| \leq C_1 \|u\| \quad \text{for all} \quad u \in X.$$

The positive definiteness of B implies

$$C_2 \|u\|^2 \leq B(u, u) = (Lu, u) \leq \|Lu\| \|u\|, \quad or$$
$$C_2 \|u\| \leq \|Lu\|, \quad \text{and analogously} \quad C_2 \|u\| \leq \|L^*u\| \quad \text{for all} \quad u \in X.$$

This proves that the symmetric and continuous bilinear form

$$S(u, w) = (L^*u, L^*w) \quad \text{is positive definite, i.e.,}$$
$$C_2^2 \|u\|^2 \leq \|L^*u\|^2 = S(u, u) \quad \text{for all} \quad u \in X,$$

and Riesz' representation theorem is applicable: For any $\ell \in X'$, there is a unique $y^* \in X$, such that

$$\ell(h) = S(y^*, h) = (LL^*y^*, h) = B(L^*y^*, h)$$
$$\text{for all} \quad h \in X,$$

proving the Lax-Milgram theorem with $y_0 = L^*y^*$.

Next we study eigenvalue problems using the direct methods in the calculus of variations. We follow the classical approach in "Methods of Mathematical Physics, Vol. 1 and 2" by R. Courant and D. Hilbert, which is based on the work of Baron J.W.S Rayleigh (1842–1919), W. Ritz (1878–1909), E. Fischer (1875–1954), H. Weyl (1885–1955), and R. Courant (1888–1972).

Let $K : X \times X \to \mathbb{R}$ be a continuous and symmetric bilinear form. Then a (global) minimizer $y \in X$ of

$$B(y) = B(y, y), \quad \text{under the constraint}$$
$$K(y) = K(y, y) = 1, \tag{3.1.26}$$

fulfills by Lagrange's multiplier rule (formally, for the time being)

$$B(y, h) = \lambda K(y, h) \quad \text{for all} \quad h \in X, \tag{3.1.27}$$

where $\lambda \in \mathbb{R}$, which follows from $\delta B(y)h = 2B(y, h)$ and $\delta K(y)h = 2K(y, h)$. Equation (3.1.27) is the weak form of a linear eigenvalue problem as will be shown in Paragraph 3.3. The existence of a minimizer is proved in the following proposition. We employ the abbreviations $B(y, y) = B(y)$ and $K(y, y) = K(y)$.

Proposition 3.1.3. *Let $B : X \times X \to \mathbb{R}$ be a bilinear form on a (real) Hilbert space fulfilling conditions (3.1.17), and let $K : X \times X \to \mathbb{R}$ be a bilinear form having the following properties:*

$$K(y,z) = K(z,y) \quad \text{for all } y,z \in X, \quad \textit{i.e., } K \text{ is symmetric,}$$
$$w\text{-}\lim_{n \to \infty} y_n = y_0 \quad \Rightarrow \quad \lim_{n \to \infty} K(y_n) = K(y_0),$$
$$\textit{i.e., } K \text{ is weakly sequentially continuous,} \tag{3.1.28}$$
$$K(y) = K(y,y) > 0 \quad \text{for all } y \neq 0.$$

Then there exists a global minimizer $u_1 \in X$ of

$$B(y) = B(y,y), \quad \text{under the constraint}$$
$$K(y) = K(y,y) = 1, \quad \textit{i.e.,} \tag{3.1.29}$$
$$B(u_1) = \inf\{B(y)|y \in X,\, K(y) = 1\} = \lambda_1 > 0.$$

The minimizer fulfills the weak eigenvalue problem

$$B(u_1,h) = \lambda_1 K(u_1,h) \quad \text{for all } h \in X. \tag{3.1.30}$$

Any minimizing sequence $(y_n)_{n \in \mathbb{N}} \subset X \cap \{K(y) = 1\}$ contains a subsequence, $(y_{n_k})_{k \in \mathbb{N}}$, that converges (strongly) to the global minimizer u_1 in X:

$$\lim_{k \to \infty} B(y_{n_k}) = \lambda_1 = B(u_1) \quad \text{and}$$
$$\lim_{k \to \infty} y_{n_k} = u_1 \quad \text{in} \quad X. \tag{3.1.31}$$

Proof. Since B is positive definite, the infimum λ_1 is nonnegative in $(3.1.29)_3$, and any minimizing sequence $(y_n)_{n \in \mathbb{N}} \subset X \cap \{K(y) = 1\}$ is bounded in X. The weak sequential compactness (3.1.13) implies the existence of a subsequence $(y_{n_k})_{k \in \mathbb{N}}$ such that $w\text{-}\lim_{k \to \infty} y_{n_k} = u_1 \in X$. Furthermore

$$B(y_{n_k}) = B(y_{n_k},y_{n_k}) = B(y_{n_k} - u_1, y_{n_k} - u_1) + 2B(y_{n_k},u_1) - B(u_1,u_1)$$
$$\geq 2B(y_{n_k},u_1) - B(u_1,u_1), \quad \text{and thus,} \tag{3.1.32}$$
$$\liminf_{k \to \infty} B(y_{n_k}) \geq \lim_{k \to \infty} 2B(y_{n_k},u_1) - B(u_1,u_1) = B(u_1),$$

where we use $B(\cdot,u_1) \in X'$, cf. the definition (3.1.7) of weak convergence. By (3.1.32), the functional B is weakly sequentially lower semi-continuous. The hypothesis $(3.1.28)_2$ on K implies $1 = \lim_{n \to \infty} K(y_{n_k}) = K(u_1)$, and therefore

$$\lambda_1 = \lim_{k \to \infty} B(y_{n_k}) \geq B(u_1) \geq \lambda_1, \quad \text{and thus,}$$
$$\lim_{k \to \infty} B(y_{n_k}) = B(u_1) = \lambda_1 = \inf\{B(y)|y \in X, K(y) = 1\}. \tag{3.1.33}$$

Since $B(u_1) = B(u_1, u_1) = \lambda_1$, we have $\lambda_1 > 0$; otherwise $\lambda_1 = 0$ and the positive definiteness of B would imply $u_1 = 0$, contradicting $K(u_1) = 1$, cf. $(3.1.28)_3$. Finally,

$$
\lim_{k \to \infty} B(y_{n_k} - u_1, y_{n_k} - u_1)
$$
$$
= \lim_{k \to \infty} B(y_{n_k}, y_{n_k}) - 2 \lim_{k \to \infty} B(y_{n_k}, u_1) + B(u_1, u_1) \qquad (3.1.34)
$$
$$
= B(u_1) - 2B(u_1) + B(u_1) = 0,
$$

which proves, by the positive definiteness of B, the (strong) convergence of $(y_{n_k})_{k \in \mathbb{N}}$ to u_1 in X.

For any $y \in X$, $y \neq 0$, we have $K(y) > 0$ and $K(y/\sqrt{K(y)}) = 1$. Therefore

$$
\begin{aligned}
B(y/\sqrt{K(y)}) &= B(y,y)/K(y,y) \geq \lambda_1, \quad \text{or} \\
B(y,y) - \lambda_1 K(y,y) &\geq 0 \quad \text{for all} \quad y \in X.
\end{aligned} \qquad (3.1.35)
$$

In other words, for all $h \in X$ and $t \in \mathbb{R}$,

$$
\begin{aligned}
&B(u_1 + th) - \lambda_1 K(u_1 + th) \\
&= B(u_1) - \lambda_1 K(u_1) + 2t(B(u_1, h) - \lambda_1 K(u_1, h)) + t^2(B(h) - \lambda_1 K(h)) \quad (3.1.36) \\
&\geq 0 = B(u_1) - \lambda_1 = B(u_1) - \lambda_1 K(u_1).
\end{aligned}
$$

Since $B(u_1 + th)$ is minimal at $t = 0$, the derivative with respect to t at $t = 0$ vanishes, i.e.,

$$
2B(u_1, h) - \lambda_1 2K(u_1, h) = \delta B(u_1)h - \lambda_1 \delta K(u_1)h = 0
$$
$$
\text{for all } h \in X. \qquad (3.1.37)
$$

Thus we rediscover Lagrange's multiplier rule in a Hilbert space.   □

Next we prove the existence of infinitely many linear independent weak eigenvectors. We assume the hypotheses stated in Proposition 3.1.3.

Proposition 3.1.4. *The weak eigenvalue problem*

$$
B(u,h) = \lambda K(u,h) \quad \text{for all } h \in X \qquad (3.1.38)
$$

has infinitely many linear independent eigenvectors $u_n \in X$ with eigenvalues λ_n, $n \in \mathbb{N}$, satisfying

$$
K(u_n, u_m) = \delta_{nm} = \begin{cases} 1 \text{ for } n = m, \\ 0 \text{ for } n \neq m, \end{cases}
$$
$$
0 < \lambda_1 \leq \lambda_2 \leq \cdots \leq \lambda_n \leq \cdots \qquad (3.1.39)
$$
$$
\text{and} \quad \lim_{n \to \infty} \lambda_n = +\infty.
$$

Proof. The proof is by induction: The base clause is the statement of Proposition 3.1.3, and for the induction hypothesis we make the following statements: Any k

"K-orthonormal" elements $u_1, \ldots, u_k$ in X fulfilling $(3.1.39)_1$ are linearly independent and for

$$U_k = span[u_1, \ldots, u_k], \quad \dim U_k = k,$$
$$W_k = \{y \in X \mid K(y, u_i) = 0 \text{ for } i = 1, \ldots, k\}, \text{we have} \qquad (3.1.40)$$
$$X = U_k \oplus W_k.$$

Indeed, for any $y \in X$, we have

$$y = \sum_{i=1}^{k} K(y, u_i) u_i + y - \sum_{i=1}^{k} K(y, u_i) u_i, \quad \text{and}$$

$$y - \sum_{i=1}^{k} K(y, u_i) u_i \in W_k. \qquad (3.1.41)$$

By the infinite dimension of X $\dim W_k = \infty$ for any $k \in \mathbb{N}$, and for any system of K-orthonormal vectors definition $(3.1.40)_2$ yields

$$X = W_0 \supset W_1 \supset W_2 \supset \cdots \supset W_k \supset W_{k+1} \supset \cdots, \qquad (3.1.42)$$

where each space W_k is closed, and therefore it is a Hilbert space. Now we are ready to formulate the induction hypothesis:

For $k = 1, \ldots, n$ there exist weak eigenvectors $u_k \in W_{k-1}$,

which are K-orthonormal in the sense of $(3.1.39)_1$,

with corresponding eigenvalues $0 < \lambda_1 \leq \lambda_2 \leq \cdots \leq \lambda_n$ satisfying $\qquad (3.1.43)$

$$B(u_k) = \inf\{B(y) \mid y \in W_{k-1}, K(y) = 1\} = \lambda_k.$$

The arguments in the proof of Proposition 3.1.3, in particular, the arguments for (3.1.33), guarantee the existence of some $u_{n+1} \in W_n$ satisfying

$$B(u_{n+1}) = \inf\{B(y) \mid y \in W_n, K(y) = 1\} = \lambda_{n+1}, \qquad (3.1.44)$$

due to the fact that the weak limit u_{n+1} of a minimizing sequence in $W_n \cap \{K(y) = 1\}$ is in W_n and in the set $\{K(y) = 1\}$ as well. The first statement follows from the definition of W_n, from the fact that $K(\cdot, u_i) \in X'$, and from the definition of weak convergence. The second statement follows from the assumed properties of K, cf. (3.1.28). Since $W_{n-1} \supset W_n$,

$$\lambda_n \leq \lambda_{n+1}, \qquad (3.1.45)$$

and the arguments for (3.1.34) show that the chosen subsequence of the minimizing sequence converges strongly to u_{n+1} in X.

The induction hypothesis $(3.1.43)_4$ and (3.1.44) together with the arguments for (3.1.35)–(3.1.37) imply

$$B(u_k, h) = \lambda_k K(u_k, h) \quad \text{for all } h \in W_{k-1}, k = 1, \ldots, n,$$
$$B(u_{n+1}, h) = \lambda_{n+1} K(u_{n+1}, h) \quad \text{for all } h \in W_n. \tag{3.1.46}$$

By the definition of $W_n \subset W_{k-1}$ for $k = 1, \ldots, n$, $(3.1.46)_1$ implies

$$B(u_k, u_{n+1}) = B(u_{n+1}, u_k) = 0 \quad \text{for } k = 1, \ldots, n, \tag{3.1.47}$$

and for any $h \in X$, $(3.1.41)_1$ and $(3.1.46)_2$ yield

$$B(u_{n+1}, h) = B(u_{n+1}, h - \sum_{i=1}^{n} K(h, u_i) u_i)$$
$$= \lambda_{n+1} K(u_{n+1}, h - \sum_{i=1}^{n} K(h, u_i) u_i) = \lambda_{n+1} K(u_{n+1}, h), \tag{3.1.48}$$

i.e., u_{n+1} is a weak eigenvector with eigenvalue λ_{n+1}. By their construction, the eigenvectors $u_1, \ldots, u_n, u_{n+1}$ are K-orthonormal, which finishes the induction step.

Assume that the sequence of eigenvalues is bounded, i.e.,

$$0 < B(u_n, u_n) = B(u_n) = \lambda_n \leq C \quad \text{for all } n \in \mathbb{N}. \tag{3.1.49}$$

The positive definiteness of B implies then that the sequence $(u_n)_{n \in \mathbb{N}}$ is bounded in X, and thus there is a subsequence $(u_{n_k})_{k \in \mathbb{N}}$ converging weakly in X, i.e., $w\text{-}\lim_{k \to \infty} u_{n_k} = u_0 \in X$. The hypotheses $(3.1.28)$ on K finally yield

$$1 = \lim_{k \to \infty} K(u_{n_k}) = K(u_0) = 1,$$
$$\lim_{k \to \infty} K(u_{n_k} - u_0) = \lim_{k \to \infty} (K(u_{n_k}) - 2K(u_{n_k}, u_0) + K(u_0)) = 0,$$
$$K(u_{n_k} - u_{n_l}) = K(u_{n_k}) - 2K(u_{n_k}, u_{n_l}) + K(u_{n_l}) = 2, \tag{3.1.50}$$
$$\lim_{l \to \infty} K(u_{n_k} - u_{n_l}) = K(u_{n_k} - u_0) = 2,$$

which is contradictory. Therefore $(3.1.39)_3$ is true. $\qquad\square$

Corollary 3.1.1. *The geometric multiplicity of each eigenvalue λ_n is finite, i.e., the dimension of the eigenspace spanned by the eigenvectors with eigenvalue λ_n is finite.*

Proof. All linear independent eigenvectors with eigenvalue λ_n can be K-orthonormalized, and the assumption of infinitely many is contradictory as shown in $(3.1.49)$ and $(3.1.50)$. $\qquad\square$

Proposition 3.1.5. *The system of all K-orthonormal weak eigenvectors $(3.1.39)_1$ is complete or forms a Schauder basis in X (cf. Definition 3.2.5): Any $y \in X$ can be developed into a "Fourier series"*

$$y = \sum_{n=1}^{\infty} c_n u_n \quad \text{where} \quad c_n = K(y, u_n) \tag{3.1.51}$$

which converges in X.

Proof. Let for arbitrary $y \in X$, let $y_N = \sum_{n=1}^{N} c_n u_n$, where the coefficients c_n are given in (3.1.51). We show that $\lim_{N \to \infty} y_N = y$ in X. Due to K-orthonormality,

$$K(y - y_N, u_n) = 0 \quad \text{for } n = 1, \ldots, N \text{ or}$$
$$y - y_N \in W_N, \tag{3.1.52}$$

cf. (3.1.40)$_2$. For any vector $\tilde{y} \in W_N, \tilde{y} \neq 0$, we have $K(\tilde{y}/\sqrt{K(\tilde{y})}) = K(\tilde{y})/K(\tilde{y}) = 1$, and by (3.1.43)$_4$ or (3.1.44),

$$B(\tilde{y}/\sqrt{K(\tilde{y})}) = B(\tilde{y})/K(\tilde{y}) \geq \lambda_{N+1}, \quad \text{or}$$
$$B(\tilde{y}) \geq \lambda_{N+1} K(\tilde{y}) \quad \text{for all} \quad \tilde{y} \in W_N. \tag{3.1.53}$$

In particular,

$$B(y - y_N) \geq \lambda_{N+1} K(y - y_N). \tag{3.1.54}$$

Using

$$B(u_n, u_m) = \lambda_n K(u_n, u_m) = \lambda_n \delta_{nm}, \tag{3.1.55}$$

one obtains

$$K(y - y_N) = K(y) - \sum_{n=1}^{N} c_n^2 = K(y) - K(y_N),$$
$$B(y - y_N) = B(y) - \sum_{n=1}^{N} \lambda_n c_n^2 = B(y) - B(y_N). \tag{3.1.56}$$

By (3.1.54), $\lambda_{N+1} > 0$, and $B(y_N) \geq 0$,

$$0 \leq K(y - y_N) \leq \frac{1}{\lambda_{N+1}} B(y - y_N) \leq \frac{1}{\lambda_{N+1}} B(y), \quad \text{and}$$
$$\lim_{N \to \infty} K(y - y_N) = 0, \tag{3.1.57}$$

where we use (3.1.39)$_3$. Therefore, in view of (3.1.56)$_1$,

$$K(y) = \lim_{N \to \infty} \sum_{n=1}^{N} c_n^2 = \sum_{n=1}^{\infty} c_n^2 < \infty. \tag{3.1.58}$$

Formula (3.1.56)$_2$ implies

$$0 \leq B(y - y_N) = B(y) - \sum_{n=1}^{N} \lambda_n c_n^2, \quad \text{or}$$

$$\lim_{N \to \infty} \sum_{n=1}^{N} \lambda_n c_n^2 = \sum_{n=1}^{\infty} \lambda_n c_n^2 \leq B(y),$$

(3.1.59)

since $\lambda_n > 0$ for all $n \in \mathbb{N}$. Convergence of this series means that the partial sums form a Cauchy sequence:

$$B(y_N - y_M) = \sum_{n=M}^{N} \lambda_n c_n^2 < \varepsilon \quad \text{provided} \quad N > M \geq N_0(\varepsilon). \qquad (3.1.60)$$

The positive definiteness of B then yields

$$C_2 \|y_N - y_M\|^2 \leq B(y_N - y_M) < \varepsilon, \quad \text{provided} \quad N > M \geq N_0(\varepsilon), \qquad (3.1.61)$$

and by completeness of the Hilbert space X, any Cauchy sequence converges to some $\tilde{y} \in X$, i.e.,

$$\lim_{N \to \infty} y_N = \tilde{y} \quad \text{in} \quad X. \qquad (3.1.62)$$

Finally, the continuity of K and $(3.1.57)_2$ yield

$$0 = \lim_{N \to \infty} K(y - y_N) = K(y - \tilde{y}), \quad \text{or} \quad \tilde{y} = y, \qquad (3.1.63)$$

where we used hypothesis $(3.1.28)_4$ on K.   □

Proposition 3.1.6. *The recursive construction described in Proposition 3.1.4 provides all weak eigenvectors (up to linear combinations in case of geometric multiplicities bigger than one) and all eigenvalues of the weak eigenvalue problem (3.1.38).*

The proof is set in Exercise 3.1.1.
The positive definiteness of B $(3.1.17)_3$ can be replaced by a "K-coercivity":

$$B(y, y) \geq C_2 \|y\|^2 - c_2 K(y, y) \quad \text{for all} \quad y \in X. \qquad (3.1.64)$$

In this case, the symmetric and continuous bilinear form

$$\tilde{B}(y, z) = B(y, z) + c_2 K(y, z) \quad \text{for} \quad y, z \in X, \qquad (3.1.65)$$

is positive definite, and the weak eigenvectors u_n with eigenvalues $\tilde{\lambda}_n$, i.e.,

$$\tilde{B}(u_n, h) = \tilde{\lambda}_n K(u_n, h) \quad \text{fulfill}$$

$$B(u_n, h) = (\tilde{\lambda}_n - c_2) K(u_n, h) \quad \text{for all} \quad h \in X, n \in \mathbb{N}. \qquad (3.1.66)$$

The "spectrum" is shifted by the constant c_2 such that not all eigenvalues $\lambda_n = \tilde{\lambda}_n - c_2$ are necessarily positive. The eigenvectors u_n for B and $\tilde{B}$ are the same, and therefore Proposition 3.1.5 is valid also if nonpositive eigenvalues are present.

The nth eigenvalue of the weak eigenvalue problem (3.1.38) is determined as follows, cf. (3.1.43), (3.1.44):

$$\lambda_n = \min\left\{\frac{B(y)}{K(y)}\middle| y \in W_{n-1}, y \neq 0\right\}, \quad \text{or}$$

$$\lambda_n = \min\left\{\frac{B(y)}{K(y)}\middle| y \in X, y \neq 0,\ K(y,u_i) = 0, i = 1,\dots,n-1\right\}, \quad (3.1.67)$$

where $u_1,\dots,u_{n-1}$ are the first $n-1$ eigenvectors.

The minimizing vector is the nth eigenvector u_n. The quotient $B(y)/K(y)$ is called **Rayleigh quotient**.

The characterization (3.1.67) of the nth eigenvalue λ_n uses the first $n-1$ eigenvectors $u_1,\dots,u_{n-1}$. The so-called **Minimax Principle** of Courant, Fischer, and Weyl characterizes the nth eigenvalue without using the first $n-1$ eigenvectors. Apart from its mathematical elegance, this principle has relevant applications as we see in Paragraph 3.3.

Proposition 3.1.7. *Let $v_1,\dots,v_{n-1}$ $(n \geq 2)$ be any vectors in a real Hilbert space X, which define a closed subspace of X:*

$$V(v_1,\dots,v_{n-1}) = \{y \in X | K(y,v_i) = 0 \ \ for \ i = 1,\dots,n-1\}. \quad (3.1.68)$$

In $V(v_1,\dots,v_{n-1})$ the Rayleigh quotient attains its minimum:

$$d(v_1,\dots,v_{n-1}) = \min\left\{\frac{B(y)}{K(y)}\middle| y \in V(v_1,\dots,v_{n-1}),\ y \neq 0\right\}. \quad (3.1.69)$$

Then the nth eigenvalue λ_n of the weak eigenvalue problem (3.1.38) is given by

$$\lambda_n = \max_{v_1,\dots,v_{n-1} \in X}\{d(v_1,\dots,v_{n-1})\}. \quad (3.1.70)$$

Proof. The arguments for (3.1.44) prove that the Rayleigh quotient attains its minimum in $V(v_1,\dots,v_{n-1})$. We determine the coefficients of the linear combination $y = \alpha_1 u_1 + \cdots + \alpha_n u_n \in X$ with the first n eigenvectors such that $y \in V(v_1,\dots,v_{n-1})$, i.e.,

$$K(y,v_i) = \sum_{k=1}^{n} \alpha_k K(u_k,v_i) = 0 \quad \text{for } i = 1,\dots,n-1, \text{and}$$

$$\sum_{k=1}^{n} \alpha_k^2 = 1. \quad (3.1.71)$$

This is possible since the homogeneous linear system $(3.1.71)_1$ with $n-1$ equations for n unknowns has nontrivial solutions $(\alpha_1, \ldots, \alpha_n) \in \mathbb{R}^n$, which can be normalized. By the K-orthonormality of the eigenvectors, cf. $(3.1.39)_1$, and by $(3.1.55)$, we obtain

$$K(y) = \sum_{k=1}^{n} \alpha_k^2 = 1 \quad \text{and}$$

$$B(y) = \sum_{k=1}^{n} \lambda_k \alpha_k^2 \leq \lambda_n \sum_{k=1}^{n} \alpha_k^2 = \lambda_n, \tag{3.1.72}$$

where we use also $(3.1.39)_2$. Therefore the minimum $d(v_1, \ldots, v_{n-1}) \leq \lambda_n$ for any choice of the vectors $v_1, \ldots, v_{n-1} \in X$. In view of $(3.1.67)$, $d(u_1, \ldots, u_{n-1}) = \lambda_n$, which proves $(3.1.70)$. $\square$

So far our results on direct methods for bilinear forms on a Hilbert space. Applications are given in Paragraph 3.3. However, we cannot apply Propositions 3.1.1, 3.1.2–3.1.4 to functionals in the setting of the first two chapters since the spaces $C^1[a,b]$ and $C^{1,pw}[a,b]$ are not reflexive and a fortiori not Hilbert spaces.

Therefore, for the direct methods of this chapter, the functionals have to be extended to a Hilbert space. In Paragraph 3.3 we investigate also the regularity that minimizers must have for the variational problem and for the Euler-Lagrange equation. This last step for a satisfactory solution of a variational problem is called "regularity theory." For this theory, like for the existence of a minimizer, a partial convexity, called "ellipticity," of the Lagrange function is crucial, cf. Proposition 1.11.4.

Remark. *All results on abstract quadratic functionals following Definition 3.1.1 are not only applicable to Sturm-Liouville boundary and eigenvalue problems as elaborated in Paragraph 3.3 (cf. (3.3.1) and (3.3.4)) but they apply also to elliptic boundary and eigenvalue problems over higher dimensional bounded domains, see for instance the standard reference: R. Courant and D. Hilbert, "Methods of Mathematical Physics, Vol.1," Interscience Publishers, New York, 1953.*

Exercises

3.1.1. Prove Proposition 3.1.6.

Hint: Show for any eigenvector u of $(3.1.38)$ with eigenvalue λ, that $\lambda = \lambda_{n_0}$ for some $n_0 \in \mathbb{N}$ and that u is a linear combination of all constructed linear independent eigenvectors with eigenvalue λ_{n_0}.

3.2   An Explicit Performance of the Direct Method in a Hilbert Space

Around the turn of the last century, Hilbert justified Dirichlet's principle by the direct method. However, a systematic application to variational problems goes back to L. Tonelli (1885–1946) in the years 1910–1930. We confine ourselves to the case that the reflexive space is a Hilbert space.

We assume that the linear space

$$L^2(a,b) = \{y|y : (a,b) \to \mathbb{R} \text{ is measurable, } \int_a^b y^2 dx < \infty\} \qquad (3.2.1)$$

is known from a course on calculus. Otherwise its definition and properties are found, e.g., in [20]. Defining

$$(y,z)_{0,2} = \int_a^b yz dx, \ \|y\|_{0,2} = \sqrt{(y,y)_{0,2}}, \qquad (3.2.2)$$

then $L^2(a,b)$ is a Hilbert space (Riesz-Fischer theorem). In this space, $y = 0$ means that $y(x) = 0$ for "almost all" $x \in (a,b)$, i.e., for all $x \in (a,b)$ except a set of measure zero. The integral is the Lebesgue integral, and the measure is the Lebesgue measure (H. Lebesgue, 1875–1941).

The functional J in (1.1.1) contains the derivative of y, and since the classical derivative is not adequate for the definition of a Hilbert space, a "weak" or "distributional" derivative is defined as follows:

Definition 3.2.1. *Assume that for* $y, z \in L^2(a,b)$, *we have*

$$\int_a^b zh + yh' dx = 0 \quad \text{for all} \quad h \in C_0^\infty(a,b). \qquad (3.2.3)$$

Then y is "weakly differentiable" with weak derivative $y' = z$.

In view of Lemma 3.2.1, the weak derivative $y' = z \in L^2(a,b)$ of $y \in L^2(a,b)$ is unique (as a function in $L^2(a,b)$), and the formula of integration by parts (1.3.5) shows that the classical derivative, also if it exists only piecewise, coincides with the weak derivative (almost everywhere). With this notion we can define a Hilbert space as follows:

Definition 3.2.2. $W^{1,2}(a,b) = \{y|y \in L^2(a,b) \text{ and } y' \in L^2(a,b) \text{ exists according to}$ (3.2.3)\} *is endowed with the scalar product and the norm*

$$(y,\bar{y})_{1,2} = (y,\bar{y})_{0,2} + (y',\bar{y}')_{0,2}, \quad \|y\|_{1,2} = \sqrt{(y,y)_{1,2}}. \qquad (3.2.4)$$

The space $W^{1,2}(a,b)$ is complete, and therefore it is a Hilbert space, which follows easily from the completeness of $L^2(a,b)$, cf. Exercise 3.2.1. $W^{1,2}(a,b)$ is called a "Sobolev space," named after S. L. Sobolev (1908–1989). Details about Sobolev spaces are found in [5], [6], [9], [25], and in many books on functional analysis, calculus of variations, and partial differential equations. We give here only those properties that we need. Apparently $C^{1,pw}[a,b] \subset W^{1,2}(a,b)$.

The next lemma extends Lemma 1.3.1.

Lemma 3.2.1. *If for $y \in L^2(a,b)$*

$$\int_a^b yh\,dx = 0 \quad \text{for all} \quad h \in C_0^\infty(a,b), \tag{3.2.5}$$

then $y(x) = 0$ for almost all $x \in (a,b)$.

Proof. We make use of the fact that the space of "test functions" $C_0^\infty(a,b)$ is dense in $L^2(a,b)$, i.e., for any $y \in L^2(a,b)$ and any $\varepsilon > 0$ there exists an $h \in C_0^\infty(a,b)$ such that $\|y - h\|_{0,2} < \varepsilon$. Then (3.2.5) implies, using Cauchy-Schwarz' inequality,

$$\|y\|_{0,2}^2 = (y,y)_{0,2} = (y-h,y)_{0,2} \le \|y-h\|_{0,2}\|y\|_{0,2}, \quad \text{and hence,}$$
$$\|y\|_{0,2} < \varepsilon, \tag{3.2.6}$$

which implies $\|y\|_{0,2} = 0$ or the claim. □

Remark. *All functions that are Lebesgue integrable are approximated by "simple" or "step functions." Apparently, in $L^2(a,b)$, a step function is approximated by test functions.*

Lemma 1.3.2 can be extended as well.

Lemma 3.2.2. *If for $y \in L^2(a,b)$*

$$\int_a^b yh'\,dx = 0 \quad \text{for all} \quad h \in C_0^\infty(a,b), \tag{3.2.7}$$

then $y(x) = c$ for almost all $x \in (a,b)$.

Proof. 1) We prove first: If

$$\int_a^b yh\,dx = 0 \quad \text{for all} \quad h \in C_0^\infty(a,b) \text{ satisfying } \int_a^b h\,dx = 0, \tag{3.2.8}$$

then $y(x) = c$ for almost all $x \in (a,b)$.

Choose any $g \in C_0^\infty(a,b)$ and some fixed $f \in C_0^\infty(a,b)$ satisfying $\int_a^b f\,dx = 1$. Then

$$h = g - \int_a^b g\,dx f \in C_0^\infty(a,b) \quad \text{satisfying} \quad \int_a^b h\,dx = 0, \tag{3.2.9}$$

and therefore by (3.2.8),

$$
\begin{aligned}
0 &= \int_a^b y h\,dx = \int_a^b y g\,dx - \int_a^b g\,dx \int_a^b y f\,dx \\
&= \int_a^b (y - \int_a^b y f\,dx) g\,dx \quad \text{for all} \quad g \in C_0^\infty(a,b),
\end{aligned}
\tag{3.2.10}
$$

which implies by Lemma 3.2.1 that $y(x) = \int_a^b y f\,dx = c$ for almost all $x \in (a,b)$.

2) We prove the general case: Choose any $g \in C_0^\infty(a,b)$ satisfying $\int_a^b g\,dx = 0$. Define $h(x) = \int_a^x g\,ds$. Then $h \in C_0^\infty(a,b)$, $h' = g$, and by (3.2.7) $\int_a^b y g\,dx = 0$. Then case 1) applies, proving the lemma. $\qquad\square$

The fundamental theorem of calculus holds also for the weak derivative and Lebesgue integral:

Lemma 3.2.3. *For $y \in W^{1,2}(a,b)$, it follows that*

$$y(x_2) - y(x_1) = \int_{x_1}^{x_2} y'\,dx \quad \text{for all} \quad a \le x_1 < x_2 \le b. \tag{3.2.11}$$

Proof. According to Definition 3.2.2 of $W^{1,2}(a,b)$, a function $y \in W^{1,2}(a,b) \subset L^2(a,b)$ can be modified on a set of measure zero, and in this sense such functions form an "equivalence class." Therefore the pointwise equality in (3.2.11) seems to make no sense. In Lemma 3.2.4 it is shown that each function in $W^{1,2}(a,b)$ has a continuous representative, and formula (3.2.11) is valid for this representative.

Defining

$$z(x) = \int_a^x y'\,ds \quad \text{for} \quad x \in [a,b] \tag{3.2.12}$$

the Cauchy-Schwarz inequality implies that z is continuous on $[a,b]$, cf. (3.2.18). For an arbitrary $h \in C_0^\infty(a,b)$ Fubini's theorem implies

$$
\begin{aligned}
\int_a^b z h'\,dx &= \int_a^b \int_a^x y' h'\,ds\,dx = \int_a^b \int_s^b y' h'\,dx\,ds \\
&= \int_a^b y'(-h)\,ds \quad \text{or} \quad \int_a^b y' h + z h'\,dx = 0.
\end{aligned}
\tag{3.2.13}
$$

By (3.2.3),

$$\int_a^b y' h + y h'\,dx = 0 \quad \text{for all} \quad h \in C_0^\infty(a,b), \tag{3.2.14}$$

which yields, in view of (3.2.13)$_2$,

$$\int_a^b (z-y)h'dx = 0 \quad \text{for all} \quad h \in C_0^\infty(a,b), \text{ or}$$

$$z(x) - y(x) = c \quad \text{for almost all } x \in (a,b), \tag{3.2.15}$$

due to Lemma 3.2.2.

Choosing the continuous representative of y, namely $y(x) = z(x) - c$, we obtain $z(x) = y(x) - y(a)$, and finally,

$$\int_{x_1}^{x_2} y'dx = \int_a^{x_2} y'dx - \int_a^{x_1} y'dx = z(x_2) - z(x_1) = y(x_2) - y(x_1). \tag{3.2.16}$$

$\square$

For the continuous representative of $y \in W^{1,2}(a,b)$, the following holds:

Lemma 3.2.4. *If $y \in W^{1,2}(a,b)$, then y is uniformly Hölder continuous with exponent $1/2$ and the following estimates hold:*

$$|y(x_2) - y(x_1)| \leq |x_2 - x_1|^{1/2} \|y'\|_{0,2} \quad \text{for all} \quad a \leq x_1 < x_2 \leq b,$$

$$\|y\|_0 = \max_{x \in [a,b]} |y(x)| \leq \frac{1}{(b-a)^{1/2}} \|y\|_{0,2} + (b-a)^{1/2} \|y'\|_{0,2}. \tag{3.2.17}$$

Proof. Relation (3.2.11) implies (3.2.17)$_1$, by virtue of the Cauchy-Schwarz inequality:

$$|y(x_2) - y(x_1)| \leq \left(\int_{x_1}^{x_2} 1 dx \right)^{1/2} \left(\int_{x_1}^{x_2} (y')^2 dx \right)^{1/2} \leq |x_2 - x_1|^{1/2} \|y'\|_{0,2}. \tag{3.2.18}$$

For a continuous function $y \in C[a,b]$, the mean value theorem for integrals reads

$$\int_a^b y dx = y(\xi)(b-a) \quad \text{for some} \quad \xi \in (a,b). \tag{3.2.19}$$

Lemma 3.2.3 yields for any $x \in [a,b]$, using also estimates like (3.2.18),

$$y(x) = y(\xi) + \int_\xi^x y'ds = \frac{1}{b-a} \int_a^b y dx + \int_\xi^x y'ds,$$

$$|y(x)| \leq \frac{(b-a)^{1/2}}{b-a} \left(\int_a^b y^2 dx \right)^{1/2} + (b-a)^{1/2} \left(\int_a^b (y')^2 dx \right)^{1/2}, \tag{3.2.20}$$

where we agree on $\int_\xi^x y'ds = -\int_x^\xi y'ds$ if $x < \xi$. Since $x \in [a,b]$ is arbitrary, (3.2.17)$_2$ is proved. $\square$

Uniformly Hölder continuous functions form a subspace of $C[a,b]$, called after O. Hölder (1859–1937):

Definition 3.2.3. $C^{1/2}[a,b] = \{y|y : [a,b] \to \mathbb{R} \text{ is uniformly Hölder continuous with exponent } 1/2\}$ *is endowed with the norm*

$$\|y\|_{1/2} = \|y\|_0 + \sup_{a \le x_1 < x_2 \le b} \frac{|y(x_2) - y(x_1)|}{|x_2 - x_1|^{1/2}}. \tag{3.2.21}$$

By the estimates of Lemma 3.2.4 and by the definition of the norm $\| \quad \|_{1/2}$ in (3.2.21), we can state for the continuous representatives of functions in $W^{1,2}(a,b)$:

Proposition 3.2.1. *The Sobolev space $W^{1,2}(a,b)$ is continuously embedded into the Hölder space $C^{1/2}[a,b]$, i.e.,*

$$W^{1,2}(a,b) \subset C^{1/2}[a,b] \quad and$$
$$\|y\|_{1/2} \le c_0 \|y\|_{1,2} \quad \text{for all } y \in W^{1,2}(a,b), \tag{3.2.22}$$

with a constant $c_0 > 0$.

The following result is due to C. Arzelà (1847–1912) and to G. Ascoli (1843–1896). It is crucial for the direct methods.

Proposition 3.2.2. *Let a sequence $(y_n)_{n \in \mathbb{N}} \subset C[a,b]$ be bounded and equicontinuous, i.e., for all $n \in \mathbb{N}$,*

$$\|y_n\|_0 = \max_{x \in [a,b]} |y_n(x)| \le C,$$

and for any $\varepsilon > 0$ there exists a $\delta(\varepsilon) > 0$ such that
$$|y_n(x_2) - y_n(x_1)| < \varepsilon, \quad \text{provided } |x_2 - x_1| < \delta(\varepsilon) \tag{3.2.23}$$
for all $x_1, x_2 \in [a,b]$.

Then $(y_n)_{n \in \mathbb{N}}$ contains a subsequence $(y_{n_k})_{k \in \mathbb{N}}$, which converges uniformly to a continuous function $y_0 \in C[a,b]$:

$$\lim_{k \to \infty} y_{n_k} = y_0 \quad in \ C[a,b], \ or$$
$$\lim_{k \to \infty} \|y_{n_k} - y_0\|_0 = 0. \tag{3.2.24}$$

A proof is given in the Appendix.

The definition (3.2.21) of the Hölder norm and the embedding (3.2.22) allows us to deduce from the Arzelà-Ascoli theorem the following proposition:

Proposition 3.2.3. *Let a sequence* $(y_n)_{n\in\mathbb{N}} \subset W^{1,2}(a,b)$ *be bounded, i.e.,*

$$\|y_n\|_{1,2} \leq C \quad \text{for all } n \in \mathbb{N}. \tag{3.2.25}$$

Then $(y_n)_{n\in\mathbb{N}}$ *contains a subsequence* $(y_{n_k})_{k\in\mathbb{N}}$, *which converges uniformly to a continuous function* $y_0 \in C[a,b]$, *cf.* (3.2.24).

The boundedness of a sequence $(y_n)_{n\in\mathbb{N}} \subset W^{1,2}(a,b)$ implies, by (3.2.22)$_2$, that it is bounded in $C[a,b]$ and, according to the definition (3.2.4) of the norm $\| \ \|_{1,2}$, it means also that the sequence of the weak derivatives $(y_n')_{n\in\mathbb{N}}$ is bounded in $L^2(a,b)$. Since $L^2(a,b)$ is a Hilbert space, we can apply the weak sequential compactness (3.1.13), and we obtain:

Proposition 3.2.4. *Let a sequence* $(y_n)_{n\in\mathbb{N}} \subset W^{1,2}(a,b)$ *be bounded. Then it contains a subsequence* $(y_{n_k})_{k\in\mathbb{N}}$ *having the following properties:*

$$\lim_{k\to\infty} y_{n_k} = y_0 \quad \text{in } C[a,b] \text{ and}$$
$$w\text{-} \lim_{k\to\infty} y_{n_k}' = z_0 \quad \text{in } L^2(a,b). \tag{3.2.26}$$

Furthermore, $y_0 \in W^{1,2}(a,b)$, *and* $y_0' = z_0$ *is the weak derivative.*

Proof. W.l.o.g let $(y_{n_k})_{k\in\mathbb{N}}$ be the subsequence, for which the Arzelà-Ascoli theorem in Proposition 3.2.3 and for which the weak sequential compactness (3.1.14) holds. According to the definition (3.1.11) of weak convergence in a Hilbert space,

$$\lim_{k\to\infty} \int_a^b y_{n_k}' h\,dx = \int_a^b z_0 h\,dx \quad \text{for all } h \in C_0^\infty(a,b). \tag{3.2.27}$$

Since y_{n_k}' is the weak derivative of y_{n_k} the convergence (3.2.26)$_1$ implies

$$\lim_{k\to\infty} \int_a^b y_{n_k}' h\,dx = -\lim_{k\to\infty} \int_a^b y_{n_k} h'\,dx = -\int_a^b y_0 h'\,dx. \tag{3.2.28}$$

Equality of the limits in (3.2.27) and in (3.2.28) proves the statements of Proposition 3.2.4.   □

Remark. *A bounded sequence* $(y_n)_{n\in\mathbb{N}}$ *in the Hilbert space* $W^{1,2}(a,b)$ *contains, by virtue of the weak sequential compactness* (3.1.13), *a subsequence* $(y_{n_k})_{k\in\mathbb{N}}$, *which converges weakly in* $W^{1,2}(a,b)$, *i.e.,* $\lim_{k\to\infty}((y_{n_k},z)_{0,2} + (y_{n_k}',z')_{0,2}) = (y_0,z)_{0,2} + (y_0',z')_{0,2}$ *for all* $z \in W^{1,2}(a,b)$. *Apparently* (3.2.26) *gives more information.*

The embeddings $W^{1,2}(a,b) \subset C^{1/2}[a,b] \subset C[a,b]$ allows us to define boundary conditions $y(a) = A$ and $y(b) = B$ for functions $y \in W^{1,2}(a,b)$. For $A = B = 0$, we define

Definition 3.2.4. $W_0^{1,2}(a,b) = \{y | y \in W^{1,2}(a,b) \text{ satisfying } y(a) = 0, y(b) = 0\}.$

The following Poincaré inequality holds:

Lemma 3.2.5. *All* $y \in W_0^{1,2}(a,b)$ *satisfy*

$$\|y\|_{0,2} \leq (b-a)\|y'\|_{0,2}. \tag{3.2.29}$$

Proof. Lemma 3.2.3 yields for $y(a) = 0$,

$$y(x) = \int_a^x y' ds, \quad \text{and hence,}$$

$$|y(x)| \leq (b-a)^{1/2} \left(\int_a^b (y')^2 dx \right)^{1/2}, \tag{3.2.30}$$

which gives the claim after squaring, integration, and taking the square root. $\square$

Remark. *The constant* $(b-a)$ *in* (3.2.29) *is not optimal: Exercise 2.1.2 gives for* $(a,b) = (0,1)$ *the constant* $1/\pi$ *instead of* $b-a = 1$. *However, this Poincaré estimate requires the existence of a global minimizer, which is proved in Paragraph 3.3.*

For functions with nonhomogeneous boundary conditions, we can prove the following estimate:

Lemma 3.2.6. *All* $y \in D = W^{1,2}(a,b) \cap \{y(a) = A, y(b) = B\}$ *satisfy*

$$\|y\|_{1,2} \leq C_1 \|y'\|_{0,2} + C_2 \tag{3.2.31}$$

with constants C_1, C_2 *that do not depend on* y.

Proof. Let $z_0 \in D$ be fixed, for instance, let z_0 be the straight line from (a,A) to (b,B). Then for $y \in D$ the function $y - z_0 = h \in W_0^{1,2}(a,b)$ and for $y = z_0 + h$ (3.2.29) implies

$$\|y\|_{0,2} \leq \|z_0\|_{0,2} + (b-a)\|h'\|_{0,2}$$
$$\leq \|z_0\|_{0,2} + (b-a)\|z_0'\|_{0,2} + (b-a)\|y'\|_{0,2} \tag{3.2.32}$$
$$= \tilde{C}_1 \|y'\|_{0,2} + \tilde{C}_2.$$

Then (3.2.31) holds for $C_1 = \sqrt{2\tilde{C}_1^2 + 1}$ and $C_2 = \sqrt{2\tilde{C}_2^2}$. $\square$

Next we perform the direct method for a functional

$$J(y) = \int_a^b F(x,y,y') dx \tag{3.2.33}$$

whose admitted functions fulfill boundary conditions $y(a) = A$ and $y(b) = B$. For this purpose, we extend the domain of definition from $C^{1,pw}[a,b] \cap \{y(a) = A, y(b) = B\}$, given in the first two chapters of this book, to $D = W^{1,2}(a,b) \cap \{y(a) = A, y(b) = B\}$.

According to Proposition 3.2.1, a function $y \in D$ is continuous on $[a,b]$, but its weak derivative is only in $L^2(a,b)$. For a continuous Lagrange function $F : [a,b] \times \mathbb{R} \times \mathbb{R} \to \mathbb{R}$, the function $F(\cdot, y, y')$ is measurable (according to a result of measure theory), but the integral (3.2.33) is not necessarily finite for each $y \in D$. Since we are only interested in minimizers we have only to guarantee that the functional is bounded from below. The functional can be unbounded from above, it can possibly attain the value $+\infty$. Since $J(y)$ is finite for $y \in C^{1,pw}[a,b]$, the infimum is finite in case J is bounded from below on D. The following proposition gives sufficient conditions for the existence of a global minimizer of J. This proposition is not the most general. For sharper results, we refer to the literature, in particular to [5].

Proposition 3.2.5. *Assume that the Lagrange function $F : [a,b] \times \mathbb{R} \times \mathbb{R} \to \mathbb{R}$ of the functional (3.2.33) is continuous and continuously partially differentiable with respect to the third variable. The hypotheses for all $x \in [a,b]$, and for all $y, y', \tilde{y}' \in \mathbb{R}$ are:*

$$F(x,y,y') \geq c_1(y')^2 - c_2 \quad \text{where} \quad c_1 > 0 \quad \text{(coercivity),}$$
$$F(x,y,\tilde{y}') \geq F(x,y,y') + F_{y'}(x,y,y')(\tilde{y}' - y') \qquad (3.2.34)$$
(partial convexity with respect to the variable y').

Then the functional (3.2.33) possesses a global minimizer $y_0 \in D = W^{1,2}(a,b) \cap \{y(a) = A, y(b) = B\}$.

Proof. We verify the hypotheses (H1), (H2), and (H3) of Paragraph 3.1.

(H1) In view of $F(x,y,y') \geq -c_2$ for all $(x,y,y') \in [a,b] \times \mathbb{R} \times \mathbb{R}$, the functional J is bounded from below.

   Therefore $m = \inf\{J(y) | y \in D\} \in \mathbb{R}$ exists, and we let $(y_n)_{n \in \mathbb{N}} \subset D$ be a minimizing sequence.

(H2) The coercivity implies for $n \geq n_0$

$$c_1 \|y_n'\|_{0,2}^2 - c_2(b-a) \leq J(y_n) \leq m+1. \qquad (3.2.35)$$

By the estimates (3.2.31) and (3.2.35), the minimizing sequence $(y_n)_{n \in \mathbb{N}}$ is bounded in $W^{1,2}(a,b)$. Proposition 3.2.4 then guarantees the existence of a subsequence $(y_{n_k})_{k \in \mathbb{N}}$ and a limit $y_0 \in W^{1,2}(a,b)$ satisfying

$$\lim_{k \to \infty} y_{n_k} = y_0 \quad \text{in } C[a,b] \text{ and}$$
$$\text{w-} \lim_{k \to \infty} y_{n_k}' = y_0' \quad \text{in } L^2(a,b). \qquad (3.2.36)$$

Due to the uniform convergence $(3.2.36)_1$, the limit fulfills the same boundary conditions as y_{n_k}, and hence $y_0 \in D$.

(H3) Let $(\tilde{y}_n)_{n \in \mathbb{N}} \subset W^{1,2}(a,b)$ be sequence converging in the sense of (3.2.36) to a limit function $\tilde{y}_0 \in W^{1,2}(a,b)$. We define

$$M_N = \{x \in [a,b] \mid (\tilde{y}_0'(x))^2 > N^2\} \quad \text{and}$$
$$K_N = [a,b] \backslash M_N. \tag{3.2.37}$$

Then for the Lebesgue measure μ,

$$\mu(M_N)N^2 < \int_{M_N} (\tilde{y}_0')^2 dx \le \|\tilde{y}_0'\|_{0,2}^2,$$
$$\mu(M_N) < \frac{1}{N^2}\|\tilde{y}_0'\|_{0,2}^2, \quad (\tilde{y}_0'(x))^2 \le N^2 \quad \text{for } x \in K_N. \tag{3.2.38}$$

By $(3.2.34)_1$, the function $\tilde{F}(x,y,y') = F(x,y,y') + c_2 \ge 0$ for all $(x,y,y') \in [a,b] \times \mathbb{R} \times \mathbb{R}$, and

$$\tilde{J}(y) = \int_a^b \tilde{F}(x,y,y')dx = J(y) + c_2(b-a). \tag{3.2.39}$$

Then the following holds:

$$\int_a^b \tilde{F}(x,\tilde{y}_n,\tilde{y}_n')dx \ge \int_{K_N} \tilde{F}(x,\tilde{y}_n,\tilde{y}_n')dx$$
$$= \int_{K_N} \tilde{F}(x,\tilde{y}_n,\tilde{y}_n') - \tilde{F}(x,\tilde{y}_n,\tilde{y}_0')dx + \int_{K_N} \tilde{F}(x,\tilde{y}_n,\tilde{y}_0')dx$$
$$\ge \int_{K_N} F_{y'}(x,\tilde{y}_n,\tilde{y}_0')(\tilde{y}_n' - \tilde{y}_0')dx + \int_{K_N} \tilde{F}(x,\tilde{y}_n,\tilde{y}_0')dx \tag{3.2.40}$$
$$= \int_{K_N} (F_{y'}(x,\tilde{y}_n,\tilde{y}_0') - F_{y'}(x,\tilde{y}_0,\tilde{y}_0'))(\tilde{y}_n' - \tilde{y}_0')dx$$
$$+ \int_{K_N} F_{y'}(x,\tilde{y}_0,\tilde{y}_0')(\tilde{y}_n' - \tilde{y}_0')dx + \int_{K_N} \tilde{F}(x,\tilde{y}_n,\tilde{y}_0')dx.$$

We investigate the last three terms separately.

$$|1^{\text{st}} \text{ term}| \le \left(\int_{K_N} (F_{y'}(x,\tilde{y}_n,\tilde{y}_0') - F_{y'}(x,\tilde{y}_0,\tilde{y}_0'))^2 dx \right)^{1/2} \|\tilde{y}_n' - \tilde{y}_0'\|_{0,2}. \tag{3.2.41}$$

For $x \in K_N$

$$|\tilde{y}_n(x)|, |\tilde{y}_0(x)| \le c, \quad \text{since } \lim_{n \to \infty} \tilde{y}_n = \tilde{y}_0 \text{ in } C[a,b],$$
$$|\tilde{y}_0'(x)| \le N \quad \text{in view of } (3.2.38)_2. \tag{3.2.42}$$

By assumption, the function $F_{y'} : [a,b] \times [-c,c] \times [-N,N] \to \mathbb{R}$ is uniformly continuous, and therefore the uniform convergence $\lim_{n \to \infty} \tilde{y}_n = \tilde{y}_0$ implies the uniform convergence $\lim_{n \to \infty} F_{y'}(\cdot,\tilde{y}_n,\tilde{y}_0') = F_{y'}(\cdot,\tilde{y}_0,\tilde{y}_0')$ on K_N. Thus the first factor in (3.2.41) converges to zero as $n \to \infty$. The second factor in (3.2.41) is bounded because $w\text{-}\lim_{n\to\infty} \tilde{y}'_n = \tilde{y}'_0$ in $L^2(a,b)$, and because weakly convergent sequences are bounded. (For the minimizing sequence, we do not need this result from functional analysis; the sequence of their weak derivatives is bounded in $L^2(a,b)$ by (3.2.35).) Thus we obtain

$$\lim_{n\to\infty} \int_{K_N} (F_{y'}(x,\tilde{y}_n,\tilde{y}'_0) - F_{y'}(x,\tilde{y}_0,\tilde{y}'_0))(\tilde{y}'_n - \tilde{y}'_0)dx = 0. \qquad (3.2.43)$$

By (3.2.42) and the continuity of $F_{y'}$

$$|F_{y'}(x,\tilde{y}_0(x),\tilde{y}'_0(x))| \le C_N \quad \text{for } x \in K_N, \text{ whence}$$
$$F_{y'}(\cdot,\tilde{y}_0,\tilde{y}'_0) \in L^2(K_N). \qquad (3.2.44)$$

Since $w\text{-}\lim_{n\to\infty} \tilde{y}'_n = \tilde{y}'_0$ in $L^2(a,b)$, the function $\tilde{y}'_0$ is the weak limit of the sequence $(\tilde{y}'_n)$ in $L^2(K_N)$ as well, and by definition of the weak limit,

$$\lim_{n\to\infty} \int_{K_N} F_{y'}(x,\tilde{y}_0,\tilde{y}'_0)(\tilde{y}'_n - \tilde{y}'_0)dx = 0. \qquad (3.2.45)$$

Thus the 2nd term in (3.2.40)$_5$ converges to zero as does the 1st term, cf. (3.2.43).

Again, (3.2.42) and the uniform continuity of $\tilde{F} : [a,b] \times [-c,c] \times [-N,N] \to \mathbb{R}$ imply the uniform convergence $\lim_{n\to\infty} \tilde{F}(\cdot,\tilde{y}_n,\tilde{y}'_0) = \tilde{F}(\cdot,\tilde{y}_0,\tilde{y}'_0)$ on K_N, whence

$$\lim_{n\to\infty} \int_{K_N} \tilde{F}(x,\tilde{y}_n,\tilde{y}'_0)dx = \int_{K_N} \tilde{F}(x,\tilde{y}_0,\tilde{y}'_0)dx. \qquad (3.2.46)$$

From (3.2.40), it then follows that

$$\liminf_{n\to\infty} \tilde{J}(\tilde{y}_n) \ge \int_{K_N} \tilde{F}(x,\tilde{y}_0,\tilde{y}'_0)dx. \qquad (3.2.47)$$

Since (3.2.47) holds for all $N \in \mathbb{N}$, the lemma of Fatou allows us to take the limit $N \to \infty$: Let χ_{K_N} be the characteristic function of the set K_N, which, in view of (3.2.37) and (3.2.38), converges pointwise almost everywhere in $[a,b]$ to the constant 1. Then

$$\liminf_{N\to\infty} \int_a^b \chi_{K_N} \tilde{F}(x,\tilde{y}_0,\tilde{y}'_0)dx \ge \int_a^b \liminf_{N\to\infty} \chi_{K_N} \tilde{F}(x,\tilde{y}_0,\tilde{y}'_0)dx$$
$$= \int_a^b \tilde{F}(x,\tilde{y}_0,\tilde{y}'_0)dx = \tilde{J}(\tilde{y}_0). \qquad (3.2.48)$$

Combining (3.2.47) and (3.2.48) yields

$$\liminf_{n\to\infty} \tilde{J}(\tilde{y}_n) \ge \tilde{J}(\tilde{y}_0), \quad \text{and from (3.2.39),}$$
$$\liminf_{n\to\infty} J(\tilde{y}_n) \ge J(\tilde{y}_0). \qquad (3.2.49)$$

Since the three hypotheses (H1), (H2), and (H3) are fulfilled, Proposition 3.1.1 guarantees the existence of a global minimizer $y_0 \in D$ of the functional J. $\qquad\square$

In Exercise 3.2.2 the coercivity $(3.2.34)_1$ is weakened to

$$F(x,y,y') \geq c_1(y')^2 - c_2|y|^q - c_3 \quad \text{where } c_1 > 0,\ 1 \leq q < 2. \tag{3.2.50}$$

It is still open whether the global minimizer $y_0 \in D \subset W^{1,2}(a,b)$ fulfills the Euler-Lagrange equation. The hypotheses of Proposition 3.2.5 do not guarantee the existence of the first variation $\delta J(y_0)h$ for all directions $h \in W_0^{1,2}(a,b)$.

Proposition 3.2.6. *Assume that the Lagrange function $F : [a,b] \times \mathbb{R} \times \mathbb{R} \to \mathbb{R}$ of the functional (3.2.33) is continuous and continuously partially differentiable with respect to the last two variables. The hypotheses for all $(x,y,y') \in [a,b] \times \mathbb{R} \times \mathbb{R}$ are:*

$$\begin{aligned}
&|F_y(x,y,y')| \leq f_1(x,y)(y')^2 + f_2(x,y), \\
&|F_{y'}(x,y,y')| \leq g_1(x,y)|y'| + g_2(x,y), \quad \text{where} \\
&f_i, g_i : [a,b] \times \mathbb{R} \to \mathbb{R} \quad \text{are continuous for } i = 1,2.
\end{aligned} \tag{3.2.51}$$

Then $J : W^{1,2}(a,b) \to \mathbb{R}$ is continuous, and there exists the first variation

$$\delta J(y)h = \int_a^b F_y(x,y,y')h + F_{y'}(x,y,y')h'dx \tag{3.2.52}$$

for all $y \in W^{1,2}(a,b)$ and all directions $h \in W^{1,2}(a,b)$.

Proof. The representation

$$\begin{aligned}
F(x,y,y') &- F(x,\tilde{y},\tilde{y}') \\
&= \int_0^1 \frac{d}{dt} F(x,\tilde{y}+t(y-\tilde{y}),\tilde{y}'+t(y'-\tilde{y}'))dt \\
&= \int_0^1 F_y(x,\tilde{y}+t(y-\tilde{y}),\tilde{y}'+t(y'-\tilde{y}'))(y-\tilde{y})dt \\
&\quad + \int_0^1 F_{y'}(x,\tilde{y}+t(y-\tilde{y}),\tilde{y}'+t(y'-\tilde{y}'))(y'-\tilde{y}')dt,
\end{aligned} \tag{3.2.53}$$

and (3.2.51) give, for $(x,y,y') \in [a,b] \times [-c,c] \times \mathbb{R}$ and for $(x,\tilde{y},\tilde{y}') \in [a,b] \times [-\tilde{c},\tilde{c}] \times \mathbb{R}$, the estimate

$$\begin{aligned}
|F(x,y,y') &- F(x,\tilde{y},\tilde{y}')| \\
&\leq c_1(|y'|^2 + |\tilde{y}'|^2)|y-\tilde{y}| + c_2(|y'| + |\tilde{y}'|)|y'-\tilde{y}'|,
\end{aligned} \tag{3.2.54}$$

where the constants c_1 and c_2 depend on c and $\tilde{c}$. By the embedding $W^{1,2}(a,b) \subset C[a,b]$, cf. Proposition 3.2.1, and by setting $\|y\|_0 \leq c_0\|y\|_{1,2} = c$, $\|\tilde{y}\|_0 \leq c_0\|\tilde{y}\|_{1,2} = \tilde{c}$, (3.2.54) gives

$$|J(y) - J(\tilde{y})|$$
$$\leq c_1(\|y\|_{1,2}^2 + \|\tilde{y}\|_{1,2}^2)\|y - \tilde{y}\|_0 + c_2(\|y\|_{1,2} + \|\tilde{y}\|_{1,2})\|y - \tilde{y}\|_{1,2}, \tag{3.2.55}$$

where the second summand is estimated by the Cauchy-Schwarz inequality. Furthermore, (3.2.55) and again the embedding $W^{1,2}(a,b) \subset C[a,b]$ show that $J : W^{1,2}(a,b) \to \mathbb{R}$ is locally Lipschitz continuous.

To prove (3.2.52) we have to show that

$$\frac{d}{dt}J(y + th)|_{t=0} = \delta J(y)h \tag{3.2.56}$$

exists and is given by (3.2.52). The proof of Proposition 1.2.1 relies on the fact that, due to uniform convergence of the difference quotient to the derivative on $[a,b]$, differentiation and integration can be interchanged. Here the possibility of that interchange is guaranteed by Lebegue's dominated convergence theorem.

For this purpose, the last two terms in (1.2.9) are estimated via (3.2.51) and the continuity of $y, h \in W^{1,2}(a,b)$, yielding $\|y\|_0, \|h\|_0 \leq c$. Finally, let $|t| \leq 1$, and we obtain

$$\left| \frac{1}{t} \int_0^t F_y(x, y(x) + sh(x), y'(x) + sh'(x)) - F_y(x, y(x), y'(x)) ds h(x) \right|$$
$$\leq c_3 + c_4(|y'(x)|^2 + |h'(x)|^2),$$
$$\left| \frac{1}{t} \int_0^t F_{y'}(x, y(x) + sh(x), y'(x) + sh'(x)) - F_{y'}(x, y(x), y'(x)) ds h'(x) \right| \tag{3.2.57}$$
$$\leq (c_5 + c_6(|y'(x)| + |h'(x)|))|h'(x)|.$$

For $y', h' \in L^2(a,b)$, both terms of (3.2.57) have integrable majorants, which do not depend on $|t| \leq 1$. They allow the interchange of the limit $t \to 0$ and integration. Following the proof of Proposition 1.2.1, the continuity of F_y and $F_{y'}$ implies that the last two terms of (1.2.9) converge to zero pointwise for each $x \in [a,b]$, and as in (1.2.15) we obtain (3.2.52). □

For the next proposition, we sharpen the growth conditions (3.2.51) on $F_y(x,y,y')$, which, however, is due to our Hilbert-space approach.

Proposition 3.2.7. *Let the Lagrange function* $F : [a,b] \times \mathbb{R} \times \mathbb{R} \to \mathbb{R}$ *of the functional*

$$J(y) = \int_a^b F(x,y,y')dx \tag{3.2.58}$$

be continuous and continuously partially differentiable with respect to the last two variables. We assume that F is coercive as in $(3.2.34)_1$ or in $(3.2.50)$, and that it is partially convex in the sense of $(3.2.34)_2$. Furthermore we assume the growth conditions for all $(x,y,y') \in [a,b] \times \mathbb{R} \times \mathbb{R}$:

$$|F_y(x,y,y')| \le f_1(x,y)|y'| + f_2(x,y),$$
$$|F_{y'}(x,y,y')| \le g_1(x,y)|y'| + g_2(x,y), \quad where$$
$$f_i, g_i : [a,b] \times \mathbb{R} \to \mathbb{R} \quad are\ continuous\ for\ i = 1,2.$$
(3.2.59)

The functional (3.2.58) *possesses a global minimizer* $y_0 \in D = W^{1,2}(a,b) \cap \{y(a) = A,\ y(b) = B\}$, *which fulfills the Euler-Lagrange equation in its weak form*

$$\int_a^b F_y(x,y_0,y_0')h + F_{y'}(x,y_0,y_0')h'dx = 0$$
(3.2.60)
$$for\ all \quad h \in W_0^{1,2}(a,b),$$

and also in its strong form

$$F_{y'}(\cdot,y_0,y_0') \in W^{1,2}(a,b) \quad and$$
$$\frac{d}{dx}F_{y'}(\cdot,y_0,y_0') = F_y(\cdot,y_0,y_0') \quad on\ (a,b).$$
(3.2.61)

Here $\frac{d}{dx}F_{y'}(\cdot,y_0,y_0')$ *is the weak derivative of* $F_{y'}(\cdot,y_0,y_0')$ *in the sense of Definition 3.2.1.*

Proof. By Proposition 3.2.5, a global minimizer y_0 of J exists, and by Definition 3.2.4, the minimizer y_0 and perturbations $y_0 + th$ for all $h \in W_0^{1,2}(a,b)$ and for all $t \in \mathbb{R}$ are in the domain of definition D. By Proposition 3.2.6 the first variation exists, and we obtain for the global minimizer

$$\frac{d}{dt}J(y_0 + th)|_{t=0} = \delta J(y_0)h = 0 \quad for\ all \quad h \in W_0^{1,2}(a,b).$$
(3.2.62)

The representation (3.2.52) and Definition 3.2.1 prove (3.2.60) and (3.2.61), provided $F_y(\cdot,y_0,y_0')$ and $F_{y'}(\cdot,y_0,y_0') \in L^2(a,b)$. This is assured by the growth conditions (3.2.59) and by $y_0 \in W^{1,2}(a,b) \subset C[a,b]$. $\square$

The minimizer $y_0 \in W^{1,2}(a,b)$ fulfills the Euler-Lagrange equation (3.2.61) with weak derivatives. It is a goal to prove enough regularity of y_0 so that the Euler-Lagrange equation is solved in the classical sense. If $y_0 \in C^1[a,b]$ this is the case, cf. Proposition 1.4.1, and if $F_{y'y'}(x,y_0(x),y_0'(x)) \ne 0$ for all $x \in [a,b]$, Exercise 1.5.1 then yields even $y_0 \in C^2[a,b]$. However, the step from $y_0 \in W^{1,2}(a,b) \subset C^{1/2}[a,b]$ (Proposition 3.2.1) to $y_0 \in C^1[a,b]$ is not simple, in general, and therefore we refer to the literature, to [5], e.g. For some special functionals, however, the required regularity comes along in a simple way, as we see in the next paragraph. Again, the ellipticity $F_{y'y'}(x,y(x),y'(x)) \ne 0$ plays a crucial role.

Besides existence and regularity, the computation of a minimizer is of interest. The theory is very helpful for this purpose, according to which a global minimizer is approximated by a minimizing sequence. Therefore the construction and computation of a minimizing sequence is important in applications.

Definition 3.2.5. *A Hilbert space X possesses a* **Schauder basis** $S = \{e_n\}_{n\in\mathbb{N}}$ *(J. Schauder, 1899–1943), if any element $y \in X$ is the limit of a unique convergent series in X:*

$$y = \sum_{n=1}^{\infty} c_n e_n = \lim_{N\to\infty} \sum_{n=1}^{N} c_n e_n, \ c_n \in \mathbb{R}. \tag{3.2.63}$$

Each separable Hilbert space possesses a Schauder basis, which can be orthonormalized. In (3.1.12) an orthonormal Schauder basis in $X = \ell^2$ is given, and Proposition 3.1.5 ascertains that a Schauder basis in a Hilbert space X consists of eigenfunctions of a weak eigenvalue problem.

Let $S = \{e_n\}_{n\in\mathbb{N}}$ be a Schauder basis in the Hilbert space $W_0^{1,2}(a,b)$ and let

$$U_N = span[e_1,\ldots,e_N] \subset W_0^{1,2}(a,b) \tag{3.2.64}$$

be the N-dimensional subspace, which is a Hilbert space for each $N \in \mathbb{N}$. For some fixed $z_0 \in D = W^{1,2}(a,b) \cap \{y(a) = A, \ y(b) = B\}$ the domain of definition can be written as $D = z_0 + W_0^{1,2}(a,b)$, and under the hypotheses of Proposition 3.2.7, there exists a global minimizer $y_0 \in D$ of $J : D \to \mathbb{R}$ and a fortiori there exists a global minimizer $y_N \in z_0 + U_N$ of $J : z_0 + U_N \to \mathbb{R}$. In view of $U_N \subset U_{N+1} \subset W_0^{1,2}(a,b)$,

$$J(y_N) \geq J(y_{N+1}) \geq J(y_0) \quad \text{for all} \quad N \in \mathbb{N}. \tag{3.2.65}$$

Proposition 3.2.8. *Under the hypotheses of Proposition 3.2.7 the sequence $(y_N)_{N\in\mathbb{N}}$ of global minimizers of the functional (3.2.58) defined on $z_0 + U_N$ is a minimizing sequence of the functional $J : z_0 + W_0^{1,2}(a,b) \to \mathbb{R}$. The coefficients $(c_1^N,\ldots,c_N^N) \in \mathbb{R}^N$ of $y_N = z_0 + \sum_{n=1}^{N} c_n^N e_n$ fulfill the N-dimensional system of equations*

$$\int_a^b \tilde{F}_y(x,c_1^N,\ldots,c_N^N)e_k + \tilde{F}_{y'}(x,c_1^N,\ldots,c_N^N)e_k' dx = 0, \quad k = 1,\ldots,N,$$

$$\tilde{F}_y(x,c_1^N,\ldots,c_N^N) = F_y(x,z_0 + \sum_{n=1}^{N} c_n^N e_n, z_0' + \sum_{n=1}^{N} c_n^N e_n'), \tag{3.2.66}$$

$$\tilde{F}_{y'}(x,c_1^N,\ldots,c_N^N) = F_{y'}(x,z_0 + \sum_{n=1}^{N} c_n^N e_n, z_0' + \sum_{n=1}^{N} c_n^N e_n').$$

Proof. By Proposition 3.2.6, $J : W^{1,2}(a,b) \to \mathbb{R}$ is continuous, and therefore for any $\varepsilon > 0$ there is a $\delta > 0$ such that $|J(y) - J(y_0)| < \varepsilon$ provided $\|y - y_0\|_{1,2} < \delta$. By the property (3.2.63) of a Schauder basis, for $\delta > 0$ there is an $N_0 \in \mathbb{N}$ and some $v_{N_0} \in U_{N_0}$ such that $\|y_0 - z_0 - v_{N_0}\|_{1,2} < \delta$. By (3.2.65),

$$0 \le J(y_N) - J(y_0) \le J(z_0 + v_{N_0}) - J(y_0) < \varepsilon$$
$$\text{for all} \quad N \ge N_0, \tag{3.2.67}$$

which proves $\lim_{N \to \infty} J(y_N) = J(y_0)$. The global minimizer $y_N \in z_0 + U_N$ fulfills
the Euler-Lagrange equation in its weak form (3.2.60) for all $h \in U_N$, which proves
(3.2.66).   □

By (3.2.36) the minimizing sequence $(y_N)_{N \in \mathbb{N}}$ contains a subsequence $(y_{N_k})_{k \in \mathbb{N}}$
such that
$$\lim_{k \to \infty} y_{N_k} = y_0 \quad \text{in } C[a,b] \text{ and}$$
$$w\text{-} \lim_{k \to \infty} y'_{N_k} = y'_0 \quad \text{in } L^2(a,b). \tag{3.2.68}$$

Under suitable hypotheses the convergence in (3.2.68) can be improved. This is
the case for quadratic functionals, which we study in the next paragraph. The
Euler-Lagrange equation is then a linear differential equation, and the regularity of a
minimizer follows easily in these cases. The system of equations (3.2.66) is a linear
system with a symmetric matrix of coefficients. For a suitable choice of a Schauder
basis $\{e_n\}_{n \in \mathbb{N}}$, it can be a diagonal matrix.

Exercises

3.2.1. Prove that the completeness of the space $L^2(a,b)$ implies the completeness
of the Sobolev space $W^{1,2}(a,b)$. Show that any Cauchy sequence in $W^{1,2}(a,b)$ converges to some limit in $W^{1,2}(a,b)$.

3.2.2. Show that Proposition 3.2.5 can be proved with the weaker coercivity
(3.2.50). Indicate where the proof has to be modified.

3.2.3. Which hypothesis for Proposition 3.2.5 is not fulfilled by the counterexample
of Weierstraß (example 3 in 1.5)?
   Show that the minimizing sequence given in Paragraph 1.5 is not bounded in
$W^{1,2}(-1,1)$.

3.3 Applications of the Direct Method

In this paragraph we prove the existence of classical solutions of boundary value
problems, in particular, of the so-called Sturm-Liouville boundary value problems.
We study the Sturm-Liouville eigenvalue problem, named after C.-F. Sturm (1803–
1855) and J. Liouville (1809–1882). The proofs are based on the direct method
in the calculus of variations in the spirit of Dirichlet's principle. The solution is a

minimizer of a related functional and it is therefore approximated by minimizing sequences. This approximation, called the Ritz or Galerkin method (W. Ritz, 1878–1909, B.G. Galerkin, 1871–1945), is the foundation for numerical methods, in particular, finite element methods, for elliptic boundary and eigenvalue problems.

1. A nonlinear boundary value problem

$$
\begin{aligned}
&y'' + f(\cdot, y) = 0 \quad \text{on} \quad [a,b], \\
&y(a) = A, \ y(b) = B, \quad \text{where} \\
&f : [a,b] \times \mathbb{R} \to \mathbb{R} \quad \text{is continuous and fulfills} \\
&|f(x,y)| \le c_2 |y|^r + c_3 \quad \text{for all } (x,y) \in [a,b] \times \mathbb{R} \\
&\text{with } 0 \le r < 1 \text{ and nonnegative constants } c_2, c_3.
\end{aligned}
\tag{3.3.1}
$$

Proposition 3.3.1. *The boundary value problem* (3.3.1) *possesses a solution* $y_0 \in C^2[a,b]$.

Proof. Let $g : [a,b] \times \mathbb{R} \to \mathbb{R}$ be a partial primitive satisfying $g_y(x,y) = f(x,y)$ for $(x,y) \in [a,b] \times \mathbb{R}$. Define for the functional $J(y) = \int_a^b F(x,y,y')dx$, with Lagrange function $F(x,y,y') = \frac{1}{2}(y')^2 - g(x,y)$, which fulfills the hypotheses $(3.2.34)_2$ and (3.2.50) for Proposition 3.2.5 and the growth conditions (3.2.59) of Proposition 3.2.7. Consequently the functional J has a global minimizer $y_0 \in D = W^{1,2}(a,b) \cap \{y(a) = A, \ y(b) = B\}$, which satisfies the Euler-Lagrange equation in its weak and in its strong version (3.2.60), (3.2.61).
Furthermore, $y_0 \in D \subset C[a,b]$ and $F_{y'}(\cdot, y_0, y_0') = y_0' \in W^{1,2}(a,b) \subset C[a,b]$. By the definition (3.2.3) of the weak derivative, (1.3.10), and the uniqueness of the weak derivative, we conclude that y_0' is the classical derivative and therefore $y_0 \in C^1[a,b]$. The Euler-Lagrange equation,

$$
\int_a^b y_0' h' - f(x,y_0)h \, dx \quad \text{for all} \quad h \in C_0^1[a,b],
\tag{3.3.2}
$$
and where $\quad y_0', f(\cdot, y_0) \in C[a,b]$,

implies by (1.3.10),

$$
\begin{aligned}
&y_0' \in C^1[a,b] \quad \text{or } y_0 \in C^2[a,b], \quad \text{and} \\
&y_0'' + f(\cdot, y_0) = 0 \quad \text{on} \quad [a,b].
\end{aligned}
\tag{3.3.3}
$$

The boundary conditions are fulfilled by $y_0 \in D$.   □

The example (3.3.8) below shows that a growth $(3.3.1)_4$ with an exponent $r = 1$ prevents the existence of a solution, in general.
For a solution $y_0 \in C^2[a,b]$ of the differential equation $(3.3.1)_1$ in the classical sense, a so-called bootstrapping increases its regularity, provided the function f is k times

continuously partially differentiable with respect to its variables: In this case $y_0 \in C^{k+2}[a,b]$.

Remark. *The differential equation* $(3.3.1)_1$ *is a special case of a general Euler-Lagrange equation (1.4.5), which is a quasilinear ordinary differential equation of second order. We assume that the Lagrange function F and accordingly the functional J depends on a real parameter* λ, *which models, for instance, a variable physical quantity. Then the Euler-Lagrange equation (1.4.5) is of the form*

$$a(\cdot, y, y', \lambda)y'' + b(\cdot, y, y', \lambda) = 0 \quad on \quad [a,b].$$

Assuming that $b(\cdot, 0, 0, \lambda) = 0$, *the boundary value problem with homogeneous boundary values* $y(a) = y(b) = 0$ *possesses the "trivial solution"* $y \equiv 0$ *for all values of the parameter* λ. *The natural question, which values of the parameter* λ *admit "nontrivial solutions," has a positive answer in the following setting: Assume that the functional* $J(y, \lambda)$ *loses its convexity when the parameter* λ *crosses a threshold* λ_0 *as sketched in Figure 3.1: For* $\lambda < \lambda_0$, *the trivial solution* $y = 0$ *is the only "stable" minimizer, which becomes "unstable" for* $\lambda > \lambda_0$, *and accordingly two new "stable nontrivial" minimizers emerge.*

This scenario of minimizers and a (local) maximizer in Figure 3.1 is reflected in a (y, λ)-*diagram of solutions of the Euler-Lagrange equation as depicted in Figure 3.2.*

The y-axis represents the (infinite-dimensional) space of admitted functions, and each point (y, λ) *represents a function for the parameter* λ.

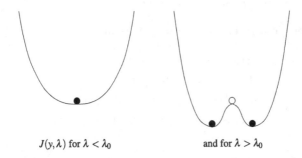

$J(y, \lambda)$ for $\lambda < \lambda_0$ and for $\lambda > \lambda_0$

Fig. 3.1 A Continuous Loss of Convexity

Figure 3.2 suggests the technical term of a "bifurcation" of solutions depending on a parameter. In this scenario a unique solution bifurcates into several solutions when the parameter exceeds a critical threshold, and this bifurcation comes along with an "exchange of stability": the trivial solution loses and the nontrivial solutions gain stability. Physicists call bifurcation also "a self-organization of new states," because it happens without external agency. If the bifurcating solutions have less symmetry than the "trivial solution," which is often the case, then bifurcation is also labeled "a spontaneous symmetry breaking." There are many applications, among which a classical one is the buckling of the so-called Euler beam: a straight beam buckles under a growing axial load at a critical threshold, thus losing its symmetry. The "trivial" straight state persists, but it is unstable beyond the critical load.

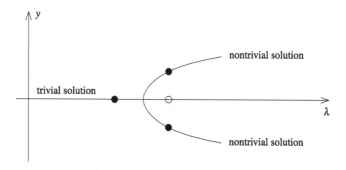

Fig. 3.2   A Bifurcation Diagram

At the "bifurcation point" $(0, \lambda_0)$, the implicit function theorem is obviously not valid. This theorem assures the existence of a unique solution in a perturbed form if the parameter is varied, and in most physical experiments it describes what one expects. The validity of the implicit function theorem is "generic" insofar as a bifurcation is an exceptional event. Under which general mathematical conditions a bifurcation occurs is a natural and interesting question. This led to the field "bifurcation theory," which is presented, e.g., in [17].

2. The Sturm-Liouville boundary value problem

$$
\begin{aligned}
&- (py')' + qy = f \quad \text{on} \quad [a,b], \\
&y(a) = 0, \; y(b) = 0, \quad \text{where} \\
&p \in C^1[a,b], \quad q, f \in C[a,b] \quad \text{and} \\
&p(x) \geq d > 0, \quad q(x) \geq 0 \quad \text{for all } x \in [a,b].
\end{aligned}
\tag{3.3.4}
$$

Proposition 3.3.2. *For all $f \in C[a,b]$, the boundary value problem (3.3.4) possesses a unique classical solution $y_0 \in C^2[a,b]$.*

Proof. On the Hilbert space $W_0^{1,2}(a,b)$, we define

$$
\begin{aligned}
B(y,z) &= \int_a^b p(x)y'z' + q(x)yz\,dx \quad \text{and} \\
\ell(y) &= \int_a^b f(x)y\,dx.
\end{aligned}
\tag{3.3.5}
$$

Then the bilinear form B fulfills the assumptions (3.1.17) of Proposition 3.1.2: Symmetry is obvious; the Cauchy-Schwarz inequality implies the continuity, by virtue of (3.3.4)$_4$ $B(y,y) \geq d\|y'\|_{0,2}^2$; positive definiteness then follows by the Poincaré inequality (3.2.29). The linear functional ℓ is continuous on $L^2(a,b)$, again, by the Cauchy-Schwarz inequality, and a fortiori it is continuous on $W_0^{1,2}(a,b)$. By the Riesz representation theorem 3.1.3, there is a unique $y_0 \in W_0^{1,2}(a,b)$ such that

$$\int_a^b p(x)y_0'h' + (q(x)y_0 - f(x))h\,dx = 0$$

$$\text{for all} \quad h \in W_0^{1,2}(a,b).$$

(3.3.6)

This Euler-Lagrange equation in a weak form yields the regularity of y_0 as in the proof of Proposition 3.3.1: By definition of the weak derivative,

$$\frac{d}{dx}(py_0') = qy_0 - f \in C[a,b] \subset L^2(a,b),$$

$$py_0' \in W^{1,2}(a,b) \subset C[a,b], \quad \text{and hence,} \quad y_0' = py_0'/p \in C[a,b],$$

$$y_0 \in C^1[a,b], \quad \text{and by (1.3.10)},$$

$$py_0' \in C^1[a,b]. \quad \text{Finally,} \quad y_0' = py_0'/p \in C^1[a,b],$$

(3.3.7)

due to $p \in C^1[a,b]$. Therefore $y_0 \in C^2[a,b]$ fulfills equation (3.3.7)$_1$ in the classical sense, and the boundary conditions are fulfilled by $y_0 \in W_0^{1,2}(a,b)$.   □

If the coefficients p, q, and f are smooth enough, more regularity of the solution y_0 can be obtained by "bootstrapping."
The conditions (3.3.4)$_4$ cannot be relaxed in an essential way:
The necessary condition of Legendre in Exercise 1.4.3 requires $p(x) \geq 0$ for all $x \in [a,b]$. The following example shows that also for $p(x) \geq d > 0$, Proposition 3.3.2 is no longer true when q is negative:

$$-y'' - \pi^2 y = f \quad \text{on} \quad [0,1],$$
$$y(0) = 0, \ y(1) = 0,$$

(3.3.8)

does not possess a solution for all $f \in C[a,b]$, and if it exists, it is not unique. Assume a solution $y \in C^2[a,b]$ of (3.3.8) for $f(x) = \sin \pi x$. Then, after two integrations by parts, we find

$$-\int_0^1 y''f + \pi^2 yf\,dx = -\int_0^1 y(f'' + \pi^2 f)dx = 0 = \int_0^1 f^2 dx > 0, \qquad (3.3.9)$$

which is a contradiction. For $f = 0$ (3.3.8) has two solutions: $y = 0$ and $y(x) = \sin \pi x$.
However, the condition $q(x) \geq 0$ is not necessary. For $p(x) \geq d > 0$, one has to assume that the homogeneous boundary value problem (3.3.4)$_{1,2}$ with $f = 0$ has only the trivial solution $y = 0$. Then the claim of Proposition 3.3.2 is true. The proof

relies on the nontrivial Riesz-Schauder theory of functional analysis, and the above-mentioned application is called the "Fredholm alternative."

Remark. *The left side of equation* (3.3.4)$_1$ *is not a general linear ordinary differ-ential equation of second order, but it is of a so-called self-adjoint form. A general equation* $-(py')' + ry' + qy$ *with an arbitrary coefficient* $r \in C[a,b]$ *can be trans-formed in a self-adjoint form by multiplication by* e^R *where* $R' = -r/p$:

$$e^R[-(py')' + ry' + qy] = -(e^R py')' + e^R qy.$$

For a computation of the unique solution $y_0 \in C^2[a,b]$ of (3.3.4), we describe the **Ritz method** of Proposition 3.2.8:
Let $S = \{e_n\}_{n \in \mathbb{N}}$ be a Schauder basis in the Hilbert space $W_0^{1,2}(a,b)$, and let $U_N = span[e_1, \ldots, e_N]$ be the N-dimensional subspace. Then

$$J(y_N) = \inf\{J(y)|y \in U_N\} \tag{3.3.10}$$

defines by Proposition 3.2.8 a minimizing sequence $(y_N)_{N \in \mathbb{N}} \subset W_0^{1,2}(a,b)$ of the functional (3.1.19), which converges by Proposition 3.1.2 strongly to the global minimizer $y_0 \in W_0^{1,2}(a,b)$ of J. This minimizer is the classical solution $y_0 \in C^2[a,b]$ of (3.3.4), cf. Propositions 3.1.2 and 3.3.2. In particular $\lim_{N \to \infty} y_N = y_0$ in $C[a,b]$ by Proposition 3.2.1, which means uniform convergence.
The coefficients of $y_N = \sum_{n=1}^N c_n^N e_n$ are determined by the linear system

$$\sum_{n=1}^N \alpha_{kn} c_n^N = \beta_k \quad , \quad k = 1, \ldots, N, \quad \text{where}$$
$$\alpha_{kn} = B(e_k, e_n), \quad \beta_k = \ell(e_k) = (f, e_k)_{0,2}, \tag{3.3.11}$$

cf. Proposition 3.2.8. The so-called stiffness matrix $(\alpha_{kn}) \in \mathbb{R}^{N \times N}$ can be reduced to a diagonal form by an appropriate choice of the Schauder basis: In case of the eigenfunctions of the corresponding Sturm-Liouville eigenvalue problem (3.3.12), the matrix (α_{kn}) is a diagonal matrix, cf. Propositions 3.3.3, 3.1.4. However, these eigenfunctions are known only in special cases.
The finite element method to solve (3.3.4) approximately relies also on the Ritz method as described before: here the "elements" e_n have "small" support such that the stiffness matrix is a "sparse," having nonzero entries only near the diagonal. There are efficient solvers for the linear system (3.3.11) in this case.

3. The Sturm-Liouville eigenvalue problem

$$-(py')' + qy = \lambda \rho y \quad \text{on} \quad [a,b],$$
$$y(a) = 0, \ y(b) = 0. \tag{3.3.12}$$

The coefficients p and q and the "weight function" ρ fulfill:

$$p \in C^1[a,b], \quad q, \rho \in C[a,b],$$
$$p(x) \geq d > 0 \quad \text{and} \quad q(x) \geq -c_2\rho(x) \quad \text{for all } x \in [a,b], \qquad (3.3.13)$$
$$\rho(x) > 0 \quad \text{for all } x \in (a,b).$$

For $\rho \equiv 1$, (3.3.12) is a common eigenvalue problem. A number $\lambda \in \mathbb{R}$ is called an eigenvalue if (3.3.12) has a nontrivial solution $y \in C^2[a,b]$, called an eigenfunction with eigenvalue λ.

If $\rho(x) \geq \delta > 0$ for all $x \in [a,b]$, then condition (3.3.13)$_2$ is fulfilled for any $q \in C[a,b]$, with some nonnegative constant c_2.

Proposition 3.3.3. *Under the hypotheses* (3.3.13), *the Sturm-Liouville eigenvalue problem* (3.3.12) *possesses infinitely many linear independent eigenfunctions* $u_n \in C^2[a,b]$ *with eigenvalues* $\lambda_n \in \mathbb{R}$, *which satify*

$$\int_a^b \rho u_n u_m dx = \delta_{nm},$$
$$\lambda_1 < \lambda_2 < \cdots < \lambda_n < \cdots, \quad \lim_{n \to \infty} \lambda_n = +\infty. \qquad (3.3.14)$$

The system $S = \{u_n\}_{n \in \mathbb{N}}$ *of the eigenfunctions provides a Schauder basis in* $W_0^{1,2}(a,b)$, *i.e., any* $y \in W_0^{1,2}(a,b)$ *can be developed into a series*

$$y = \sum_{n=1}^{\infty} c_n u_n \quad \text{where} \quad c_n = \int_a^b \rho y u_n dx, \qquad (3.3.15)$$

which converges in $W_0^{1,2}(a,b)$.

Proof. On the Hilbert space $W_0^{1,2}(a,b)$ we define the bilinear forms

$$B(y,z) = \int_a^b p(x)y'z' + q(x)yz\,dx,$$
$$K(y,z) = \int_a^b \rho(x)yz\,dx, \qquad (3.3.16)$$

and we note that they are symmetric, continuous, and also that B is K-coercive according to (3.1.64).

Furthermore $K(y) = K(y,y) > 0$ for $y \neq 0$. It is true that K is weakly sequentially continuous in the sense of (3.1.28)$_2$, but the proof is not simple. Therefore we show only the property of K that we use in the proofs of the propositions in Paragraph 3.1.

Let $(y_n)_{n \in \mathbb{N}}$ be a minimizing sequence, which is bounded in $W_0^{1,2}(a,b)$. By (3.1.13) or (3.1.14) and Proposition 3.2.4, this sequence contains a subsequence $(y_{n_k})_{k \in \mathbb{N}}$, which converges weakly in $W_0^{1,2}(a,b)$ as well as in the sense (3.2.26) to a function $y_0 \in W_0^{1,2}(a,b)$. The uniform convergence (3.2.26)$_1$ then implies $\lim_{k \to \infty} K(y_{n_k}) = K(y_0)$.

Therefore all results of Paragraph 3.1 on the weak eigenvalue problem,

$$B(u,h) = \lambda K(u,h) \quad \text{for all} \quad h \in W_0^{1,2}(a,b), \tag{3.3.17}$$

hold for B and K defined in (3.3.16). In particular, for $u \in W_0^{1,2}(a,b)$,

$$\int_a^b p(x)u'h' + (q(x) - \lambda \rho(x))uh\,dx = 0$$
$$\text{for all} \quad h \in W_0^{1,2}(a,b), \tag{3.3.18}$$

and we may copy the arguments in (3.3.7) for the regularity of y_0: Each weak eigen-function $u \in W_0^{1,2}(a,b)$ satisfying (3.3.18) for some eigenvalue λ is in $C^2[a,b]$ and fulfills (3.3.12) in the classical sense. If the coefficients p, q, and the weight function ρ are "smooth" a "bootstrapping" yields as much regularity of the eigenfunction as p, q, and ρ allow.

The claim $\lambda_n < \lambda_{n+1}$ for all $n \in \mathbb{N}$ means that all eigenvalues have a geometric multiplicity one. Assume $\lambda_n = \lambda_{n+1}$. By the homogeneous boundary condition at $x = a$ there exist $(\alpha, \beta) \neq (0,0) \in \mathbb{R}^2$ such that

$$\alpha u_n(a) + \beta u_{n+1}(a) = 0, \quad \text{and}$$
$$\alpha u'_n(a) + \beta u'_{n+1}(a) = 0. \tag{3.3.19}$$

For the linear differential equation of second order

$$y'' + \frac{p'}{p}y' + \frac{\lambda_n \rho - q}{p}y = 0 \tag{3.3.20}$$

the initial values $y(a)$ and $y'(a)$ determine a unique solution, which, by the trivial ini-tial values (3.3.19), is the trivial solution $\alpha u_n + \beta u_{n+1} = 0$. Therefore $\alpha K(u_n, u_n) + \beta K(u_{n+1}, u_n) = \alpha = 0$ (cf. (3.2.39)$_1$), and $\alpha K(u_n, u_{n+1}) + \beta K(u_{n+1}, u_{n+1}) = \beta = 0$, which contradicts the choice $(\alpha, \beta) \neq (0,0)$. □

Remark. *Each of the eigenvalues λ_n not only have geometric multiplicity one, but by symmetry also algebraic multiplicity one: Assume that*

$$-(pu')' + qu - \lambda_n \rho u = u_n \tag{3.3.21}$$

has a solution $u \in C^2[a,b] \cap \{u(a) = 0,\ u(b) = 0\}$. Then

$$B(u,u_n) + \lambda_n K(u,u_n) = \|u_n\|_{0,2}^2,$$
$$B(u_n,u) + \lambda_n K(u_n,u) = 0, \tag{3.3.22}$$

and hence, $u_n = 0$ by the symmetry of B and K. This contradiction shows that there is no generalized eigenfunction proving algebraic simplicity of the eigenvalue.

According to (3.1.67), the eigenvalues are determined as follows:

$$\lambda_n = \min\left\{\frac{B(y)}{K(y)}\Big| y \in W_0^{1,2}(a,b), y \neq 0, K(y,u_i) = 0,\ i = 1,\dots,n-1\right\},$$

where $u_1,\dots,u_{n-1}$ are the first $n-1$ eigenfunctions. (3.3.23)

The minimum is attained in the nth eigenfunction u_n,
which can be normalized to $K(u_n) = 1$.

The quotient $B(y)/K(y)$ is called **Rayleigh quotient**.
For $n = 1$ (3.3.23) implies a **Poincaré inequality**:

$$\lambda_1 \int_a^b \rho y^2 dx \leq \int_a^b p(y')^2 + q y^2 dx \quad \text{for all} \quad y \in W_0^{1,2}(a,b). \tag{3.3.24}$$

The eigenvalues can also be determined by a **minimax principle**:

Proposition 3.3.4. *Let the functions $v_1,\dots,v_{n-1} \in L^2(a,b)$ for $n \geq 2$ define the following closed subspace of $W_0^{1,2}(a,b)$:*

$$V(v_1,\dots,v_{n-1}) = \{y \in W_0^{1,2}(a,b)| K(y,v_i) = 0 \quad \text{for} \quad i = 1,\dots,n-1\}, \quad (3.3.25)$$

and let

$$d(v_1,\dots,v_{n-1}) = \min\left\{\frac{B(y)}{K(y)}\Big| y \in V(v_1,\dots,v_{n-1}), y \neq 0\right\}. \tag{3.3.26}$$

Then

$$\lambda_n = \max_{v_1,\dots,v_{n-1} \in L^2(a,b)} \{d(v_1,\dots,v_{n-1})\}. \tag{3.3.27}$$

The **proof** is the same as that for Proposition 3.1.7.

The next propositions are consequences of the minimax principle: We establish the monotonicity of the eigenvalues of the Sturm-Liouville eigenvalue problem in dependence on the weight function ρ and on the length of the interval (a,b).

Proposition 3.3.5. *Assume in $(3.3.13)_2$ that $q(x) \geq 0$ for all $x \in [a,b]$. For two continuous weight functions ρ_1, ρ_2 of the Sturm-Liouville eigenvalue problem (3.3.12), suppose that*

$$0 < \delta \leq \rho_1(x) \leq \rho_2(x) \quad \text{for all} \quad x \in [a,b]. \tag{3.3.28}$$

Then the nth eigenvalues $\lambda_n = \lambda_n(\rho)$ of (3.3.12) satisfy

$$\lambda_n(\rho_1) \geq \lambda_n(\rho_2) \quad \text{for all} \quad n \in \mathbb{N}. \tag{3.3.29}$$

Proof. Define $K_i(y,z) = \int_a^b \rho_i yz\, dx$, $K_i(y) := K_i(y,y)$, and let $d_i(v_1,\dots,v_{n-1})$ be the minimum (3.3.26) of the Rayleigh quotient $B(y)/K_i(y)$, $i = 1,2$, in $V(v_1,\dots,v_{n-1})$ for arbitrary functions $v_1,\dots,v_{n-1} \in L^2(a,b)$. Since

$$K_1(y, v_k) = K_2(y, \frac{\rho_1}{\rho_2} v_k), \quad k = 1, \ldots, n-1,$$

$$\frac{B(y)}{K_1(y)} \geq \frac{B(y)}{K_2(y)} \quad \text{for} \quad y \in W_0^{1,2}(a,b), y \neq 0, \tag{3.3.30}$$

definitions (3.3.27) and (3.3.23), with the first $n-1$ eigenfunctions $u_1^2, \ldots, u_{n-1}^2$ determined with weight function ρ_2, imply

$$\lambda_n(\rho_1) \geq d_1 \left(\frac{\rho_2}{\rho_1} u_1^2, \ldots, \frac{\rho_2}{\rho_1} u_{n-1}^2 \right) \geq d_2(u_1^2, \ldots, u_{n-1}^2) = \lambda_n(\rho_2), \tag{3.3.31}$$

which is the claim (3.3.29). In $(3.3.30)_2$ we use $B(y) > 0$ for $y \neq 0$, guaranteed by the assumption on q. $\qquad\square$

If $\rho_1(x_0) < \rho_2(x_0)$ for at least one $x_0 \in (a,b)$, then $\lambda_1(\rho_1) > \lambda_1(\rho_2) > 0$, cf. Exercise 3.3.4.

Proposition 3.3.6. *For two intervals of the Sturm-Liouville eigenvalue problem* (3.3.12), *suppose that*

$$(a_1, b_1) \subsetneq (a_2, b_2). \tag{3.3.32}$$

Then the nth eigenvalues $\lambda_n = \lambda_n(a,b)$ of (3.3.12) satisfy

$$\lambda_n(a_1, b_1) > \lambda_n(a_2, b_2) \quad \text{for all} \quad n \in \mathbb{N}. \tag{3.3.33}$$

Proof. If we extend any $y \in W_0^{1,2}(a_1, b_1)$ by zero off the interval (a_1, b_1), then $W_0^{1,2}(a_1, b_1) \subset W_0^{1,2}(a_2, b_2)$, cf. Exercise 3.3.3. We denote the bilinear forms (3.3.16) over the interval (a_i, b_i) by B_i and K_i, $i = 1,2$.

Let $u_1^2, \ldots, u_{n-1}^2$ be the first $n-1$ eigenfunctions over (a_2, b_2). Then by (3.3.23) and (3.3.27),

$$\lambda_n(a_2, b_2)$$
$$= \min \left\{ \frac{B_2(y)}{K_2(y)} \,\Big|\, 0 \neq y \in W_0^{1,2}(a_2, b_2), K_2(y, u_k^2) = 0, k = 1, \ldots, n-1 \right\}$$
$$\leq \min \left\{ \frac{B_1(y)}{K_1(y)} \,\Big|\, 0 \neq y \in W_0^{1,2}(a_1, b_1), K_1(y, u_k^2) = 0, k = 1, \ldots, n-1 \right\} \tag{3.3.34}$$
$$\leq \lambda_n(a_1, b_1),$$

which is due to the fact that in $(3.3.34)_3$ B_1 can be replaced by B_2, and K_1 can be replaced by K_2 if $y \in W_0^{1,2}(a_1, b_1)$ is extended by zero off (a_1, b_1).

Assume that $\lambda_n(a_2, b_2) = \lambda_n(a_1, b_1)$. Then there exists a nonzero function $y_n \in W_0^{1,2}(a_1, b_1)$, which, after extension by zero, minimizes the Rayleigh quotient $B_2(y)/K_2(y)$ in $W_{n-1} = \{y \in W_0^{1,2}(a_2, b_2) \mid K_2(y, u_k^2) = 0, k = 1, \ldots, n-1\}$.

The proof of Proposition 3.1.4 then shows that y_n is a weak eigenfunction, and Proposition 3.3.3 finally shows that $y_n \in C^2[a_2,b_2]$ is a classical eigenfunction of (3.3.12) with eigenvalue $\lambda_n(a_2,b_2)$. Since (a_1,b_1) is properly contained in (a_2,b_2) an extension by zero implies that there is some $x_0 \in (a_2,b_2)$ such that $y_n(x_0) = y_n'(x_0) = 0$. As mentioned already in the proof of Proposition 3.3.3, the uniqueness of the initial value problem (3.3.20) with $y(x_0) = y'(x_0) = 0$ implies $y_n(x) = 0$ for all $x \in [a_2,b_2]$, contradicting $y_n \neq 0$. This proves $\lambda_n(a_1,b_1) > \lambda_n(a_2,b_2)$. $\quad\square$

Here is an example:

$$-y'' = \lambda y \quad \text{on} \quad [a,b]$$
$$y(a) = 0, \quad y(b) = 0, \tag{3.3.35}$$

has the eigenfunctions

$$u_n(x) = \sin n\pi \frac{x-a}{b-a} \quad \text{with eigenvalues} \quad \lambda_n = \left(\frac{n\pi}{b-a}\right)^2, \quad n \in \mathbb{N}. \tag{3.3.36}$$

In this example the nth eigenfunction u_n has precisely $n-1$ simple zeros in (a,b). The simplicity follows from the fact that $y(x_0) = y_0'(x_0) = 0$ for some $x_0 \in (a,b)$ implies $y = 0$. This property concerning the number of simple zeros of the eigenfunctions of a Sturm-Liouville eigenvalue problem holds more generally:

Proposition 3.3.7. *The nth eigenfunction u_n of a Sturm-Liouville eigenvalue problem (3.3.12) has at most $n-1$ simple zeros in (a,b).*

Proof. Assume that the eigenfunction $u_n \in C^2[a,b] \cap \{y(a) = 0, y(b) = 0\}$ has m simple zeros $a = x_0 < x_1 < \cdots < x_m < x_{m+1} = b$ where $m \geq n$.

For $i = 1, \ldots, n$, we define

$$w_i(x) = \begin{cases} c_i u_n(x) & \text{for } x \in [x_{i-1}, x_i], \\ 0 & \text{for } x \notin [x_{i-1}, x_i]. \end{cases} \tag{3.3.37}$$

By Exercise 3.3.3, $w_i \in W_0^{1,2}(x_{i-1}, x_i) \subset W_0^{1,2}(a, x_n)$. Choose the constants $c_i \neq 0$ such that $K(w_i) = 1$. Then by (3.3.17) and the definition (3.3.37) of the functions w_i, we have

$$B(w_i, w_j) = \lambda_n K(w_i, w_j) = \lambda_n \delta_{ij}, \quad i, j = 1, \ldots, n,$$
$$\text{with the } n^{\text{th}} \text{ eigenvalue} \quad \lambda_n = \lambda_n(a,b). \tag{3.3.38}$$

For arbitrary $v_1, \ldots, v_{n-1} \in L^2(a, x_n)$ there exist n coefficients α_i to determine $y = \alpha_1 w_1 + \cdots + \alpha_n w_n \in W_0^{1,2}(a, x_n)$ such that

$$\int_a^{x_n} \rho y v_k dx = \sum_{i=1}^n \alpha_i K(w_i, v_k) = 0 \quad \text{for} \quad k = 1, \ldots, n-1 \quad \text{and}$$
$$\sum_{i=1}^n \alpha_i^2 = 1. \tag{3.3.39}$$

Then $K(y) = 1$, $B(y) = \lambda_n$, and an application of the minimax principle of Proposition 3.3.4 gives

$$\lambda_n(a, x_n) \leq \lambda_n = \lambda_n(a, b). \tag{3.3.40}$$

The assumption $(a, x_n) \subsetneq (a, b)$ and (3.3.40) contradict the monotonicity (3.3.33) of Proposition 3.3.6.   □

Proposition 3.3.7 implies that the first eigenfunction u_1 has no zero in (a, b), which means that it has one sign in (a, b). Typically one chooses a positive sign of u_1 in (a, b). For all eigenfunctions the following proposition holds:

Proposition 3.3.8. *The nth eigenfunction u_n of the Sturm-Liouville eigenvalue problem (3.3.12) has precisely $n - 1$ simple zeros in (a, b). Between two consecutive zeros of u_n, there is precisely one simple zero of u_{n+1}, where the zeros $u_n(a) = 0$ and $u_n(b) = 0$ are taken into account.*

Proof. Assume that between two consecutive zeros $x_i < x_{i+1}$ of u_n there is no zero of u_{n+1}. Then there are two zeros $\tilde{x}_j < \tilde{x}_{j+1}$ of u_{n+1} with $(x_i, x_{i+1}) \subset (\tilde{x}_j, \tilde{x}_{j+1}) \subset (a, b)$, and u_n as well as u_{n+1} has one sign in (x_i, x_{i+1}) and $(\tilde{x}_j, \tilde{x}_{j+1})$, respectively. Therefore λ_n and λ_{n+1} are the first eigenvalues of the eigenvalue problem over $[x_i, x_{i+1}]$ and $[\tilde{x}_j, \tilde{x}_{j+1}]$, respectively; due to K-orthogonality, all other eigenfunctions have to change sign in the underlying interval. According to Proposition 3.3.6, $\lambda_n \geq \lambda_{n+1}$, contradicting $\lambda_n < \lambda_{n+1}$, cf. (3.3.14)$_2$. Thus u_{n+1} has a zero between two zeros of u_n.

We have shown that if u_n has the zeros $a = x_0 < x_1 < \cdots < x_m < x_{m+1} = b$, then u_{n+1} has at least $m + 1$ zeros in (a, b).

We know that u_1 has no zero, and thus, u_2 has at least one zero in (a, b). The induction hypothesis that u_n has at least $n - 1$ zeros in (a, b) implies that u_{n+1} has at least n zeros in (a, b). Thus this is true for all $n \in \mathbb{N}$, and in combination with Proposition 3.3.7, we have proved Proposition 3.3.8.   □

We remark that Proposition 3.3.8 excludes that u_n and u_{n+1} have a common zero in (a, b).

More applications of the direct method are found in the literature cited in the reference list.

Exercises

3.3.1. According to (3.3.23), the nth eigenfunction $u_n \in C^2[a, b]$ is the global minimizer of B defined on $C_0^1[a, b]$ under the isoperimetric constraints

$$K(y) = 1,$$
$$K(y, u_k) = 0 \quad \text{for} \quad k = 1, \dots, n-1.$$

Assuming the regularity $u_k \in C^2[a, b]$ for $k = 1, \dots, n-1$, Proposition 2.1.3 gives that u_n satisfies the Euler-Lagrange equation

$$\frac{d}{dx}(2pu'_n) = 2qu_n + 2\tilde{\lambda}_n \rho u_n + \sum_{k=1}^{n-1} \tilde{\lambda}_k \rho u_k \quad \text{on } [a, b],$$

with Lagrange multipliers $\tilde{\lambda}_n, \tilde{\lambda}_1, \dots, \tilde{\lambda}_{n-1}$, provided the constraints are not critical for u_n. Prove:

a) The isoperimetric constraints are not critical for u_n.
b) $\tilde{\lambda}_k = 0$ for $k = 1, \dots, n-1$.
c) $\tilde{\lambda}_n = -B(u_n)$.

3.3.2. Show that the system $\{u_n\}_{n \in \mathbb{N}}$ of all eigenfunctions of (3.3.12) is also complete or a Schauder basis in $L^2(a, b)$.

Hint: $\{u_n\}_{n \in \mathbb{N}}$ is complete in $L^2(a, b) \Leftrightarrow$ the closure of span$[u_n | n \in \mathbb{N}]$ in $L^2(a, b)$ is $L^2(a, b)$, cf. [25], V.4.

3.3.3. For any $y \in W_0^{1,2}(a, b)$, show that its extension by zero off $[a, b]$ defines a function $\tilde{y} \in W_0^{1,2}(c, d)$ for all $c \le a < b \le d$.

3.3.4. Under the assumptions of Proposition 3.3.5, show that the additional assumption $\rho_1 \ne \rho_2$ implies $\lambda_1(\rho_1) > \lambda_1(\rho_2)$.

Appendix

First we prove some facts about manifolds, some of which are probably known from courses on calculus. The advantage of a revision here is that we introduce the terminology, which we use in applications of the calculus of variations. We prove Lagrange's multiplier rule, Liouville's theorem on volume preservation, the weak sequential compactness of bounded sets in a Hilbert space, and the theorem of Arzelà-Ascoli.

For a continuously totally differentiable mapping

$$\Psi : \mathbb{R}^n \to \mathbb{R}^m, \quad \text{where} \quad m < n, \tag{A.1}$$

we investigate the nonempty zero set

$$M = \{x \in \mathbb{R}^n | \Psi(x) = 0\} \subset \mathbb{R}^n \tag{A.2}$$

under the assumption that the Jacobian matrix

$$D\Psi(x) = \left(\frac{\partial \Psi_i}{\partial x_j}(x) \right)_{\substack{i=1,\ldots,m \\ j=1,\ldots,n}} \in \mathbb{R}^{m \times n} \tag{A.3}$$

has the maximal rank m for all $x \in M$. If the linear spaces $\mathbb{R}^n$ and $\mathbb{R}^m$ are endowed with the canonical bases, we can identify the Jacobian matrix with a linear transformation $D\Psi(x) \in L(\mathbb{R}^n, \mathbb{R}^m)$. Then the following statements are equivalent:

$$\begin{aligned} &\text{Range} D\Psi(x) = \mathbb{R}^m, \quad \text{or } D\Psi(x) \text{ is surjective,} \\ &\text{dimKernel} D\Psi(x) = n - m > 0 \quad \text{for all} \quad x \in M. \end{aligned} \tag{A.4}$$

The transposed matrix represents the dual or adjoint transformation, which is characterized by the Euclidean scalar product in $\mathbb{R}^m$ and $\mathbb{R}^n$ as follows:

$$(D\Psi(x)y, z)_{\mathbb{R}^m} = (y, D\Psi(x)^* z)_{\mathbb{R}^n} \quad \text{for all} \quad y \in \mathbb{R}^n, \quad z \in \mathbb{R}^m. \tag{A.5}$$

© Springer International Publishing AG 2018
H. Kielhöfer, *Calculus of Variations*, Texts in Applied Mathematics 67,
https://doi.org/10.1007/978-3-319-71123-2

The transposed matrix $D\Psi(x)^*$ has the same rank m as the matrix $D\Psi(x)$, and $D\Psi(x)^* \in L(\mathbb{R}^m, \mathbb{R}^n)$ is injective, which is easily proved by (A.5): $\mathrm{Range}D\Psi(x) = \mathbb{R}^m$ implies $\mathrm{Kernel}D\Psi(x)^* = \{0\}$.

For a subspace $U \subset \mathbb{R}^n$ we define the orthogonal complement by $U^\perp = \{y \in \mathbb{R}^n | (y, u) = 0 \text{ for all } u \in U\}$. Then $\mathbb{R}^n$ is the direct sum $\mathbb{R}^n = U \oplus U^\perp$, which we prove in (A.45)–(A.48). In particular,

$$
\begin{aligned}
(\mathrm{Range}D\Psi(x)^*)^\perp &= \{y \in \mathbb{R}^n | (y, D\Psi(x)^* z)_{\mathbb{R}^n} = 0 \text{ for all } z \in \mathbb{R}^m\} \\
&= \{y \in \mathbb{R}^n | (D\Psi(x)y, z)_{\mathbb{R}^m} = 0 \text{ for all } z \in \mathbb{R}^m\} \\
&= \{y \in \mathbb{R}^n | D\Psi(x)y = 0\} = \mathrm{Kernel}D\Psi(x), \\
\mathbb{R}^n &= \mathrm{Kernel}D\Psi(x) \oplus \mathrm{Range}D\Psi(x)^*,
\end{aligned}
\tag{A.6}
$$

and the direct sum is orthogonal. We define for $x \in M$

$$
\begin{aligned}
T_x M &= \mathrm{Kernel}D\Psi(x), \quad \dim T_x M = n - m, \\
N_x M &= \mathrm{Range}D\Psi(x)^* = (T_x M)^\perp, \quad \dim N_x M = m
\end{aligned}
\tag{A.7}
$$

and therefore $\mathbb{R}^n = T_x M \oplus N_x M$ for all $x \in M$.

Below we fix $x_0 \in M$ and $x \in \mathbb{R}^n$ are arbitrary. Then there is a unique decomposition $x - x_0 = y + z$, where $y \in T_{x_0}M$ and $z \in N_{x_0}M$. Defining

$$
\begin{aligned}
F(y, z) &= \Psi(x_0 + y + z) \quad \text{yields a mapping} \\
F &: T_{x_0}M \times N_{x_0}M \to \mathbb{R}^m, \quad \text{satisfying} \\
F(0, 0) &= 0.
\end{aligned}
\tag{A.8}
$$

The mapping is continuously totally differentiable with respect to its two variables, and in particular

$$
D_z F(0, 0) = D\Psi(x_0)|_{N_{x_0}M} : N_{x_0}M \to \mathbb{R}^m \quad \text{is bijective.}
\tag{A.9}
$$

Observe that $N_{x_0}M$ is the orthogonal complement of $T_{x_0}M = \mathrm{Kern}D\Psi(x_0)$. Property (A.9) allows the application of the implicit function theorem: There exists a neighborhood $B_r(0) = \{y \in T_{x_0}M | \|y\| < r\}$ and a continuously differentiable mapping

$$
\begin{aligned}
\varphi &: B_r(0) \subset T_{x_0}M \to N_{x_0}M, \quad \text{satisfying} \quad \varphi(0) = 0 \quad \text{and} \\
F(y, \varphi(y)) &= 0 \quad \text{for all } y \in B_r(0).
\end{aligned}
\tag{A.10}
$$

Moreover, all zeros of F in a neighbourhood of $(0, 0) \in T_{x_0}M \times N_{x_0}M = \mathbb{R}^n$ are given by $(y, \varphi(y))$. By definition (A.8), this implies that the graph of φ shifted by the vector x_0 coincides with the set M in a neighborhood $U(x_0)$ of x_0 in $\mathbb{R}^n$:

$$
\{x_0 + y + \varphi(y) | y \in B_r(0) \subset T_{x_0}M\} = M \cap U(x_0).
\tag{A.11}
$$

We now show that the affine subspace $x_0 + T_{x_0}M$ is tangent to the set M in x_0. By virtue of $F(y, \varphi(y)) = 0$ for all $y \in B_r(0)$, differentiation via the chain rule at $(0,0)$ yields

$$D_y F(0,0) + D_z F(0,0) D\varphi(0) = 0 \quad \text{in } L(T_{x_0}M, \mathbb{R}^m). \qquad (A.12)$$

According to definition (A.8),

$$D_y F(0,0) = D\Psi(x_0)|_{T_{x_0}M} = 0, \quad \text{due to } T_{x_0}M = \text{Kernel} D\Psi(x_0),$$

$$D_z F(0,0) = D\Psi(x_0)|_{N_{x_0}M} : N_{x_0}M \to \mathbb{R}^m \quad \text{is bijective, and hence,} \qquad (A.13)$$

$$D\varphi(0) = 0 \quad \text{in } L(T_{x_0}M, N_{x_0}M).$$

The derivatives of φ in all directions of the subspace $T_{x_0}M$ vanish in $y = 0$, and therefore the term "**tangent space of M in x_0**" is justified. The orthogonal space $N_{x_0}M$ is called "**normal space of M in x_0**".

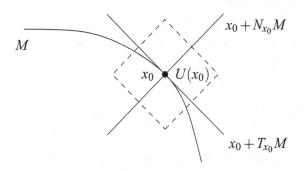

Fig. A.1 Tangent and Normal Space

We define on the neighborhood $U(x_0) = \{x = x_0 + y + z | y \in T_{x_0}M, \|y\| < r, z \in N_{x_0}M, \|z\| < r\}$, where $\| \quad \|$ is the Euclidean norm, the mapping $H(x) = H(x_0 + y + z) = x + \varphi(y)$. By (A.11), $H((x_0 + T_{x_0}M) \cap U(x_0)) = M \cap U(x_0)$. This mapping is also called a "local straightening," and due to $DH(x_0) = E + D\varphi(0) = E$ (which is the identity matrix), $H : U(x_0) \to H(U(x_0))$ is a diffeomorphism, cf. Figure A.2.

The set M defined by (A.2), with the maximal rank condition (A.3) of the Jacobian matrix, is a so-called **continuously differentiable manifold of dimension** $n - m$.

Any direct and orthogonal decomposition $\mathbb{R}^n = U \oplus V$ defines orthogonal projections, whose properties are the following:

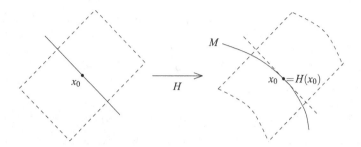

Fig. A.2 A Local Straightening

The unique decomposition $x = y + z$ for $x \in \mathbb{R}^n$, $y \in U$, $z \in V$
defines via $Px = y$ and $Qx = z$,
linear transformations $P, Q \in L(\mathbb{R}^n, \mathbb{R}^n)$, satisfying:

$$P = I - Q, \quad Q = I - P, \quad P^2 = P, \quad Q^2 = Q, \tag{A.14}$$

$$\text{Range} P = U, \quad \text{Kernel} P = V, \quad \text{Range} Q = V, \quad \text{Kernel} Q = U,$$

$$(Px, \tilde{x}) = (x, P\tilde{x}) = (Px, P\tilde{x}),$$

$$(Qx, \tilde{x}) = (x, Q\tilde{x}) = (Qx, Q\tilde{x}) \quad \text{for } x, \tilde{x} \in \mathbb{R}^n.$$

The symmetries are consequences of the orthogonality of the two subspaces:
$(Px, \tilde{x}) = (Px, P\tilde{x} + Q\tilde{x}) = (Px, P\tilde{x}) = (Px + Qx, P\tilde{x}) = (x, P\tilde{x})$ and analog relations
for $(Qx, \tilde{x})$.

P is called the **orthogonal projection** on U along V, and Q is called the orthogonal projection on V along U.

For the direct and orthogonal decomposition $\mathbb{R}^n = T_x M \oplus N_x M$ for $x \in M$, the orthogonal projections are referred to as $P(x)$ and $Q(x)$, where $P(x)$ is the projection on $T_x M$ along $N_x M$ and $Q(x)$ is the projection on $N_x M$ along $T_x M$.

Next we investigate the differentiability of these projections. Assume that

$$\Psi \in C^k(\mathbb{R}^n, \mathbb{R}^m) \quad \text{for} \quad k \geq 1, \tag{A.15}$$

i.e., Ψ is k times continuously partially differentiable. Then the components of the Jacobian matrix $D\Psi(x)$ and also of the transposed matrix $D\Psi(x)^*$ are $(k-1)$ times continuously partially differentiable with respect to $x \in \mathbb{R}^n$. By assumption, $D\Psi(x)^* \in L(\mathbb{R}^m, \mathbb{R}^n)$ is injective, and thus, it maps the canonical basis in $\mathbb{R}^m$ onto a basis in $N_x M$, which is $(k-1)$ times continuously partially differentiable with respect to $x \in \mathbb{R}^n$. When this basis is orthonormalized according to E. Schmidt (1876–1959), we obtain an orthonormal basis $\{b_1(x),\dots,b_m(x)\} \subset N_xM$,

which is $(k-1)$ times continuously partially differentiable

with respect to $x \in M$, and the orthogonal projection

$Q(x) : \mathbb{R}^n \to N_xM$ along T_xM is defined by

(A.16)

$$Q(x)\tilde{x} = \sum_{i=1}^{m} (\tilde{x}, b_i(x))b_i(x) \quad \text{for} \quad \tilde{x} \in \mathbb{R}^n.$$

As summarized in (A.14)

$$P(x) = I - Q(x) : \mathbb{R}^n \to T_xM$$
is the orthogonal projection on T_xM along N_xM.

(A.17)

By their construction, both projections Q and P are $(k-1)$ times continuously partially differentiable with respect to $x \in M$. (The differentiability holds in a neighborhood of $x \in M$ in $\mathbb{R}^n$, due to the fact that the maximal rank condition of the Jacobian matrix holds also in a neighborhood of $x \in M$ in $\mathbb{R}^n$.)

We show that locally (where $\| \quad \|$ is the Euclidean norm),

$$T_xM = P(x)T_{x_0}M, \quad N_xM = Q(x)N_{x_0}M$$
for $\|x - x_0\| < \delta$, where $0 < \delta$ is sufficiently small.

(A.18)

If $P(x) : T_{x_0}M \to T_xM$ is injective, then it is also surjective. Assume $P(x)y = 0$ for $y \in T_{x_0}M$. Since $P(x_0)y = y$ and $\|y\| = \|P(x_0)y - P(x)y\| \le \|P(x_0) - P(x)\|\|y\| < \varepsilon\|y\|$, due to the continuity of the projection, we conclude that $y = 0$. The argument for $Q(x) : N_{x_0}M \to N_xM$ is the same.

Let $\{a_1,\dots,a_{n-m}\}$ be an orthonormal basis in $T_{x_0}M$. By (A.18), $\{P(x)a_1,\dots,P(x)a_{n-m}\}$ is a basis in T_xM for all $\|x - x_0\| < \delta$, which we orthonormalize. Hence, we obtain

an orthonormalized basis $\{a_1(x),\dots,a_{n-m}(x)\} \subset T_xM$,

which is $(k-1)$ times continuously partially differentiable

with respect to $x \in M$, and

$P(x) : \mathbb{R}^n \to T_xM$ along N_xM is defined by

(A.19)

$$P(x)\tilde{x} = \sum_{i=1}^{n-m} (\tilde{x}, a_i(x))a_i(x) \quad \text{for} \quad \tilde{x} \in \mathbb{R}^n.$$

The sets $\{T_xM | x \in M\}$ and $\{N_xM | x \in M\}$ are called tangent and normal bundle, respectively, and they can be given the structure of a manifold.

Next we establish the **multiplier rule of Lagrange**. Let

$$f : \mathbb{R}^n \to \mathbb{R} \quad \text{be continuously totally differentiable,} \qquad \text{(A.20)}$$

and assume that $x_0 \in M$ minimizes f locally under the constraint $\Psi(x) = 0$, i.e.,

$$f(x_0) \leq f(x) \quad \text{for all } x \in M \text{ satisfying} \quad \|x - x_0\| < d \tag{A.21}$$

where $d > 0$ is some constant. Then, by (A.11), the curve $\{x(t) = x_0 + at + \varphi(at) | t \in (-\varepsilon, \varepsilon)\} \subset M$ for any $a \in T_{x_0}M$ (with ε depending on a), and by (A.21), $g : (-\varepsilon, \varepsilon) \to \mathbb{R}$, defined by $g(t) = f(x_0 + at + \varphi(at))$, is locally minimal at $t = 0$. Consequently, (where $(\ ,\ )$ is the Euclidean scalar product),

$$\begin{aligned} g'(0) &= (\nabla f(x_0), \dot{x}(0)) = (\nabla f(x_0), a + D\varphi(0)a) \\ &= (\nabla f(x_0), a) = 0 \quad \text{where we use (A.13)}_3. \end{aligned} \tag{A.22}$$

Since $a \in T_{x_0}M$ is arbitrary, the gradient $\nabla f(x_0) \in (T_{x_0}M)^{\perp} = N_{x_0}M = \text{Range}D\Psi(x_0)^*$, cf. (A.7). By convention (cf. (A.3)), the columns of $D\Psi(x_0)^* \in \mathbb{R}^{n \times m}$ are the gradients of Ψ_i, $i = 1, \ldots, m$, i.e.,

$$\begin{aligned} D\Psi(x_0)^* &= (\nabla\Psi_1(x_0) \cdots \nabla\Psi_m(x_0)), \quad \text{and} \\ \nabla f(x_0) &\in \text{Range}D\Psi(x_0)^* = N_{x_0}M \quad \text{means} \\ \nabla f(x_0) &+ \sum_{i=1}^{m} \lambda_i \nabla\Psi_i(x_0) = 0 \quad \text{for some} \quad \lambda = (\lambda_1, \ldots, \lambda_m) \in \mathbb{R}^m. \end{aligned} \tag{A.23}$$

The constants $\lambda_1, \ldots, \lambda_m$ are called **Lagrange multipliers**.

Since $D\Psi(x_0)^* \in L(\mathbb{R}^m, \mathbb{R}^n)$ has maximal rank m and $\dim\text{Range}D\Psi(x_0)^* = \dim N_{x_0}M = m$, the linear transformation

$$D\Psi(x_0)^* : \mathbb{R}^m \to N_{x_0}M \quad \text{is an isomorphism,} \tag{A.24}$$

and thus, the Lagrange multipliers in (A.23)$_3$ are uniquely determined.

Next we prove **Liouville's theorem**, cf. (2.5.61). The system

$$\dot{x} = f(x), \quad f : \mathbb{R}^n \to \mathbb{R}^n, \tag{A.25}$$

where f is a continuously totally differentiable vector field, generates a flow $\varphi(t, z)$, i.e.,

$$\frac{\partial}{\partial t}\varphi(t, z) = f(\varphi(t, z)), \quad \varphi(0, z) = z \in \mathbb{R}^n. \tag{A.26}$$

Differentiation of (A.26) with respect to z (which is possible by the continuously differentiable dependence of the flow on the initial condition) yields ($D = D_z$)

$$\begin{aligned} \frac{\partial}{\partial t}D\varphi(t, z) &= Df(\varphi(t, z))D\varphi(t, z), \quad \text{where} \\ D\varphi(0, z) &= E \quad (= \text{the identity matrix}). \end{aligned} \tag{A.27}$$

The Jacobian matrix of $\varphi(t, .)$ has the columns

$$D\varphi(t, z) = (\varphi_{z_1}(t, z) \cdots \varphi_{z_n}(t, z)). \tag{A.28}$$

Since the determinant is linear with respect to each column, differentiation with respect to t gives, in view of (A.27),

$$\frac{\partial}{\partial t} \det D\varphi(t,z) = \sum_{i=1}^{n} \det(\varphi_{z_1}(t,z) \cdots \frac{\partial}{\partial t} \varphi_{z_i}(t,z) \cdots \varphi_{z_n}(t,z))$$

$$= \sum_{i=1}^{n} \det(\varphi_{z_1}(t,z) \cdots Df(\varphi(t,z))\varphi_{z_i}(t,z) \cdots \varphi_{z_n}(t,z)) \qquad (A.29)$$

$$= \mathrm{trace} Df(\varphi(t,z)) \det(\varphi_{z_1}(t,z) \cdots \varphi_{z_i}(t,z) \cdots \varphi_{z_n}(t,z))$$

$$= \mathrm{div} f(\varphi(t,z)) \det D\varphi(t,z),$$

where we use a result of linear algebra on the trace of a matrix. The differential equation (A.29) for $\det D\varphi(t,z)$ has the solution

$$\det D\varphi(t,z) = \det D\varphi(0,z) \exp \int_0^t \mathrm{div} f(\varphi(s,z)) ds \qquad (A.30)$$

$$= 1 \exp 0 = 1,$$

where we use $(A.27)_2$ and $\mathrm{div} f(x) = 0$ for all $x \in \mathbb{R}^n$. Let Ω be measurable set in $\mathbb{R}^n$. Then the change-of-variable formula for Lebesgue integrals over measurable sets yields

$$\mu(\varphi(t,\Omega)) = \int_{\varphi(t,\Omega)} 1 dz = \int_{\Omega} |\det D\varphi(t,z)| dz = \int_{\Omega} 1 dz = \mu(\Omega), \qquad (A.31)$$

which is Liouville's theorem (2.5.61). If $\mathrm{div} f(x)$ does not vanish, then formulas (A.30) and (A.31) describe how a flow changes a volume.

Next we prove the **sequential weak compactness in a Hilbert space**, cf. (3.1.13).

Let $(y_n)_{n\in\mathbb{N}}$ be a bounded sequence in a Hilbert space X, i.e., $\|y_n\| \leq C$ for all $n \in \mathbb{N}$. Then the numbers

$$\alpha_{nm} = (y_n, y_m) \in \mathbb{R}, \quad n, m \in \mathbb{N}, \qquad (A.32)$$

are bounded in $\mathbb{R} : |\alpha_{nm}| \leq C^2$ for all $n, m \in \mathbb{N}$. By the Bolzano-Weierstraß theorem one can select subsequences of each previous sequence such that

$$(\alpha_{n_k^1, m})_{k\in\mathbb{N}} \quad \text{converges for} \quad m = 1,$$

$$(\alpha_{n_k^2, m})_{k\in\mathbb{N}} \quad \text{converges for} \quad m = 1, 2, \text{ and so on,} \qquad (A.33)$$

$$(\alpha_{n_k^i, m})_{k\in\mathbb{N}} \quad \text{converges for} \quad m = 1, \ldots, i.$$

This construction yields a diagonal sequence

$$(\alpha_{n_k^k, m})_{k\in\mathbb{N}} \quad \text{which converges for all } m \in \mathbb{N}. \qquad (A.34)$$

With the notation $\alpha_{n_k^k,m} = \alpha_{km}$, we obtain a sequence such that

$$\lim_{k\to\infty} \alpha_{km} = \lim_{k\to\infty} (y_k, y_m) \quad \text{exists in } \mathbb{R} \text{ for each } m \in \mathbb{N}. \tag{A.35}$$

Note that $(y_k)_{k\in\mathbb{N}}$ is a subsequence of the original sequence $(y_n)_{n\in\mathbb{N}}$.

Let $U := \text{cl}_X(\text{span}[y_n, n \in \mathbb{N}])$, the closure of the subspace of X spanned by the vectors $\{y_1, y_2, \ldots\}$. Then for any $u \in U$ and $\varepsilon > 0$, there are some $N \in \mathbb{N}$ and some $\tilde{y} \in \text{span}[y_1, \ldots, y_N] = U_N$ such that

$$\|u - \tilde{y}\| < \frac{\varepsilon}{4C}, \tag{A.36}$$

where C bounds the sequence $\|y_n\|$. Then for the subsequence $(y_k)_{k\in\mathbb{N}}$

$$\begin{aligned}
|(y_k, u) - (y_l, u)| &= |(y_k - y_l, u)| \\
&\leq |(y_k - y_l, \tilde{y})| + |(y_k, u - \tilde{y})| + |(y_l, u - \tilde{y})| \\
&\leq |(y_k - y_l, \tilde{y})| + 2C\frac{\varepsilon}{4C}.
\end{aligned} \tag{A.37}$$

By (A.35), the sequence $((y_k, \tilde{y}))_{k\in\mathbb{N}}$ is convergent, and hence,

$$|(y_k - y_l, \tilde{y})| < \frac{\varepsilon}{2} \quad \text{for} \quad k, \ell \geq k_0(\varepsilon, \tilde{y}). \tag{A.38}$$

The arguments thus far demonstrate that

$$((y_k, u))_{k\in\mathbb{N}} \quad \text{is a Cauchy sequence for each } u \in U, \tag{A.39}$$

and therefore it is a convergent sequence in $\mathbb{R}$.

The Hilbert space is decomposed as

$$X = U \oplus U^{\perp}, \tag{A.40}$$

where $U^{\perp}$ is the orthogonal complement of the closed subspace U, cf. (A.45)–(A.48) below. If $y = u + w$, $u \in U$, $w \in U^{\perp}$, then apparently, since $y_k \in U$,

$$(y_k, y) = (y_k, u) \quad \text{for all} \quad k \in \mathbb{N}. \tag{A.41}$$

By (A.39)

$$\lim_{k\to\infty} (y_k, y) = \ell(y) \in \mathbb{R} \quad \text{for each} \quad y \in X. \tag{A.42}$$

The abovedefined transformation $\ell : X \to \mathbb{R}$ is linear, and by $|(y_k, y)| \leq C\|y\|$ for all $k \in \mathbb{N}$, the estimate

$$|\ell(y)| \leq C\|y\| \tag{A.43}$$

holds, proving continuity of ℓ or $\ell \in X'$. By Riesz' representation theorem (3.1.9), there exists a unique $z \in X$ such that $\ell(y) = (y, z) = (z, y)$ for all $y \in X$. Finally, by (A.42),

$$\lim_{k \to \infty} (y_k, y) = (z, y) \quad \text{for all } y \in X \quad \text{or}$$
$$w - \lim_{k \to \infty} y_k = z. \tag{A.44}$$

We have proven that the bounded sequence $(y_n)_{n \in \mathbb{N}}$ contains a weakly convergent subsequence $(y_k)_{k \in \mathbb{N}}$.

We now establish the decomposition (A.40) via Riesz' representation theorem: The orthogonal complement

$$U^\perp = \{w \in X \,|\, (w, u) = 0 \text{ for all } u \in U\}, \tag{A.45}$$

is a closed subspace of X satisfying $U \cap U^\perp = \{0\}$. We define on $U \subset X$

$$\ell : U \to \mathbb{R} \quad \text{by } \ell(u) = (u, y), \quad \text{where } y \in X \text{ is arbitrary.} \tag{A.46}$$

Then $\ell \in U'$, and since the closed subspace $U \subset X$ is a Hilbert space as well, there exists a unique $z \in U$ such that

$$\ell(u) = (u, z), \quad \text{or}$$
$$(u, y) = (u, z) \quad \text{for all} \quad u \in U. \tag{A.47}$$

Then this vector $z \in U$ yields the decomposition

$$y = z + y - z, \quad \text{satisfying}$$
$$(y - z, u) = 0 \quad \text{for all } u \in U, \quad \text{meaning} \tag{A.48}$$
$$y - z \in U^\perp.$$

Finally we prove the **Arzelà-Ascoli Theorem**, cf. Proposition 3.2.2.

Let $S = \{x_m \,|\, m \in \mathbb{N}\} \subset [a, b]$ be a countable dense subset, for instance the rational numbers. Then the numbers

$$\alpha_{nm} = y_n(x_m) \in \mathbb{R}, \quad n, m \in \mathbb{N}, \tag{A.49}$$

are bounded in $\mathbb{R}$, i.e., $|\alpha_{nm}| \leq C$ for all $n, m \in \mathbb{N}$. By the arguments in (A.32)–(A.35), there exists a subsequence $(y_k)_{k \in \mathbb{N}}$ of the sequence $(y_n)_{n \in \mathbb{N}}$ such that

$$\lim_{k \to \infty} y_k(x_m) \quad \text{exists in } \mathbb{R} \text{ for each } m \in \mathbb{N}. \tag{A.50}$$

For any $x \in [a, b]$ and for any $\varepsilon > 0$ there is an $x_m \in S$ such that

$$|x - x_m| < \delta\left(\frac{\varepsilon}{3}\right), \tag{A.51}$$

where δ is taken from the equicontinuity (3.2.23). Then the following estimates hold:

$$|y_k(x) - y_l(x)|$$
$$\le |y_k(x) - y_k(x_m)| + |y_k(x_m) - y_l(x_m)| + |y_l(x_m) - y_l(x)| \qquad \text{(A.52)}$$
$$< \frac{\varepsilon}{3} + |y_k(x_m) - y_l(x_m)| + \frac{\varepsilon}{3} < \varepsilon \quad \text{for} \quad k, l \ge k_0(\varepsilon, m),$$

where we use the convergence (A.50). The index k_0 depends on x_m and consequently on x. However, this does not alter the following result, namely that $(y_k(x))_{k\in\mathbb{N}}$ is a Cauchy sequence and therefore a convergent sequence for each $x \in [a,b]$:

$$\lim_{k \to \infty} y_k(x) = y_0(x) \quad \text{for each} \quad x \in [a,b]. \qquad \text{(A.53)}$$

The limit function y_0 is uniformly continuous on $[a,b]$:

$$|y_0(x) - y_0(\tilde{x})|$$
$$\le |y_0(x) - y_k(x)| + |y_k(x) - y_k(\tilde{x})| + |y_k(\tilde{x}) - y_0(\tilde{x})| \qquad \text{(A.54)}$$
$$< \frac{\varepsilon}{3} + \frac{\varepsilon}{3} + \frac{\varepsilon}{3} = \varepsilon \quad \text{for} \quad k \ge k_1(\varepsilon, x, \tilde{x}) \text{ and } |x - \tilde{x}| < \delta\left(\frac{\varepsilon}{3}\right),$$

where we use the equicontinuity (3.2.23) and the convergence (A.53). (The dependence of k_1 on x and $\tilde{x}$ is not relevant.)

We show the uniform convergence (3.2.24): For $\varepsilon > 0$ there exist finitely many points $\tilde{x}_1, \ldots, \tilde{x}_N$ in $[a,b]$ such that for each $x \in [a,b]$ there exists an $\tilde{x}_m$ satisfying

$$|x - \tilde{x}_m| < \delta\left(\frac{\varepsilon}{3}\right). \qquad \text{(A.55)}$$

Then

$$|y_k(x) - y_0(x)|$$
$$\le |y_k(x) - y_k(\tilde{x}_m)| + |y_k(\tilde{x}_m) - y_0(\tilde{x}_m)| + |y_0(\tilde{x}_m) - y_0(x)| \qquad \text{(A.56)}$$
$$\le \frac{\varepsilon}{3} + |y_k(\tilde{x}_m) - y_0(\tilde{x}_m)| + \frac{\varepsilon}{3} < \varepsilon \quad \text{for} \quad k \ge k_2(\varepsilon),$$

where we use (3.2.23), (A.54), and the convergence (A.53). For all points $\tilde{x}_1, \ldots, \tilde{x}_N$ the number k_2 depends only on ε.

Solutions of the Exercises

0.1 $\|B - A\| = \|x(t_b) - x(t_a)\| = \|\int_{t_a}^{t_b} \dot{x}(t)dt\| \leq \int_{t_a}^{t_b} \|\dot{x}(t)\|dt.$

0.2 Choose the coordinates $A = (0, y_1)$, $P = (x, 0)$, $B = (x_2, y_2)$. Then the length of the line segments are: $AP = (x^2 + y_1^2)^{1/2}$, $PB = ((x_2 - x)^2 + y_2^2)^{1/2}$. The running times are:

$$T_1 + T_2 = \frac{1}{v_1}(x^2 + y_1^2)^{1/2} + \frac{1}{v_2}((x_2 - x)^2 + y_2^2)^{1/2}.$$

Differentiation with respect to x gives

$$\frac{x}{v_1(x^2 + y_1^2)^{1/2}} - \frac{x_2 - x}{v_2((x_2 - x)^2 + y_2^2)^{1/2}} = 0, \quad \text{or} \quad \frac{\sin \alpha_1}{v_1} = \frac{\sin \alpha_2}{v_2}.$$

1.1.1 The function y is continuous on $[0, 1]$, but it has corners in all points $\frac{1}{n}$. Therefore there is no finite partition of $[0, 1]$ for which y fulfills Definition 1.1.1.

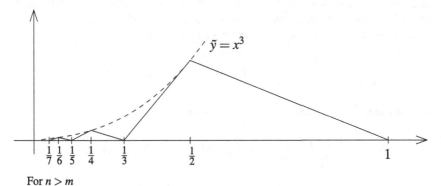

For $n > m$

Graph of the Function of Exercise 1.1.1

© Springer International Publishing AG 2018
H. Kielhöfer, *Calculus of Variations*, Texts in Applied Mathematics 67,
https://doi.org/10.1007/978-3-319-71123-2

For $n > m$

$$y_m(x) - y_n(x) = \begin{cases} y(x) & \text{for } x \in [\frac{1}{2n+1}, \frac{1}{2m+1}], \\ 0 & \text{for } x \in [0,1] \setminus [\frac{1}{2n+1}, \frac{1}{2m+1}], \end{cases}$$

and therefore $\|y_m - y_n\|_{1,pw,[0,1]} \leq \frac{1}{(2m+2)^3} + \frac{2m+3}{(2m+2)^2} < \varepsilon$ for $n > m \geq n_0(\varepsilon)$.
Assume that there is some limit $y_0 \in C^{1,pw}[0,1]$. Then, according to Definition
1.1.2, $\lim_{n \to \infty} \|y_n - y_0\|_{0,[0,1]} = 0$. Since $(y_n - y)(x) = -y(x)$ for $x \in [0, \frac{1}{2n+1}]$ and
$(y_n - y)(x) = 0$ for $x \in [\frac{1}{2n+1}, 1]$, we obtain $\|y_n - y\|_{0,[0,1]} = \frac{1}{(2n)^3}$, and therefore $\lim_{n \to \infty} y_n = y$ in $C[0,1]$. Uniqueness of the limit in $C[0,1]$ implies $y_0 = y \notin$
$C^{1,pw}[0,1]$, which is a contradiction.

1.1.2 We show the continuity in some $y_0 \in C^{1,pw}[a,b]$. By Definition 1.1.1, the
functions y_0 and y_0' are bounded on $[a,b]$; the latter is bounded on all intervals
$[x_{i-1}, x_i], i = 1, \ldots, m$. Therefore their respective ranges are compact intervals $[-c,c]$
and $[-c',c']$. The continuous function F is uniformly continuous on the compact set
$[a,b] \times [-c,c] \times [-c',c']$. Hence for $y \in C^{1,pw}[a,b]$,

$$|F(x,y(x),y'(x)) - F(x,y_0(x),y_0'(x))| < \varepsilon \quad \text{for all} \quad x \in [a,b],$$
$$\text{if } \|y - y_0\|_{1,pw,[a,b]} < \delta(\varepsilon).$$

Then, by definition 1.1.3,

$$|J(y) - J(y_0)| < \varepsilon(b-a) \quad \text{if } \|y - y_0\|_{1,pw} < \delta(\varepsilon).$$

Observe that $(x, y_0(x), y_0'(x)) \in [a,b] \times [-c,c] \times [-c',c']$ if c and c' are large enough
and $\delta = \delta(\varepsilon)$ is small enough.

1.2.1 We prove estimate (1.2.17), which entails continuity. The functions y and h
have w.l.o.g. the same partition, and for $x \in [x_{i-1}, x_i]$, $(x, y(x), y'(x)) \in [x_{i-1}, x_i] \times$
$[-c,c] \times [-c',c']$. Hence, due to continuity of F_y and of $F_{y'}$,

$$\begin{aligned}|F_y(x,y(x),y'(x))| &\leq \tilde{C} \\ |F_{y'}(x,y(x),y'(x))| &\leq \tilde{C}'\end{aligned} \quad \text{for all} \quad x \in [x_{i-1}, x_i], \ i = 1, \ldots, m,$$

with constants $\tilde{C} = \tilde{C}(y)$ and $\tilde{C}' = \tilde{C}'(y)$. We then obtain

$$\begin{aligned}|\delta J(y)h| &= \left| \sum_{i=1}^{m} \int_{x_{i-1}}^{x_i} F_y(x,y(x),y'(x))h(x) + F_{y'}(x,y(x),y'(x))h'(x)dx \right| \\ &\leq \sum_{i=1}^{m} \int_{x_{i-1}}^{x_i} |F_y(x,y(x),y'(x))||h(x)| + |F_{y'}(x,y(x),y'(x))||h'(x)|dx \\ &\leq \sum_{i=1}^{m} (x_i - x_{i-1})(\tilde{C}\|h\|_{0,[a,b]} + \tilde{C}'\|h'\|_{0,[x_{i-1},x_i]}) \\ &\leq (b-a)\max\{\tilde{C},\tilde{C}'\}\|h\|_{1,pw} = C(y)\|h\|_{1,pw}.\end{aligned}$$

1.2.2 Follow the proof of Proposition 1.2.1, and show that the second derivative with respect to t and integration can be interchanged. Then the formula for the second variation follows from the second derivative of $F(x, y+th, y'+th')$ with respect to t in $t = 0$.

1.2.3 The proof is essentially the same as for Proposition 1.2.1, cf. Exercise 1.2.1.

1.4.1 Let $[x_0 - \frac{1}{n}, x_0 + \frac{1}{n}] \subset I$ for $n \geq n_0$. Then for the saw tooth h_n of height 1,

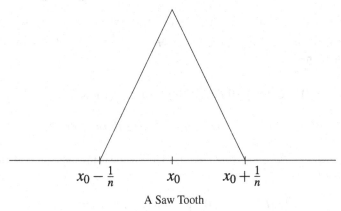

$$x_0 - \frac{1}{n} \qquad x_0 \qquad x_0 + \frac{1}{n}$$

A Saw Tooth

properties a), b), c) hold.

1.4.2 According to the hypotheses and Lemma 1.3.3, an integration by parts is allowed:

$$\delta^2 J(y)(h,h) = \int_a^b F_{yy}h^2 + 2F_{yy'}hh' + F_{y'y'}(h')^2 dx$$

$$= \int_a^b F_{yy}h^2 - \frac{d}{dx}(F_{yy'}h)h + F_{yy'}hh' + F_{y'y'}(h')^2 dx$$

$$= \int_a^b (F_{yy} - \frac{d}{dx}F_{yy'})h^2 + F_{y'y'}(h')^2 dx.$$

1.4.3 Assume that $F_{y'y'}(x_0, y(x_0), y'(x_0)) < 0$ for some $x_0 \in [x_{i-1}, x_i] \subset [a,b]$. By continuity of $F_{y'y'}(\cdot, y, y')$ on $[x_{i-1}, x_i]$, there is a compact interval $I \subset [x_{i-1}, x_i] \cap (a,b)$ such that $F_{y'y'}(x, y(x), y'(x)) \leq -c_1 < 0$ for all $x \in I$. W.l.o.g we can assume that P, defined in Exercise 1.4.2, is continuous on I as well. Hence, $P(x, y(x), y'(x)) \leq c_2$ on I. Then by (1.4.11) using a sequence $(h_n)_{n \in \mathbb{N}}$ described in Exercise 1.4.1,

$$0 \leq \int_a^b Ph_n^2 + Q(h_n')^2 dx \leq c_2 \int_I h_n^2 + \int_I F_{y'y'}(h_n')^2 dx$$

$$\leq c_2 \int_a^b h_n^2 dx - c_1 \int_a^b (h_n')^2 dx < 0 \quad \text{for} \quad n \geq n_1 \geq n_0,$$

due to the fact that the first term converges to 0 and the second term goes to $-\infty$. This contradiction proves that the assumption is wrong.

1.4.4 Define $g(t) = J(y + th)$, where $h \in C_0^{1,pw}[a,b]$ such that $y + th \in D$ for all $t \in \mathbb{R}$. According to Exercise 1.2.2, $g''(t) = \delta^2 J(y + th)(h,h) \geq 0$ for all $t \in \mathbb{R}$, by assumption. The formula given in the hint (easy to verify) gives

$$J(y + h) = J(y) + \delta J(y)h + \int_0^1 (1 - t)\delta^2 J(y + th)(h,h)\,dt$$

$$\geq J(y) \quad \text{for all} \quad h \in C_0^{1,pw}[a,b].$$

For an arbitrary $\tilde{y} \in D$, we set $\tilde{y} = y + \tilde{y} - y = y + h$, where $h = \tilde{y} - y \in C_0^{1,pw}[a,b]$.

1.4.5 We define $\tilde{h} = h/\|h\|$, where $\|h\| = \|h\|_{1,pw}$, and $g(t) = J(y + t\tilde{h})$. Then by Taylor's theorem

$$g(t) = g(0) + g'(0)t + \frac{1}{2}g''(0)t^2 + r(t), \quad \text{and we have}$$

$$J(y + t\tilde{h}) = J(y) + \delta J(y)t\tilde{h} + \frac{1}{2}\delta^2 J(y)(t\tilde{h}, t\tilde{h}) + \tilde{R}(y, \tilde{h}; t).$$

By the estimates (1.2.17) and from Exercise 1.2.3 while using the uniform continuity of F on compact sets, we obtain

$$\lim_{t \to 0} \tilde{R}(y, \tilde{h}; t) = 0 \quad \text{uniformly for} \quad \|\tilde{h}\| = 1.$$

In the sequel, the function $y \in D$ is arbitrary but fixed. Furthermore

$$\frac{d}{dt}\tilde{R}(y, \tilde{h}; t) = \delta J(y + t\tilde{h})\tilde{h} - \delta J(y)\tilde{h} - t\delta^2 J(y)(\tilde{h}, \tilde{h}),$$

$$\frac{d^2}{dt^2}\tilde{R}(y, \tilde{h}; t) = \delta^2 J(y + t\tilde{h})(\tilde{h}, \tilde{h}) - \delta^2 J(y)(\tilde{h}, \tilde{h}).$$

Using the same estimates for $\delta J(y + t\tilde{h})$ and $\delta^2 J(y + t\tilde{h})$, and the uniform continuity of $F_y, F_{y'}, F_{yy}, F_{yy'}, F_{y'y'}$ on compact sets, we obtain

$$\lim_{t \to 0} \frac{d}{dt}\tilde{R}(y, \tilde{h}; t) = 0 \quad \text{uniformly for} \quad \|\tilde{h}\| = 1,$$

$$\lim_{t \to 0} \frac{d^2}{dt^2}\tilde{R}(y, \tilde{h}; t) = 0 \quad \text{uniformly for} \quad \|\tilde{h}\| = 1.$$

The mean value theorem yields

$$\tilde{R}(y, \tilde{h}; t)/t^2 = \frac{1}{t}\frac{d}{dt}\tilde{R}(y, \tilde{h}; \tau) = \frac{\tau}{t}\frac{d}{dt}\tilde{R}(y, \tilde{h}; \tau)/\tau = \frac{\tau}{t}\frac{d^2}{dt^2}\tilde{R}(y, \tilde{h}; \sigma),$$

where $0 < |\sigma| < |\tau| < |t|$.

By the above result,

$$\lim_{t \to 0} \tilde{R}(y, \tilde{h}; t)/t^2 = 0 \quad \text{uniformly for} \quad \|\tilde{h}\| = 1.$$

Setting $t = \|h\|$ gives $t\tilde{h} = h$, $R(y,h) = \tilde{R}(y,\tilde{h}; \|h\|)$, and finally

$$\lim_{\|h\|\to 0} R(y,h)/\|h\|^2 = \lim_{\|h\|\to 0} \tilde{R}(y,h/\|h\|; \|h\|)/\|h\|^2 = 0.$$

1.4.6 For arbitrary $\tilde{y} \in D$, we set $\tilde{y} = y + \tilde{y} - y = y + h$, where $h \in C_0^{1,pw}[a,b]$. Then by Exercise 1.4.5,

$$J(\tilde{y}) = J(y) + \delta J(y)h + \frac{1}{2}\delta^2 J(y)(h,h) + R(y,h), \quad \text{and thus,}$$

$$J(\tilde{y}) - J(y) \geq \frac{1}{2}C\|h\|^2_{1,pw} - |R(y,h)|$$

$$= (\frac{1}{2}C - |R(y,h)|/\|h\|^2_{1,pw})\|h\|^2_{1,pw} > 0,$$

provided $|R(y,h)|/\|h\|^2_{1,pw} < \frac{1}{2}C$, which, due to Exercise 1.4.5, holds for $\|h\|_{1,pw} < d$. According to Definition 1.4.1, the function y is a local minimizer of J.

1.4.7

a) $J(0) = 0$ and $J(y) = \int_0^1 (y')^2(1+y')dx \geq 0$ provided $\|y\|_{1,pw} < 1$.

b)

$$J(y_{n,b}) = \frac{1}{bn^2}\left(1 + \frac{1}{bn}\right) + \frac{1}{n^2(1-b)}\left(1 - \frac{1}{n(1-b)}\right)$$

$$= \frac{1}{n^2}\left(\frac{1}{b(1-b)} + \frac{1}{b^2n} - \frac{1}{n(1-b)^2}\right) = \frac{1}{n^2 b(1-b)}\left(1 + \frac{1-2b}{b(1-b)n}\right) < 0$$

if $\frac{1}{2} < b < 1$ and $\frac{b(1-b)}{2b-1} < \frac{1}{n}$.

This holds for $b = b_n$ close to 1. Finally, $\|y_{n,b}\|_0 = \frac{1}{n} < d$ for $n \geq n_0$.

1.4.8 All solutions of the Euler-Lagrange equation are piecewise straight lines having slopes $y' = \pm\sqrt{c_1}$, $c_1 \geq 0$, cf. (1.4.7), (1.4.8). By $\delta^2 J(y)(h,h) = 6\int_0^1 y'(h')^2 dx$ (cf. Exercise 1.2.2), the necessary condition (1.4.11) for a local minimizer is only fulfilled for $y' \geq 0$ on $[0,1]$. The boundary conditions allow only $y' = 0$, and therefore, $y = 0$. The analogous argument for $-J$ shows that the only possible local maximizer is $y = 0$. Let

$$y_0(x) = \begin{cases} 3x & \text{for } x \in [0,\frac{1}{3}], \\ 1 - \frac{3}{2}(x - \frac{1}{3}) & \text{for } x \in [\frac{1}{3}, 1]. \end{cases}$$

Then $y_0 \in D$ and $J(y_0) = 27/4$. Since $J(\alpha y_0) = \alpha^3 J(y_0)$, the range of the functional J on D is $\mathbb{R}$. Furthermore, $\|\alpha y_0\|_{1,pw,[0,1]} = |\alpha|\|y_0\|_{1,pw} = 4|\alpha| < d$ for $|\alpha| < d/4$, and the functional J is positive and negative in any neighbourhood of $y = 0$. Since $J(0) = 0$, the function $y = 0$ is neither a local minimizer nor maximizer of J on D.

1.5.1 By assumption, the function $f = F_{y'}(\cdot, y, y')$ is continuously differentiable on the interval $[x_{i-1}, x_i]$. We set $G(x, z) = F_{y'}(x, y(x), z) - f(x)$, and then $G : [x_{i-1}, x_i] \times \mathbb{R} \to \mathbb{R}$ is continuously partially differentiable with respect to both variables, by assumption on F and because $y \in C^1[x_{i-1}, x_i]$. We know that $G(x_0, y'(x_0)) = 0$ and $G_z(x_0, y'(x_0)) = F_{y'y'}(x_0, y(x_0), y'(x_0)) \neq 0$, by assumption. The implicit function theorem then guarantees the existence of a unique continuously differentiable function $z(x)$ that fulfills the equation $G(x, z(x)) = 0$ in a neighborhood of x_0 and $z(x_0) = y'(x_0)$. By uniqueness, $z(x) = y'(x)$ since $G(x, y'(x)) = 0$ as well. Continuous differentiability of $z = y'$ implies two times continuous differentiability of y in a neighborhood of x_0.

1.5.2

a) $J(y) = \int_0^1 y' dx = y(1) - y(0) = 1$ for all admitted functions $y \in D$. Any admitted function satisfies the Euler-Lagrange equation.

b) $J(y) = \frac{1}{2} \int_0^1 \frac{d}{dx}(y^2) dx = \frac{1}{2}((y(1))^2 - (y(0))^2) = \frac{1}{2}$ for all admitted functions, and each of them fulfill the Euler-Lagrange equation.

c) $J(y) = \frac{1}{2} \int_0^1 x \frac{d}{dx}(y^2) dx = -\frac{1}{2} \int_0^1 y^2 dx + \frac{1}{2}$ for all admitted functions. Choosing the admitted functions $y_n(x) = x^n$, we obtain $\int_0^1 y_n^2 dx = \frac{1}{2n+1}$. Hence the supremum of J is $\frac{1}{2}$. The supremum is not a maximum, there is no minimizer, and no admitted function satisfies the Euler-Lagrange equation.

1.5.3

a) $2\frac{d}{dx}y' = 2$, or $y'' = 1$. Only the solution $y(x) = \frac{1}{2}(x^2 + x)$ fulfills the boundary conditions. Since $\delta^2 J(\tilde{y})(h,h) = 2\int_0^1 (h')^2 dx \geq 0$ for all admitted functions $\tilde{y}$, the solution is a global minimizer, cf. Exercise 1.4.4.

b) $2\frac{d}{dx}(y' + y) = 2y'$, or $y'' = 0$. Only the solution $y(x) = -\frac{1}{3}x + \frac{2}{3}$ fulfills the boundary conditions. Since $\delta^2 J(\tilde{y})(h,h) = 2\int_{-1}^2 (h')^2 + 2hh' dx = 2\int_{-1}^2 (h')^2 + \frac{d}{dx}(h^2) dx = 2\int_{-1}^2 (h')^2 dx \geq 0$ for all admitted functions $\tilde{y}$ and for all $h \in C_0^{1,pw}[-1, 2]$, the solution is a global minimizer.

c) $2\frac{d}{dx}(y' + x) = 0$ or $y'' = -1$. Only the solution $y(x) = \frac{1}{2}(x - x^2)$ fulfills the boundary conditions. Since $\delta^2 J(\tilde{y})(h,h) = 2\int_0^1 (h')^2 dx \geq 0$ for all admitted functions $\tilde{y}$, the solution is a global minimizer.

d) $2\frac{d}{dx}(y' + y) = 2(y' + y)$ or $y'' = y$. Only the solution $y(x) = \sinh x / \sinh 2$ fulfills the boundary conditions. Since $\delta^2 J(\tilde{y})(h,h) = \int_0^2 2(h')^2 + 4hh' + 2h^2 dx = 2\int_0^2 (h' + h)^2 dx \geq 0$ for all admitted functions $\tilde{y}$, the solution is a global minimizer.

1.8.1 A multiple application of L'Hospital's rule gives $f(0) = 0$ and $f'(0) = \frac{1}{3}$. The asymptotic behavior $\lim_{\tau \to 2\pi} f(\tau) = +\infty$ is obvious.

We show $f'(\tau) > 0$ for $\tau \in (0, 2\pi)$ in different subintervals. We compute the derivative and define

$$f'(\tau) = \frac{2(1 - \cos \tau) - \tau \sin \tau}{(1 - \cos \tau)^2} := \frac{g(\tau) - h(\tau)}{(1 - \cos \tau)^2}.$$

First of all, $g'(\tau) - h'(\tau) = \sin\tau - \tau\cos\tau = \cos\tau(\tan\tau - \tau) > 0$ for $\tau \in (0, \frac{\pi}{2})$. Therefore, since $g(0) = h(0) = 0$, $g(\tau) - h(\tau) > 0$ and thus, $f'(\tau) > 0$ for $\tau \in (0, \frac{\pi}{2})$. We see also that $g'(\tau) - h'(\tau) > 0$ for $\tau \in [\frac{\pi}{2}, \pi]$, and hence, $g(\tau) - h(\tau) > 0$ and $f'(\tau) > 0$ for $\tau \in [\frac{\pi}{2}, \pi]$. Since $g(\tau) > 0$ and $h(\tau) < 0$ for $\tau \in (\pi, 2\pi)$, it follows that $g(\tau) - h(\tau) > 0$, and therefore, $f'(\tau) > 0$ for $\tau \in (\pi, 2\pi)$.

1.8.2 The running time on the line $\tilde{y}(x) = \frac{2}{\pi}x$ from $x = 0$ to $x = b$ is, according to (1.8.4),

$$T = \frac{1}{\sqrt{2g}} \int_0^b \sqrt{\frac{1 + (\frac{2}{\pi})^2}{\frac{2}{\pi}x}}\,dx = \sqrt{\frac{b\pi}{g}\left(1 + \frac{4}{\pi^2}\right)}.$$

The running time on the cycloid is $\sqrt{\frac{b\pi}{g}}$; according to (1.8.19), $\tau_b = \pi$ and $r = \frac{B}{2} = \frac{b}{\pi}$. Thus the ratio of the running times is $T : T_{min} = \sqrt{1 + \frac{4}{\pi^2}}$.

1.9.1

a) The Euler-Lagrange equation is $\frac{d}{dx}2y' = 1$ piecewise on $[0,1]$. Since $2y' \in C^{1,pw}[0,1] \subset C[0,1]$, cf. $(1.4.3)_1$, the solution is given by $2y'(x) = x + c_1$ for all $x \in [0,1]$. Hence $y(x) = \frac{1}{4}x^2 + \frac{1}{2}c_1x + c_2$, noting that $y \in C[0,1]$ as well.

b) The natural boundary conditions are $y'(0) = 0$ and $y'(1) = 0$. No solution fulfills the natural boundary conditions, because $y'(0) = 0$ implies $y'(1) = \frac{1}{2}$.

c) $y(x) = \frac{1}{4}x^2 + \frac{3}{4}x$ is the only solution.

d) Without boundary conditions, no solution is locally extremal because the natural boundary conditions cannot be fulfilled. With boundary conditions, Proposition 1.4.3 is applicable because the Lagrange function is convex. Another argument is the following: The second variation $\delta^2 J(\tilde{y})(h,h) = \int_0^1 (h')^2 dx \geq 0$ for all $\tilde{y} \in D = C^{1,pw}[0,1] \cap \{y(0) = 0,\ y(1) = 1\}$, and thus, according to Exercise 1.4.4, each solution of the Euler-Lagrange equation in D is a global minimizer.

1.9.2 The Euler-Lagrange equation reads

$$\frac{d}{dx}2y' = \frac{1}{1+y^2} \quad \text{piecewise on} \quad [a,b].$$

The regularity $(1.4.3)_1$ implies $y' \in C[a,b]$. Hence $y \in C^1[a,b]$, and since $F_{y'y'}(y,y') = 2$, Exercise 1.5.1 implies that $y \in C^2(a,b)$. Therefore

$$2y'' = \frac{1}{1+y^2} > 0 \quad \text{on all} \quad (a,b).$$

The natural boundary conditions on minimizers are $y'(a) = 0$ and $y'(b) = 0$. Rolle's theorem implies the existence of some $x \in (a,b)$ such that $y''(x) = 0$, which, however, is excluded by the Euler-Lagrange equation. Hence, there is no (local or global) minimizer, even though the functional is bounded from below by $-\frac{\pi}{2}(b-a)$.

1.10.1 Since $\dot{x}(t) > 0$ on $[t_{i-1}, t_i]$, x is strictly monotone and maps $[t_{i-1}, t_i]$ one-to-one onto $[x_{i-1}, x_i]$. Since $x(t_a) = a = x(t_0)$ and $x(t_b) = b = x(t_m)$, we obtain a partition $a = x_0 < x_1 < \cdots < x_m = b$ of the interval $[a, b]$. Let $\psi_i : [x_{i-1}, x_i] \to [t_{i-1}, t_i]$ be the continuously differentiable inverse function of the function x, i.e., $\psi_i(x) = t$ if $x(t) = x$. Setting $\tilde{y}(x) = y(\psi_i(x))$, we obtain $\tilde{y}(x) = y(t)$ and $\tilde{y} \in C^1[x_{i-1}, x_i]$. Since $y(\psi_i(x_i)) = y(t_i) = y(\psi_{i+1}(x_i))$ for $i = 1, \ldots, m-1$, $\tilde{y} \in C[a, b]$, and altogether $\tilde{y} \in C^{1,pw}[a, b]$.

1.10.2

a) Let $h \in (C_0^{1,pw}[t_a, t_b])^n$. Then

$$\delta J(x)h = \int_{t_a}^{t_b} (DF(x)h, \dot{x}) + (F(x), \dot{h})\, dt$$

$$= \int_{t_a}^{t_b} (h, DF(x)^* \dot{x}) - \left(\frac{d}{dt} F(x), h\right) dt$$

$$= \int_{t_a}^{t_b} ((DF(x)^* - DF(x))\dot{x}, h)\, dt,$$

where $DF(x)$ is the Jacobian matrix of F in x, and $DF(x)^*$ is the transposed matrix.

b) The system of Euler-Lagrange equations reads

$$(DF(x) - DF(x)^*)\dot{x} = 0 \quad \text{for all} \quad t \in [t_a, t_b],$$

or $\dot{x} \in \text{Kernel}(DF(x) - DF(x)^*)$. That system does not have solutions in any D if the kernel is $\{0\}$.

c) In the case $DF(x) = DF(x)^*$, $\delta J(x) = 0$, and any $x \in D$ solves the Euler-Lagrange equations piecewise. The symmetry of the Jacobian matrix implies that the vector field F possesses a potential $f : \mathbb{R}^n \to \mathbb{R}$, i.e., $F(x) = \nabla f(x)$. Then, for all $x \in D$, by Lemma 1.3.3 with $h \equiv 1$, we have

$$J(x) = \int_{t_a}^{t_b} (\nabla f(x), \dot{x})\, dt = \int_{t_a}^{t_b} \frac{d}{dt} f(x)\, dt = f(x(t_b)) - f(x(t_a)) = f(B) - f(A),$$

which means that the curvilinear integral is path independent.

1.10.3 The Euler-Lagrange equations read

$$\frac{d}{dt} \Phi_{\dot{x}}(x, \dot{x}) = \Phi_x(x, \dot{x}) \quad \text{on} \quad [t_a, t_b].$$

By the assumed regularity of the local minimizer solving the Euler-Lagrange system, we may differentiate:

$$\frac{d}{dt}(\Phi(x,\dot{x}) - (\dot{x},\Phi_{\dot{x}}(x,\dot{x})))$$

$$= (\Phi_x(x,\dot{x}),\dot{x}) + (\Phi_{\dot{x}}(x,\dot{x}),\ddot{x}) - (\ddot{x},\Phi_{\dot{x}}(x,\dot{x})) - (\dot{x},\frac{d}{dt}(\Phi_{\dot{x}}(x,\dot{x}))$$

$$= (\Phi_x(x,\dot{x}) - \frac{d}{dt}\Phi_{\dot{x}}(x,\dot{x}),\dot{x}) = 0,$$

which proves the claim.

1.10.4 By $(1.10.22)_1$, $L_{\dot{x}}(x,\dot{x}) = m\dot{x} \in (C^1[t_a,t_b])^3$, and hence $x \in (C^2[t_a,t_b])^3$. By the chain rule, we have

$$\frac{d}{dt}E(x,\dot{x}) = m(\ddot{x},\dot{x}) + (\text{grad}V(x),\dot{x}) = 0 \quad \text{for} \quad t \in [t_a,t_b],$$

which proves constant energy.

1.10.5 Follow the proof of Proposition 1.2.1, where differentiation and integration are interchanged. This is possible for the second derivative as well and we obtain:

$$\frac{d}{ds}J(x+sh) = \int_{t_a}^{t_b}(\Phi_x(t,x+sh,\dot{x}+s\dot{h}),h) + (\Phi_{\dot{x}}(t,x+sh,\dot{x}+s\dot{h}),\dot{h})dt,$$

$$\frac{d^2}{ds^2}J(x+sh)\big|_{s=0}$$

$$= \int_{t_a}^{t_b}(D_x^2\Phi(t,x,\dot{x})h,h) + 2(D_xD_{\dot{x}}\Phi(t,x,\dot{x})\dot{h},h) + (D_{\dot{x}}^2\Phi(t,x,\dot{x})\dot{h},\dot{h})dt,$$

with the matrices

$$D_x^2\Phi(t,x,\dot{x}) = \left(\frac{\partial^2}{\partial x_i\partial x_j}\Phi(t,x,\dot{x})\right)_{\substack{i=1,\dots,n \\ j=1,\dots,n}},$$

$$D_xD_{\dot{x}}\Phi(t,x,\dot{x}) = \left(\frac{\partial^2}{\partial x_i\partial \dot{x}_j}\Phi(t,x,\dot{x})\right)_{\substack{i=1,\dots,n \\ j=1,\dots,n}},$$

$$D_{\dot{x}}^2\Phi(t,x,\dot{x}) = \left(\frac{\partial^2}{\partial \dot{x}_i\partial \dot{x}_j}\Phi(t,x,\dot{x})\right)_{\substack{i=1,\dots,n \\ j=1,\dots,n}}, \quad i = \text{row index}, \quad j = \text{column index}.$$

1.10.6

a) The second variation of the action reads

$$\delta^2J(x)(h,h) = \int_{t_a}^{t_b}m\|\dot{h}\|^2 - (D^2V(x)h,h)dt.$$

b) The assumption $(D^2V(x)h,h) \leq 0$ implies

$$\delta^2J(\hat{x})(h,h) \geq 0 \quad \text{for all} \quad \hat{x} \in (C^1[t_a,t_b])^3, \quad h \in (C_0^1[t_a,t_b])^3.$$

Let the path $x \in (C^2[t_a, t_b])^3$ satisfy the system (1.10.27), implying

$$\delta J(x)h = 0 \quad \text{for all} \quad h \in (C_0^1[t_a, t_b])^3.$$

Let a path $\hat{x} \in (C^1[t_a, t_b])^3$ fulfill the same boundary conditions as x, i.e., $\hat{x} - x = h \in (C_0^1[t_a, t_b])^3$. Setting

$$g(s) = J(x + sh), \quad \text{we obtain}$$

$$J(\hat{x}) = g(1) = g(0) + g'(0) + \int_0^1 (1-s)g''(s)ds$$

$$= J(x) + \delta J(x)h + \int_0^1 (1-s)\delta^2 J(x+sh)(h,h)ds$$

$$\geq J(x), \quad \text{due to both } \delta J(x)h = 0 \text{ and } \delta^2 J(x+sh)(h,h) \geq 0 \text{ for } s \in [0,1].$$

1.11.1 The curve $\{(x, y(x)) \mid x \in [a, b]\}$ satisfies the Euler-Lagrange equations (1.10.9), for the Lagrange function (1.11.5), in all admitted parametrizations, and thus also when parametrized by x. This means

$$\frac{d}{dx}(F(y, y') - y' F_{y'}(y, y')) = F_x(y, y') = 0,$$

piecewise on $[a, b]$. Constancy follows by the second Weierstraß-Erdmann corner condition.

1.11.2 The function W has local minima at $z = -1$ and at $z = \frac{1}{2}$, having values $W(-1) = -\frac{1}{3}$ and $W(\frac{1}{2}) = -\frac{5}{96}$. At $z = 0$, the function W has a local maximum. Since $W(-1)$ is a global minimum, a global minimizer of the functional J is given by $y = -x$ and $y' = -1$. All parallel lines $y(x) = -x + c$ are global minimizers, too.

1.11.3 The minimizers found in Exercise 1.11.2 do not satisfiy the boundary conditions, the line $y = 0$ is not a minimizer, and thus, global minimizers must have corners. The slopes are inserted in Figure 1.20 and denoted $c_1^1 = \alpha$ and $c_1^2 = \beta$. We study a "prototype" y having only one corner.

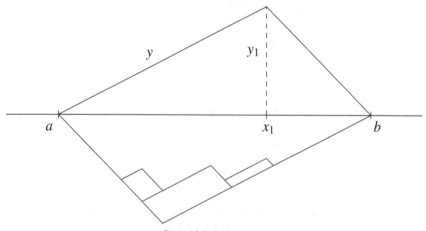

Global Minimizers

The negative slope is $c_1^1 = \frac{y_1}{x_1 - b}$, and the positive slope is $c_1^2 = \frac{y_1}{x_1 - a}$. The tangent depicted in Figure 1.20 has the following property:

$$W(z) \geq W(c_1^1) + \frac{W(c_1^2) - W(c_1^1)}{c_1^2 - c_1^1}(z - c_1^1) \quad \text{for all} \quad z \in \mathbb{R}.$$

Let $\hat{y} \in C^{1,pw}[a,b] \cap \{y(a) = 0, y(b) = 0\}$. Then $\int_a^b \hat{y}' dx = 0$, and therefore,

$$
\begin{aligned}
J(\hat{y}) = \int_a^b W(\hat{y}') dx &\geq \left(W(c_1^1) - \frac{W(c_1^2) - W(c_1^1)}{c_1^2 - c_1^1} c_1^1 \right)(b-a) \\
&= \left(W(c_1^1)\frac{c_1^2}{c_1^2 - c_1^1} - W(c_1^2)\frac{c_1^1}{c_1^2 - c_1^1} \right)(b-a).
\end{aligned}
$$

For the "prototype," we obtain

$$
\begin{aligned}
J(y) = \int_a^b W(y') dx &= W(c_1^2)(x_1 - a) + W(c_1^1)(b - x_1) \\
&= \left(W(c_1^1)\frac{b - x_1}{b - a} - W(c_1^2)\frac{a - x_1}{b - a} \right)(b - a) \\
&= \left(W(c_1^1)\frac{c_1^2}{c_1^2 - c_1^1} - W(c_1^2)\frac{c_1^1}{c_1^2 - c_1^1} \right)(b - a),
\end{aligned}
$$

which proves $\int_a^b W(\hat{y}') dx \geq \int_a^b W(y') dx$ for all $\hat{y} \in C^{1,pw}[a,b] \cap \{y(a) = 0, y(b) = 0\}$. The functional J has the same value for all saw-tooth functions having the slopes c_1^1 and c_1^2, and fulfilling the boundary conditions $y(a) = 0$ and $y(b) = 0$. Therefore they are all global minimizers.

2.1.1

$$
\sum_{i=1}^m \lambda_i \delta K_i(y)h = 0 \quad \text{for all } h \in D_0 \Leftrightarrow
$$

$$(\lambda, \delta K(y)h) = 0 \quad \text{for all } h \in D_0, \text{ where}$$

$$\lambda = (\lambda_1, \ldots, \lambda_m), \quad \delta K(y)h = (\delta K_1(y)h, \ldots, \delta K_m(y)h)$$

and $(\ ,\ )$ is the scalar product in $\mathbb{R}^m$.

If $\delta K(y)$ is surjective, then $\lambda = 0$. On the other hand, if the scalar product is zero for $\lambda = 0$ only, then $\delta K(y)$ is surjective.

2.1.2 Assume (i); $\delta K(y)h = 2 \int_0^1 yh dx$, and y is not critical for K, due to $\delta K(y)y = 2$. An application of Proposition 2.1.1 gives (up to the factor 2)

$$y'' = \lambda y \quad \text{piecewise on} \quad [0,1].$$

Since y is continuous on $[0,1]$, the minimizer satisfies $y \in C^2[0,1]$ and the Euler-Lagrange equation on all of $[0,1]$. The Lagrange multiplier λ is determined as follows:

$$\int_0^1 y''y\,dx = -\int_0^1 (y')^2 dx = \lambda \int_0^1 y^2 dx = \lambda,$$

and hence $-\lambda = \min\{J(y)\}$ under the constraint $K(y) = 1$. Since for all $h \in D = C_0^{1,pw}[0,1]$, one obtains $K(h/\alpha) = 1$ for $\alpha^2 = \int_0^1 h^2 dx > 0$; the inequality $J(h/\alpha) \geq -\lambda$ holds, which proves the Poincaré inequality.

Assume (ii). For $\tilde{y} \in D$ satisfying $K(\tilde{y}) = 1$, the difference $\tilde{y} - y = h \in C_0^{1,pw}[0,1]$. Then for $\tilde{y} = y + h$,

$$J(\tilde{y}) = J(y) + 2\int_0^1 y'h'\,dx + \int_0^1 (h')^2 dx$$

$$= J(y) - 2\int_0^1 y''h\,dx + \int_0^1 (h')^2 dx$$

$$= J(y) - 2\lambda \int_0^1 yh\,dx + \int_0^1 (h')^2 dx,$$

$$1 = K(y) + 2\int_0^1 yh\,dx + \int_0^1 h^2 dx, \quad \text{and thus,}$$

$$-2\int_0^1 yh\,dx = \int_0^1 h^2 dx \quad (\text{since } K(y) = 1).$$

Therefore, by the Poincaré inequality,

$$J(\tilde{y}) = J(y) + \lambda \int_0^1 h^2 dx + \int_0^1 (h')^2 dx \geq J(y),$$

which proves (i).

Explicitly, $y(x) = \sqrt{2}\sin\pi x$ and $\lambda = -\pi^2$. The other candidates $y_n(x) = \sqrt{2}\sin n\pi x$ and $\lambda = -n^2\pi^2$ are excluded for $n \geq 2$, due to the observation that the functions $y_n \in C_0^{1,pw}[0,1]$ do not satisfy the Poincaré inequality with λ_n for $n \geq 2$.

2.1.3 For $m = \pm 1$, the constants $y = \pm 1$ satisfy the constraints and $J(y) = 0$. Therefore they are global minimizers.
Assume $m \neq \pm 1$. According to Exercise 2.1.1, a function $y \in C^{1,pw}[0,1]$ is not critical for $K = (K_1, K_2)$ if

$$2\lambda_1 y + \lambda_2 = 0 \quad \text{on} \quad [0,1]$$

holds for $(\lambda_1, \lambda_2) = (0,0)$ only. Thus, only constant functions y can be critical. However, a constant cannot fulfill both constraints. Therefore, a nonconstant (local) minimizer y must satisfy the Euler-Lagrange equation

$$2y'' = 2\lambda_1 y + \lambda_2 \quad \text{piecewise on} \quad [0,1],$$

and the natural boundary conditions

$$y'(0) = 0 \quad \text{and} \quad y'(1) = 0.$$

From $(2.1.5)_1$ we have $y' \in C[0,1]$. Hence $y \in C^2[0,1]$, and the Euler-Lagrange equation is satisfied on all of $[0,1]$. We obtain

$$2\int_0^1 y'' dx = 0 = 2\lambda_1 \int_0^1 y dx + \lambda_2 = 2\lambda_1 m + \lambda_2, \quad \text{or} \quad \lambda_2 = -2\lambda_1 m.$$

The Euler-Lagrange equation with the natural boundary conditions admits the (non-constant) solutions

$$y_n(x) = a\cos n\pi x + m, \quad \lambda_1 = -n^2\pi^2, \ n \in \mathbb{N},$$

which fulfill the constraints for $\frac{1}{2}a^2 + m^2 = 1$. This implies $m^2 < 1$, and in this case $J(y_n) = n^2\pi^2(1 - m^2)$, which is minimal for $n = 1$ only.
If a global minimizer exists for $m^2 < 1$, then it is given by

$$y_1(x) = \sqrt{2(1-m^2)}\cos\pi x + m \quad , \text{with} J(y_1) = \pi^2(1-m^2).$$

In this case, one obtains, for $m = 0$ and for any $h \in C^{1,pw}[0,1] \cap \{\int_0^1 h dx = 0\}$ and $\alpha^2 = \int_0^1 h^2 dx > 0$,

$$K_1(h/\alpha) = 1, \quad K_2(h/\alpha) = 0, \quad J(h/\alpha) \geq \pi^2,$$

which implies a Poincaré inequality

$$\pi^2 \int_0^1 h^2 dx \leq \int_0^1 (h')^2 dx \quad \text{for all} \quad h \in C^{1,pw}[0,1] \cap \{\int_0^1 h dx = 0\}.$$

Note that the existence of a global minimizer of J under the given constraints is assumed.

2.2.1 Let $(x,y) \in (C^1[t_a, t_b])^2$ satisfy $(x(t_a), y(t_a)) = (0, A)$ and $(x(t_b), y(t_b)) = (b, 0)$, where b needs to be determined. By (2.2.5), the isoperimetric constraint is

$$K(x,y) = \frac{1}{2}\int_{t_a}^{t_b} y\dot{x} - x\dot{y} dt = S,$$

where it must be observed that the curve runs from $(0, A)$ to $(b, 0)$. (The curvilinear integrals along the axes are zero.) The surface of revolution to be minimized is

$$J(x,y) = 2\pi \int_{t_a}^{t_b} y\sqrt{\dot{x}^2 + \dot{y}^2} dt,$$

cf. (1.6.1) and (1.11.5). We omit the factor 2π, and we apply Proposition 2.1.5. Since

$$\delta K(x,y)(h_1, h_2) = -\int_{t_a}^{t_b} \dot{y}h_1 - \dot{x}h_2 dt$$

for $(h_1, h_2) \in (C_0^1[t_a, t_b])^2)$, (after integration by parts) nonconstant curves $(x, y) \in (C^1[t_a, t_b])^2$ are not critical for the isoperimetric constraint. The Euler-Lagrange equations read

$$\frac{d}{dt}\left(\frac{y\dot{x}}{\sqrt{\dot{x}^2 + \dot{y}^2}} + \frac{1}{2}\lambda y\right) = -\frac{1}{2}\lambda \dot{y},$$

$$\frac{d}{dt}\left(\frac{y\dot{y}}{\sqrt{\dot{x}^2 + \dot{y}^2}} - \frac{1}{2}\lambda x\right) = \sqrt{\dot{x}^2 + \dot{y}^2} + \frac{1}{2}\lambda \dot{x}.$$

Due to the invariance (2.1.37), the Euler-Lagrange equations are invariant with respect to admitted reparametrizations. We parametrize by the arc length and obtain $\dot{x}^2 + \dot{y}^2 = 1$ and $[t_a, t_b] = [0, L]$, cf. (2.6.6). Then the above equations become

$$(1) \quad y(\dot{x} + \lambda) = c_1,$$
$$(2) \quad y\dot{y} - \lambda x = t + c_2,$$

for $t \in [0, L]$. Since $y(L) = 0$, the constant $c_1 = 0$, and the natural boundary condition $y(L)(\dot{x}(L) + \frac{1}{2}\lambda) = 0$ is satisfied. Since we require $y(t) > 0$ for $t \in [0, L]$, equation (1) implies $\dot{x} + \lambda = 0$. Thus from by $x(0) = 0$, we have

$$x(t) = -\lambda t.$$

The relation $\dot{x}^2 + \dot{y}^2 = 1$ implies $\dot{y}^2 = 1 - \lambda^2$. Since $y(0) = A$ and $y(L) = 0$, we then have

$$y(t) = -\sqrt{1 - \lambda^2}\, t + A \quad \text{and} \quad \sqrt{1 - \lambda^2}\, L = A.$$

From equation (2), $c_2 = -A\sqrt{1 - \lambda^2}$, and the constraint yields

$$\frac{1}{2}\int_0^L y\dot{x} - x\dot{y}\, dt = -\frac{1}{2}\lambda AL = S.$$

The curve is a straight line, which begins at $(0, A)$ and meets the x-axis at

$$b = x(L) = -\lambda L = \frac{2S}{A}.$$

This straight line segment is the only curve that satisfies all necessary conditions on a minimizer, and therefore it is the minimizer, provided it exists.

2.4.1 The curve $\{(x, y(x)) | x \in [a, b]\}$ satisfies the Euler-Lagrange equation (2.1.35), regardless of its (admitted) parametrization, cf. (1.11.1)–(1.11.5). Since F and G do not depend explicitly on x, the derivatives with respect to $\tilde{x}$ vanish: $\Phi_{\tilde{x}} = \Psi_{\tilde{x}} = 0$. The first equation of $(2.1.35)_2$ yields, when parametrized by x,

$$\frac{d}{dx}(F(y, y') + \lambda G(y, y') - y'(F_{y'}(y, y') + \lambda G_{y'}(y, y'))) = 0,$$

piecewise on $[a, b]$, and the continuity $(2.4.5)_2$ implies (2.4.6).

2.5.1 The trajectory fulfills the system (2.5.32). Multiplying the three equations by $\dot{x}_k, \dot{y}_k, \dot{z}_k$, respectively, and summing up yield

$$\frac{d}{dt} \sum_{k=1}^{N} \frac{1}{2} m_k (\dot{x}_k^2 + \dot{y}_k^2 + \dot{z}_k^2) + \frac{d}{dt} V(x) = \sum_{i=1}^{m} \lambda_i \frac{d}{dt} \Psi_i(x) = 0,$$

due to $\Psi_i(x(t)) = 0, \quad i = 1, \ldots, m,$ on $[t_a, t_b].$

This proves the constancy of the total energy along the trajectory.

2.5.2 For $T = \frac{1}{2} m (\dot{x}_1^2 + \dot{x}_2^2 + \dot{x}_3^2)$, $V = mgx_3$, and $\Psi(x_1, x_2, x_3) = x_1 + x_3 - 1 = 0$ the system (2.5.32) becomes

$$m\ddot{x}_1 = \lambda,$$
$$m\ddot{x}_2 = 0,$$
$$m\ddot{x}_3 = -mg + \lambda.$$

Subtracting the third equation from the first one, integrating twice yields $(x_1 - x_3)(t)$. Using $x_1 + x_3 = 1$ and the initial conditions $x_1(0) = 0$, $\dot{x}_1(0) = 0$, $x_3(0) = 1$, $\dot{x}_3(0) = 0$, one obtains unique solutions $x_1(t)$ and $x_3(t)$. The second equation gives, together with the initial conditions $x_2(0) = 0$ and $\dot{x}_2(0) = v_2$, the unique solution $x_2(t)$:

$$x_1(t) = \frac{1}{4} gt^2, \quad x_2(t) = v_2 t, \quad x_3(t) = -\frac{1}{4} gt^2 + 1.$$

At $t = t_1 = \frac{2}{\sqrt{g}}$, the point mass m has reached the plane $x_3 = 0$. The running time does not depend on v_2. The time of the free fall $x_3(t) = -\frac{1}{2} gt^2 + 1$ from $x_3 = 1$ to $x_3 = 0$ is $t_2 = \sqrt{\frac{2}{g}}$.

2.5.3 For $T = \frac{1}{2} m (\dot{x}_1^2 + \dot{x}_2^2 + \dot{x}_3^2)$, $V = mgx_3$, and $\Psi(x_1, x_2, x_3) = x_1^2 + x_2^2 + x_3^2 - \ell^2 = 0$, the system (2.5.32) becomes

$$m\ddot{x}_1 = 2\lambda x_1,$$
$$m\ddot{x}_2 = 2\lambda x_2,$$
$$m\ddot{x}_3 = -mg + 2\lambda x_3.$$

For spherical coordinates one obtains, with $r = \ell$,

$$\dot{x}_1 = \ell \cos\theta \cos\varphi \dot{\theta} - \ell \sin\theta \sin\varphi \dot{\varphi},$$
$$\dot{x}_2 = \ell \cos\theta \sin\varphi \dot{\theta} + \ell \sin\theta \cos\varphi \dot{\varphi},$$
$$\dot{x}_3 = -\ell \sin\theta \dot{\theta},$$

which gives

$$T = \frac{1}{2}m\ell^2(\dot{\theta}^2 + \sin^2\theta\,\dot{\phi}^2), \quad V = mg\ell\cos\theta.$$

The Euler-Lagrange equations for $L(\theta, \phi, \dot{\theta}, \dot{\phi}) = T(\dot{\theta}, \dot{\phi}) - V(\theta)$ read

$$\ddot{\theta} = \sin\theta\cos\theta\,\dot{\phi}^2 + \frac{g}{l}\sin\theta,$$

$$\frac{d}{dt}\sin^2\theta\,\dot{\phi} = 0, \quad \text{or} \quad \sin^2\theta\,\dot{\phi} = c.$$

The second relation is the conservation law given in Exercise 2.5.4.

2.5.4 From the system describing the spherical pendulum, cf. Exercise 2.5.3, one deduces

$$m(\ddot{x}_1 x_2 - \ddot{x}_2 x_1) = 0 \quad \text{and} \quad \frac{d}{dt}(\dot{x}_1 x_2 - \dot{x}_2 x_1) = 0.$$

By formula (2.2.5),

$$\int_{t_a}^{t} x_1\dot{x}_2 - \dot{x}_1 x_2\,dt = 2F(t),$$

where $F(t)$ is the area depicted in the following figure:

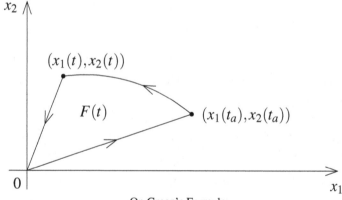

On Green's Formula

  The curvilinear integrals along the line segments from 0 to $(x_1(t_a), x_2(t_a))$ and from $(x_1(t), x_2(t))$ to 0 are zero because the vector (x_1, x_2) is orthogonal to the outer normal vector $(\dot{x}_2, -\dot{x}_1)$. From $x_1\dot{x}_2 - \dot{x}_1 x_2 = c$, differentiation gives

$$\frac{d}{dt}F(t) = \frac{c}{2},$$

which means constancy of the area velocity.

2.6.1 The geodesic parametrized by arc length satisfies the following system:

$$\ddot{x}_1 = 2\lambda x_1, \qquad x_1^2 + x_2^2 = 1,$$
$$\ddot{x}_2 = 2\lambda x_2, \qquad \dot{x}_1^2 + \dot{x}_2^2 + \dot{x}_3^2 = 1,$$
$$\ddot{x}_3 = 0.$$

Differentiation of $x_1^2 + x_2^2 = 1$ two times gives

$$\dot{x}_1^2 + \dot{x}_2^2 + x_1\ddot{x}_1 + x_2\ddot{x}_2 = \dot{x}_1^2 + \dot{x}_2^2 + 2\lambda(x_1^2 + x_2^2) = 0 \quad \text{and}$$
$$2\lambda = -\dot{x}_1^2 - \dot{x}_2^2 = \dot{x}_3^2 - 1.$$

By the third equation, $x_3(s) = c_1 s + c_0$, and $x_3(0) = 0$ implies $c_0 = 0$. If L denotes the length of the geodesic, we have $x_3(L) = c_1 L = 1$ or $c_1 = L^{-1}$. By the geometry of the cylinder we have apparently $L > 1$ or $c_1 < 1$, which is confirmed by the differential equations

$$\ddot{x}_1 = (c_1^2 - 1)x_1, \quad \ddot{x}_2 = (c_1^2 - 1)x_2,$$

which are solvable under the constraint $x_1^2 + x_2^2 = 1$ for $c_1^2 - 1 < 0$ only. The general solutions are

$$x_1(s) = a_1 \cos\sqrt{1 - c_1^2}\, s + b_1 \sin\sqrt{1 - c_1^2}\, s,$$
$$x_2(s) = a_2 \cos\sqrt{1 - c_1^2}\, s + b_2 \sin\sqrt{1 - c_1^2}\, s.$$

The boundary conditions give:
$$x_1(0) = a_1 = 1, \ x_2(0) = a_2 = 0, \ x_1(L) = \cos\sqrt{1 - c_1^2}L + b_1 \sin\sqrt{1 - c_1^2}L = -1,$$

$$x_2(L) = b_2 \sin\sqrt{1 - c_1^2}L = 0.$$
The condition $x_1^2 + x_2^2 = 1$ implies by differentiation, $x_1\dot{x}_1 + x_2\dot{x}_2 = 0$, which at $s = 0$ gives $b_1 = 0$. The relation $\dot{x}_1^2 + \dot{x}_2^2 = 1 - c_1^2$ determines, at $s = 0$, the coefficient $b_2 = \pm 1$, which implies, from $x_2(L) = \pm \sin\sqrt{1 - c_1^2}L = 0$, that $\sqrt{1 - c_1^2}L = \pi$. Finally, since $c_1 = L^{-1}$, we obtain $L = \sqrt{1 + \pi^2}$. We find two geodesics from $A = (1,0,0)$ to $B = (-1,0,1)$:

$$\left\{ x(s) = \left(\cos\frac{\pi}{\sqrt{1 + \pi^2}}s, \ \pm\sin\frac{\pi}{\sqrt{1 + \pi^2}}s, \ \frac{1}{\sqrt{1 + \pi^2}}s \right) \,\Big|\, s \in [0, \sqrt{1 + \pi^2}] \right\}.$$

2.6.2 The Euler-Lagrange equations for geodesics parametrized by arc length read in Cartesian coordinates

$$\ddot{x}_1 = 2\lambda x_1, \qquad x_1^2 + x_2^2 = x_3^2, \quad x_3 \geq 0,$$
$$\ddot{x}_2 = 2\lambda x_2, \qquad \dot{x}_1^2 + \dot{x}_2^2 + \dot{x}_3^2 = 1,$$
$$\ddot{x}_3 = -2\lambda x_3.$$

Differentiation of $\|x(s)\|^2$ two times yields

$$\frac{d}{ds}\|x\|^2 = 2(x,\dot{x}), \quad \frac{d^2}{ds^2}\|x\|^2 = 2\|\dot{x}\|^2 + 2(x,\ddot{x}) = 2 + 2\lambda(x_1^2 + x_2^2 - x_3^2) = 2.$$

In generalized coordinates, parametrized by the arc length, $(2.6.19)_1$ yields, for $f(r) = r$, $f'(r) = 1$, $f''(r) = 0$,

$$2\ddot{r} = r\dot{\varphi}^2 \quad \text{and} \quad 2\dot{r}^2 + r^2\dot{\varphi}^2 = 1.$$

This gives for $r(s)^2$

$$\frac{d}{ds}r^2 = 2r\dot{r}, \quad \frac{d^2}{ds^2}r^2 = 2\dot{r}^2 + 2r\ddot{r} = 1 - r^2\dot{\varphi}^2 + r^2\dot{\varphi}^2 = 1.$$

Since $\|x\|^2 = 2r^2$, the two differential equations coincide.

Any geodesic, which is not a meridian, fulfills Clairaut's relation (2.6.62). Hence $r \geq c_1 > 0$ for some constant c_1. The differential equation for r^2 is solved via

$$r(s)^2 = \frac{1}{2}s^2 + r_1 s + r_0^2, \quad \text{where} \quad r(0)^2 = r_0^2 > 0, \frac{d}{ds}r(s)^2|_{s=0} = r_1.$$

The estimate $r^2 \geq c_1^2$ implies $\frac{1}{2}r_1^2 + c_1^2 \leq r_0^2$. Finally,

$$\lim_{s \to \pm\infty} r(s)^2 = \infty.$$

2.7.1 Let $x \in (C^1[t_a, t_b])^n$ satisfy $\|\dot{x}\|^2 - 1 = 0$ and (2.7.23). Then

$$2\frac{d}{dt}(\lambda\dot{x}) = 0 \quad \text{or} \quad \lambda\dot{x} = c.$$

Furthermore,

$$\lambda^2 = \lambda^2\|\dot{x}\|^2 = \|c\|^2.$$

If $\lambda = \pm\|c\| \neq 0$, then $\ddot{x} = 0$, and hence x is a straight line. Conversely, if x is not a straight line, then $\lambda = 0$, and according to Definition 2.7.3, x is normal.

2.7.2 Let $x \in (C^1[t_a, t_b])^n$ satisfy $x(t_a) = A$, $\tilde{\Psi}(x, \dot{x}) = D\Psi(x)\dot{x} = 0$, and (2.7.3). Then

$$\frac{d}{dt}\sum_{i=1}^{m}\lambda_i\nabla\Psi_i(x) = \nabla\left(\sum_{i=1}^{m}\lambda_i(\nabla\Psi_i(x), \dot{x})\right) = \nabla\left(\sum_{i=1}^{m}\lambda_i\frac{d}{dt}\Psi_i(x)\right)$$

$$= \sum_{i=1}^{m}\lambda_i\frac{d}{dt}\nabla\Psi_i(x), \quad \text{and hence} \quad \sum_{i=1}^{m}\dot{\lambda}_i\nabla\Psi_i(x) = 0.$$

Since $D\Psi(x)$ has the maximal rank m for $x \in M = \{x \in \mathbb{R}^n | D\Psi(x)\dot{x} = \frac{d}{dt}\Psi(x) = 0\} = \{x \in \mathbb{R}^n | \Psi(x) = \Psi(x(t_a)) = \Psi(A)\}$, the last equation implies that $\dot{\lambda}_i = 0$ or

$\lambda_i = c_i$ for $i = 1, \ldots, m$. These constants do not necessarily vanish. Hence, according to Definition 2.7.3, x is not normal.

2.7.3 We use the following identities for $\tilde{\Psi}(x, \dot{x}) = D\Psi(x)\dot{x} = \frac{d}{dt}\Psi(x)$:
$\nabla\Psi_i(x) = \tilde{\Psi}_{i,\dot{x}}(x, \dot{x}) = (\frac{d}{dt}\Psi_i(x))_{\dot{x}}$, $\tilde{\Psi}_{i,x}(x, \dot{x}) = \frac{d}{dt}\nabla\Psi_i(x)$. Then

$$\frac{d}{dt}(\Phi + \sum_{i=1}^{m} \tilde{\lambda}_i \tilde{\Psi}_i)_{\dot{x}} = (\Phi + \sum_{i=1}^{m} \tilde{\lambda}_i \tilde{\Psi}_i)_x \Leftrightarrow$$

$$\frac{d}{dt}\Phi_{\dot{x}} + \frac{d}{dt}\sum_{i=1}^{m}\tilde{\lambda}_i\nabla\Psi_i = \Phi_x + \sum_{i=1}^{m}\tilde{\lambda}_i\frac{d}{dt}\nabla\Psi_i \Leftrightarrow$$

$$\frac{d}{dt}\Phi_{\dot{x}} = \Phi_x - \sum_{i=1}^{m}\dot{\tilde{\lambda}}_i\nabla\Psi_i.$$

The equivalence of the two systems holds with $\lambda_i = -\dot{\tilde{\lambda}}_i$.
Remark: The functions $\tilde{\lambda}_i$ may be replaced by $\tilde{\lambda}_i + c_i$ with constants c_i without changing system (2.7.21):

$$\frac{d}{dt}\sum_{i=1}^{m}c_i\tilde{\Psi}_{i,\dot{x}} = \frac{d}{dt}\sum_{i=1}^{m}c_i\nabla\Psi_i = \sum_{i=1}^{m}c_i\frac{d}{dt}\nabla\Psi_i = \sum_{i=1}^{m}c_i\tilde{\Psi}_{i,x}.$$

2.7.4

a) Let the function $(y, z) \in (C^1[a, b])^{n+m}$ satisfy the nonholonomic constraint and (2.7.23) on $[a, b]$, i.e.,

$$-\frac{d}{dx}(\sum_{i=1}^{m}\lambda_i G_i)_{y'} = -(\sum_{i=1}^{m}\lambda_i G_i)_y,$$

$$\frac{d}{dx}\lambda_i = 0, \quad i = 1, \ldots, m.$$

Thus, the functions λ_i are constant on $[a, b]$, and the first n-dimensional system becomes

$$\sum_{i=1}^{m}\lambda_i(G_{i,y} - \frac{d}{dx}G_{i,y'}) = 0 \quad \text{on} \quad [a, b] \Leftrightarrow$$

$$\sum_{i=1}^{m}\lambda_i\int_a^b(G_{i,y}, h) + (G_{i,y'}, h')dx = 0, \text{ which means}$$

$$\sum_{i=1}^{m}\lambda_i\delta K_i(y)h = 0 \quad \text{for all} \quad h \in (C_0^{1,pw}[a, b])^n,$$

cf. Proposition 1.4.2. The function (y, z) is normal if and only if this implies $\lambda_1 = \cdots = \lambda_m = 0$. This is precisely the condition that y is not critical for the isoperimetric constraints, cf. Exercise 2.1.1.

b) Since the Lagrange function F does not depend on z and z', the last m equations of (2.7.21) are identical to the last m equations of (2.7.23), which imply that all

λ_i are constant, cf. part a). If the local minimizer (y,z) is normal with respect to the nonholonomic constraint, then, according to Corollary 2.7.1, one can choose $\lambda_0 = 1$. In this case (2.7.21) is converted to (2.1.20).

2.7.5

a) For $G(x,y,z,y',z') = y' - z$, one obtains

$$D_pG(x,y,z,p_1,p_2) = (1,0),$$

and hence D_pG has rank $m = 1$.

b) The boundary value problem (2.7.10) reads

$$y' - z = 0, \quad y(a) = A_1, \quad z(a) = A_2, \quad y(b) = \tilde{B}_1, \quad z(b) = \tilde{B}_2.$$

Since $z \in C^1[a,b]$ is constrained only by the values $z(a)$ and $z(b)$, the equation

$$y(b) = \int_a^b z(\xi)d\xi + A_1 = \tilde{B}_1$$

can be solved for any $\tilde{B}_1$.

c) Let $y' - z = 0$ and suppose that (2.7.23) is satisfied, i.e.,

$$\frac{d}{dx}\lambda = 0 \quad \text{and} \quad 0 = -\lambda.$$

Therefore any solution of $y' - z = 0$ is normal.

d) System (2.7.21) reads, for $\lambda_0 = 1$, cf. Corollary 2.7.1:

$$\frac{d}{dx}(F_{y'} + \lambda) = F_y, \quad \frac{d}{dx}F_{z'} = -\lambda, \text{ and hence,}$$

$$\frac{d^2}{dx^2}F_{y''} - \frac{d}{dx}F_{y'} = -F_y \quad \text{on} \quad [a,b].$$

Here $F_{y''} = F_{y''}(\cdot,y,y',y'')$, $F_{y'} = F_{y'}(\cdot,y,y',y'')$, and $F_y = F_y(\cdot,y,y',y'')$.

2.7.6 The constraint $G(y,y') = 0$ means geometrically that the vector y' is orthogonal to the vector $g(y)$. Specifically

$$y' = \alpha g^{\perp}(y) \quad \text{where} \quad g^{\perp}(y) = (-g_2(y), g_1(y)),$$

and $\alpha = \alpha(y,y')$ is a continuous scalar function.

a) System (2.7.23) reads (g is a column):

$$\frac{d}{dx}(\lambda g) = \lambda(\nabla g_1 y_1' + \nabla g_2 y_2') = \lambda Dg^* y',$$

where $Dg^* = Dg(y)^*$ is the transposed Jacobian matrix of the vector field g, and y' is a column. Furthermore,

$$\lambda' g + \lambda Dg y' = \lambda Dg^* y', \quad \text{or} \quad \lambda' g = \lambda (Dg^* - Dg) y'.$$

Inserting the column $y' = \alpha g^\perp(y)$ gives

$$\lambda' g = \lambda (Dg^* - Dg)\alpha g^\perp = \lambda (g_{2,y_1} - g_{1,y_2})\alpha g = \lambda (\text{rot} g)\alpha g, \quad \text{and hence}$$
$$\lambda' = \alpha (\text{rot} g)\lambda,$$

due to $g(y) \neq 0$. For any continuous scalar function $\alpha(\text{rot} g)$ of x this linear differential equation possesses a nontrivial solution $\lambda = \lambda(x)$. Therefore any solution y of $G(y,y') = 0$ is not normal.

b) Let $y \in (C^2[a,b])^2$ satisfy $G(y,y') = 0$ and $y(a) = A$, $y(b) = B$. Then

$$y' = \alpha g^\perp(y) \quad \text{where} \quad \alpha = \alpha(y,y') = \frac{y_2'}{g_1(y_1,y_2)} = -\frac{y_1'}{g_2(y_1,y_2)},$$

and because $g(y) \neq 0$, at least one expression is defined. By assumption on y, the function $\alpha \in C^1[a,b]$. Therefore, the solution of the initial value problem

$$w' = \alpha g^\perp(w), \quad w(a) = A,$$

is unique, i.e., $w = y$ on $[a,b]$. Define

$$\beta(x) = \int_a^x \alpha(y(s),y'(s))ds,$$

and let z be the unique solution of the initial value problem

$$z' = g^\perp(z), \quad z(0) = A.$$

Then $w(x) = z(\beta(x))$ solves

$$w'(x) = z'(\beta(x))\alpha(y(x),y'(x)) = g^\perp(z(\beta(x)))\alpha, \quad \text{or}$$
$$w' = \alpha g^\perp(w), \quad w(a) = z(0) = A.$$

Hence $w = y$ or $y(x) = z(\beta(x))$. In particular,

$$y(b) = z(\beta(b)) \in \{z \in \mathbb{R}^2 | z = z(x), x \in \mathbb{R}\},$$

which implies that the boundary value problem (2.7.10) is not solvable for all $\tilde{B}$ in a full neighborhood of B; any endpoint of y has to be on the curve described by z.

2.8.1 As a first step, we choose $h_1,\dots,h_m \in (C_0^{1,pw}[t_a,t_b])^n$ satisfying (2.1.21), and we define, for arbitrary $h \in (C_0^{1,pw}[t_a,t_b])^n$, the functions (2.1.22). Then, by (2.1.26) and (2.1.27), the Lagrange multipliers do not depend on h, and the Euler-Lagrange equations (2.1.35) are established.

As a second step, we define (2.1.22) with the admitted perturbation $h(s, \cdot)$ constructed in (2.8.7), (2.8.8) with $h_1, \ldots, h_m \in (C_0^{1,pw}[t_a, t_b])^n$ as in the first step. Then we obtain (2.1.26) with the same Lagrange multipliers (2.1.27). There is one difference: In $(2.1.26)_1$, the function h has to be replaced by $\frac{\partial}{\partial s}h(0, \cdot) = \eta/7y$. Specifically, (2.1.26) gives, cf. (2.1.28),

$$\int_{t_a}^{t_b} ((\Phi + \sum_{i=1}^{m} \lambda_i \Psi_i)_x, \eta y) + ((\Phi + \sum_{i=1}^{m} \lambda_i \Psi_i)_{\dot{x}}, \dot{\eta} y) dt$$

$$= \int_{t_a}^{t_b} ((\Phi + \sum_{i=1}^{m} \lambda_i \Psi_i)_x - \frac{d}{dt}(\Phi + \sum_{i=1}^{m} \lambda_i \Psi_i)_{\dot{x}}, \eta y) dt + (\Phi + \sum_{i=1}^{m} \lambda_i \Psi_i)_{\dot{x}}, \eta y)|_{t_a}^{t_b} = 0.$$

In view of the Euler-Lagrange equation (2.1.35), the integral vanishes, and choosing $\eta(t_a) = 1$, $\eta(t_b) = 0$, $y \in T_{x(t_a)}M_a$, $\|y\| \leq 1$, we obtain the transversality

$$(\Phi + \sum_{i=1}^{m} \lambda_i \Psi_i)_{\dot{x}}(x(t_a), \dot{x}(t_a)) \in N_{x(t_a)}M_a,$$

with the same Lagrange multipliers as in the Euler-Lagrange equations.

2.8.2 Defining $\tilde{\Psi} = (\Psi, \Psi_a) : \mathbb{R}^n \to \mathbb{R}^{m+m_a}$, the set M_a is given by $\tilde{\Psi}(x) = 0$, and by the maximal rank of the Jacobian matrix, M_a is an $(n - (m + m_a))$-dimensional manifold contained in M, cf. the Appendix.

Following (2.8.5), let $x(t_a) + sy + \varphi(sy) \in M_a$, where $y \in T_{x(t_a)}M_a$, $\varphi(sy) \in N_{x(t_a)}M_a$, $\|y\| \leq 1$ and $s \in (-r, r)$. We choose $h \in (C^2[t_a, t_b])^n$ satisfying

$$h(t_a) = y \in T_{x(t_a)}M_a, \quad h(t_b) = 0,$$

and we obtain as in (2.5.10)

$$a(t) = P(x(t))h(t) \in T_{x(t)}M \quad \text{for} \quad t \in [t_a, t_b],$$
$$a(t_a) = y, \quad \text{since} \quad T_{x(t_a)}M_a \subset T_{x(t_a)}M, \quad \text{and} \quad a(t_b) = 0.$$

We define a function H as in (2.5.12), where we insert the above function a. Following (2.5.12)–(2.5.21), we construct a perturbation $h : (-\varepsilon_0, \varepsilon_0) \times [t_a, t_b] \to \mathbb{R}^n$ satisfying

$$x(t) + h(s, t) \in M \quad \text{for} \quad (s, t) \in (-\varepsilon_0, \varepsilon_0) \times [t_a, t_b],$$
$$h(s, t) = sa(t) + b(s, t), \quad a(t) \in T_{x(t)}M, \quad b(s, t) \in N_{x(t)}M.$$

Since $x(t_a) + sy + \varphi(sy) \in M_a \subset M$,

$$\Psi(x(t_a) + sa(t_a) + \varphi(sy)) = 0, \quad \text{and}$$
$$\Psi(x(t_a) + sa(t_a) + b(s, t_a)) = 0.$$

Since $b(s, t_a) \in N_{x(t_a)}M \subset N_{x(t_a)}M_a$, and $\varphi(sy) \in N_{x(t_a)}M_a$ is locally unique, we conclude

$$b(s,t_a) = \varphi(sy) = \varphi(sa(t_a)), \quad \text{and}$$
$$h(s,t_a) \in M_a \quad \text{for all} \quad s \in (-\varepsilon_0, \varepsilon_0), 0 < \varepsilon_0 \leq r.$$

To summarize, $h(s,\cdot)$ is a perturbation, which is admitted for the holonomic constraint as well as for the boundary condition. Furthermore, $J(x + h(s,\cdot))$ is locally minimal at $s = 0$ (observe $(h(0,\cdot) = 0)$). We follow (2.5.23):

$$\frac{d}{ds}J(x + h(s,\cdot))|_{s=0} = 0$$

$$= \int_{t_a}^{t_b} (\Phi_x(x,\dot{x}),a) + (\Phi_{\dot{x}}(x,\dot{x}),\dot{a})dt$$

$$= \int_{t_a}^{t_b} (\Phi_x(x,\dot{x}) - \frac{d}{dt}\Phi_{\dot{x}}(x,\dot{x}),a)dt + (\Phi_{\dot{x}}(x,\dot{x}),a)|_{t_a}^{t_b}$$

$$= -(\Phi_{\dot{x}}(x(t_a),\dot{x}(t_a)),y), \quad \text{where} \quad y = a(t_a) \in T_{x(t_a)}M_a.$$

The argument is that in view of the Euler-Lagrange equation (2.5.8) and the orthogonality of $\sum_{i=1}^{m} \lambda_i(t)\nabla\Psi_i(x(t)) \in N_{x(t)}M$ to $a(t) \in T_{x(t)}M$, the integral vanishes. Since $y \in T_{x(t_a)}M_a$ is restricted only by $\|y\| \leq 1$, we obtain the transversality

$$\Phi_{\dot{x}}(x(t_a),\dot{x}(t_a)) \in N_{x(t_a)}M_a,$$

which is weaker than the natural boundary condition (2.5.67).

2.8.3 The Euler-Lagrange equation reads $\frac{d}{dx}2x^3y' = 0$, and thus, $y(x) = \frac{c_1}{x^2} + c_2$. The condition $y(1) = 0$ implies $c_1 + c_2 = 0$. By transversality,

$$\frac{4c_1^2}{x^3} + \left(\frac{4}{x^3} - \frac{2c_1}{x^3}\right)4c_1 = 0, \quad \text{and hence} \quad c_1 = 4.$$

This gives $y(x) = 4\left(\frac{1}{x^2} - 1\right)$ and $(x_b, y(x_b)) = (\sqrt{2}, -2)$.

2.8.4 The Euler-Lagrange equation $y'' = y$ has the solutions $y(x) = c_1e^x + c_2e^{-x}$. The condition $y(0) = 1$ implies $c_1 + c_2 = 1$. The transversality on the line $\psi(x) = 2$ (hence $\psi'(x) \equiv 0$) reads $y^2 - (y')^2 = 0$, which implies $c_1c_2 = 0$. Since $y(x) = e^x$ cannot fulfill $y(x_b) = 2$ for some positive x_b, only $y(x) = e^x$ remains, and $x_b = \ln 2$.

2.8.5 The Euler-Lagrange equation reads

$$\frac{d}{dx}(y' + y + 1) = y' + 1.$$

Since $y' + y + 1 \in C^{1,pw}[0,1]$ and $y + 1 \in C^{1,pw}[0,1]$, it follows that $y' \in C^{1,pw}[0,1]$ and

$$y'' + y' = y' + 1, \quad \text{or} \quad y'' = 1 \text{ piecewise on } [0,1].$$

Solutions $y \in C^{1,pw}[0,1] \subset C[0,1]$ satisfying $y' \in C^{1,pw}[0,1] \subset C[0,1]$ are

$$y(x) = \frac{1}{2}x^2 + c_1 x + c_2.$$

Transversality in this case means precisely the natural boundary condition, in particular

$$y'(0) + y(0) + 1 = 0, \quad \text{and} \quad y'(1) + y(1) + 1 = 0.$$

This gives the solution $y(x) = \frac{1}{2}x^2 - \frac{3}{2}x + \frac{1}{2}$.

2.8.6 (i) The solution $y(x) = 4\left(\frac{1}{x^2} - 1\right)$ of Exercise 2.8.2 is a global minimizer:
Continuously differentiable perturbations $y + h$ satisfying $h(1) = 0$ and $y(x_0) + h(x_0) = \frac{2}{x_0^2} - 3$, or $\frac{2}{x_0^2} + h(x_0) = 1$ for some $x_0 > 1$, are admitted. Then

$$J(y+h) = \int_1^{x_0} x^3 \left(-\frac{8}{x^3} + h'(x)\right)^2 dx$$

$$= \int_1^{\sqrt{2}} x^3 \frac{64}{x^6} dx + \int_{\sqrt{2}}^{x_0} x^3 \frac{64}{x^6} dx - 16h(x_0) + \int_1^{x_0} x^3 (h'(x))^2 dx$$

$$= J(y) + 16\left(1 - \frac{2}{x_0^2} - h(x_0)\right) + \int_1^{x_0} x^3 (h'(x))^2 dx$$

$$> J(y), \quad \text{since} \quad 1 - \frac{2}{x_0^2} - h(x_0) = 0 \text{ holds for} \quad x_0 > 1.$$

(ii)   The solution $y(x) = e^x$ of Exercise 2.8.4 is a global minimizer:
Continuously differentiable perturbations $y + h$ satisfying $h(0) = 0$ and $y(x_0) + h(x_0) = 2$, or $e^{x_0} + h(x_0) = 2$ for some $x_0 > 0$, are admitted. Then

$$J(y+h) = \int_0^{x_0} (y+h)^2 + (y'+h')^2 dx$$

$$= \int_0^{\ln 2} y^2 + (y')^2 dx + \int_{\ln 2}^{x_0} y^2 + (y')^2 dx + 2\int_0^{x_0} (y - y'')h dx + 2y'(x_0)h(x_0)$$

$$+ \int_0^{x_0} h^2 + (h')^2 dx = J(y) + e^{2x_0} - 4 + 2e^{x_0}h(x_0) + \int_0^{x_0} h^2 + (h')^2 dx$$

$$= J(y) - (h(x_0))^2 + \int_0^{x_0} h^2 + (h')^2 dx,$$

where we use $e^{x_0} = 2 - h(x_0)$. By Example 5 in Paragraph 1.5,

$$\int_0^{x_0} h^2 + (h')^2 dx \geq \text{Min}\left\{\int_0^{x_0} \tilde{h}^2 + (\tilde{h}')^2 dx \,|\, \tilde{h}(0) = 0, \tilde{h}(x_0) = h(x_0)\right\}$$

$$= \int_0^{x_0} c_1^2 (e^x - e^{-x})^2 + c_1^2 (e^x + e^{-x})^2 dx$$

$$= c_1^2 (e^{2x_0} - e^{-2x_0}) = h(x_0)c_1 (e^{x_0} + e^{-x_0}) \quad (c_1 (e^{x_0} - e^{-x_0}) = h(x_0))$$

$$= (h(x_0))^2 \coth x_0 > (h(x_0))^2 \quad \text{for} \quad x_0 > 0.$$

Therefore $J(y+h) \geq J(y)$ for all admitted perturbations h.

(iii) Any $y+h$, where $h \in C^{1,pw}[0,1]$, is an admitted perturbation of the solution $y(x) = \frac{1}{2}x^2 - \frac{3}{2}x + \frac{1}{2}$ of Exercise 2.8.5:

$$J(y+h) = \int_0^1 \frac{1}{2}(y'+h')^2 + (y+h)(y'+h') + y' + h' + y + h\,dx$$

$$= J(y) + \int_0^1 (-y'' - y' + y' + 1)h\,dx + (y'+y+1)h|_0^1 + \int_0^1 \frac{1}{2}(h')^2 + hh'\,dx$$

$$= J(y) + \frac{1}{2}((h(1))^2 - (h(0))^2 + \int_0^1 (h')^2\,dx),$$

where the other terms vanish by virtue of the Euler-Lagrange equation and the natural boundary conditions.

Choosing $h_1(x) = \varepsilon(x + \frac{1}{2})$, we obtain

$$J(y+h_1) = J(y) + \frac{3}{2}\varepsilon^2 > J(y),$$

and for $h_2(x) = \varepsilon(-x + \frac{3}{2})$,

$$J(y+h_2) = J(y) - \frac{1}{2}\varepsilon^2 < J(y).$$

Therefore the solution of Exercise 2.8.5 is not a local minimizer.

2.9.1 According to Proposition (2.5.1), the local minimizer x solves

$$\frac{d}{dt}\Phi_{\dot{x}}(x,\dot{x}) = \Phi_x(x,\dot{x}) + \sum_{i=1}^m \lambda_i \nabla \Psi_i(x) \quad \text{on} \quad [t_a, t_b].$$

Due to the invariance of Φ, differentiation with respect to x and s yields

$$\nabla \Psi_i(h^s(x))Dh^s(x) = \nabla \Psi_i(x), \quad \nabla \Psi_i(h^s(x))\frac{\partial}{\partial s}h^s(x) = 0,$$

where here and in the sequel, $\nabla \Psi_i$ are row vectors and the product of $\nabla \Psi_i$ and $\frac{\partial}{\partial s}h^s$ is the Euclidean scalar product in $\mathbb{R}^n$. By the computations (2.9.11),

$$\frac{d}{dt}\Phi_{\dot{x}}(x,\dot{x}) - \Phi_x(x,\dot{x}) - \sum_{i=1}^m \lambda_i \nabla \Psi_i(x) = 0$$

$$= \left(\frac{d}{dt}\Phi_{\dot{x}}(h^s(x), \frac{d}{dt}h^s(x)) - \Phi_x(h^s(x), \frac{d}{dt}h^s(x)) - \sum_{i=1}^m \lambda_i \nabla \Psi_i(h^s(x))\right) Dh^s(x).$$

Thus, $h^s(x)$ solves the Euler-Lagrange system with the same Lagrange multipliers for all $s \in (-\delta, \delta)$. Using (2.9.16) and (2.9.17), we obtain

$$\frac{d}{dt}\left(\Phi_{\dot{x}}(h^s(x), Dh^s(x)\dot{x})\frac{\partial}{\partial s}h^s(x)\right)$$

$$= \frac{d}{dt}\Phi_{\dot{x}}(h^s(x), \frac{d}{dt}h^s(x))\frac{\partial}{\partial s}h^s(x) + \Phi_{\dot{x}}(h^s(x), Dh^s(x)\dot{x})\frac{d}{dt}\frac{\partial}{\partial s}h^s(x)$$

$$= \left(\Phi_x(h^s(x), \frac{d}{dt}h^s(x)) + \sum_{i=1}^m \lambda_i \nabla \Psi_i(h^s(x))\right)\frac{\partial}{\partial s}h^s(x)$$

$$+ \Phi_{\dot{x}}(h^s(x), Dh^s(x)\dot{x})\frac{\partial}{\partial s}Dh^s(x)\dot{x} = 0,$$

where we use (2.9.15) and the invariance of Ψ.

Remark *It is not necessary that the curve x be a local minimizer under the holonomic constraint. It suffices that x solves the Euler-Lagrange system (2.5.8).*

2.10.1 For $\beta = 0$ and for $\dot{r} = 0$ for m_1 at a distance R from m_2, formula (2.10.17) gives $E = -k/R$. For $\dot{r} < 0$ the square root in (2.10.17) is negative. Let $G(r)$ be a primitive of $-1/F(r)$. Then

$$\frac{d}{dt}G(r) = G'(r)\dot{r} = (-1/F(r))F(r) = -1, \quad \text{or } G(r) = c - t.$$

For $t = 0$, we have $G(R) = c$, and for $t = T$, we obtain $G(0) = G(R) - T$. This gives the following estimate:

$$T = G(R) - G(0) = \int_0^R G'(r)dr = \sqrt{\frac{m}{2k}}\int_0^R \frac{1}{\sqrt{\frac{1}{r} - \frac{1}{R}}}dr$$

$$= \sqrt{\frac{m}{2k}}\int_0^R \sqrt{\frac{r}{1 - \frac{r}{R}}}dr = \sqrt{\frac{m}{2k}}\sqrt{R}\int_0^R \sqrt{\frac{\frac{r}{R}}{1 - \frac{r}{R}}}dr$$

$$= \sqrt{\frac{m}{2k}}R^{3/2}\int_0^1 \sqrt{\frac{s}{1-s}}dx < 2\sqrt{\frac{m}{2k}}R^{3/2}.$$

2.10.2 Take $\beta = 0, E = 0$, and the negative square root in (2.10.17). Then, for $r = R$, the velocity is

$$\dot{r} = -\sqrt{\frac{2k}{m}\frac{1}{\sqrt{R}}},$$

and the computation in Exercise 2.10.1 gives the falling time

$$T = \sqrt{\frac{m}{2k}}\int_0^R \sqrt{r}dr = \frac{2}{3}\sqrt{\frac{m}{2k}}R^{3/2}.$$

3.1.1 Let $u \in X$ be a weak eigenvector with eigenvalue λ. Then by the symmetry of B and K, we have

$$B(u, u_n) = \lambda K(u, u_n), \quad \text{and}$$
$$B(u_n, u) = \lambda_n K(u_n, u) \quad \text{for all } n \in \mathbb{N}.$$
$$\text{Hence, } (\lambda - \lambda_n) K(u, u_n) = 0.$$

If $\lambda \neq \lambda_n$ for all $n \in \mathbb{N}$, then $c_n = K(u, u_n) = 0$ for all $n \in \mathbb{N}$. Proposition (3.1.5) then implies $u = 0$, which is not a weak eigenvector. Therefore $\lambda = \lambda_{n_0}$ for some $n_0 \in \mathbb{N}$. By the above argument $K(u, u_n) = 0$ for all $n \neq n_0$, and again Proposition 3.1.5 and Corollary 3.1.1 imply that u is a linear combination of the weak eigenvectors with eigenvalue λ_{n_0}.

3.2.1 According to Definition 3.2.2 of the norm in $W^{1,2}(a,b)$, a sequence $(y_n)_{n \in \mathbb{N}}$ is a Cauchy sequence in $W^{1,2}(a,b)$ if $(y_n)_{n \in \mathbb{N}}$ as well as $(y_n')_{n \in \mathbb{N}}$ are Cauchy sequences in $L^2(a,b)$. Both sequences have a limit, y_0 and z_0 in $L^2(a,b)$, respectively. According to Definition 3.2.1,

$$(y_n', h)_{0,2} = -(y_n, h')_{0,2} \quad \text{for all} \quad h \in C_0^\infty(a,b),$$

and for all $n \in \mathbb{N}$. Taking the limits, the continuity of the scalar product (Cauchy-Schwarz inequality) implies

$$(z_0, h)_{0,2} = -(y_0, h') \quad \text{for all} \quad h \in C_0^\infty(a,b).$$

Hence $z_0 = y_0'$ in the weak sense. By Definition 3.2.2, $y_0 \in W^{1,2}(a,b)$, and $\lim_{n \to \infty} \|y_n - y_0\|_{1,2} = 0$.

3.2.2 By (3.2.50), we deduce

$$J(y) \geq c_1 \|y'\|_{0,2}^2 - c_2 \int_a^b |y|^q dx - c_3(b-a)$$
$$\geq c_1 \|y'\|_{0,2}^2 - c_2(b-a)^{1-(q/2)} \left(\int_a^b y^2 dx \right)^{q/2} - c_3(b-a), \quad \text{and}$$

in view of Hölder's inequality,

$$J(y) \geq \frac{c_1}{2C_1^2} \|y\|_{0,2}^2 - c_2(b-a)^{1-(q/2)} \|y\|_{0,2}^q - \frac{c_1 C_2^2}{C_1^2} - c_3(b-a) \quad \text{(using (3.2.31))}$$
$$\geq -c_4, \quad \text{due to } c_1 > 0 \text{ and } q < 2.$$

Inserting a minimizing sequence, we obtain

$$c_1 \|y_n'\|_{0,2}^2 - c_2 \int_a^b |y_n|^q dx - c_3(b-a) \leq m+1 \quad \text{or}$$
$$m+1+c_3(b-a) \geq c_1 \|y_n'\|_{0,2}^2 - c_2(b-a)^{1-(q/2)} \|y_n\|_{0,2}^q$$
$$\geq \frac{c_1}{2C_1^2} \|y_n\|_{1,2}^2 - c_2(b-a)^{1-(q/2)} \|y_n\|_{1,2}^q - \frac{c_1 C_2^2}{C_1^2}, \quad \text{where again we use (3.2.31).}$$

Hence, since $c_1 > 0$ and $q < 2$, the sequence is bounded in $W^{1,2}(a,b)$: $\|y_n\|_{1,2} \le C_3$ for all $n \in \mathbb{N}$.

Again, by virtue of (3.2.50),

$$\tilde{J}(y) = J(y) + c_2 \int_a^b |y|^q dx + c_3(b-a) \ge 0 \quad \text{for all} \quad y \in D,$$

and $\lim_{n \to \infty} \tilde{y}_n = \tilde{y}_0$ in $C[a,b]$. This is the same as uniform convergence in $[a,b]$, which implies

$$\lim_{n \to \infty} c_2 \int_a^b |\tilde{y}_n|^q dx + c_3(b-a) = c_2 \int_a^b |\tilde{y}_0|^q dx + c_3(b-a).$$

Thus, for nonnegative $\tilde{F}$, cf. (3.2.49)$_1$, we have

$$\liminf_{n \to \infty} \tilde{J}(\tilde{y}_n) \ge \tilde{J}(\tilde{y}_0),$$

which implies finally the lower semi-continuity of J:

$$\liminf_{n \to \infty} J(\tilde{y}_n) = \liminf_{n \to \infty} \tilde{J}(\tilde{y}_n) - \lim_{n \to \infty} c_2 \int_a^b |\tilde{y}_n|^q dx - c_3(b-a)$$

$$\ge \tilde{J}(\tilde{y}_0) - c_2 \int_a^b |\tilde{y}_0|^q dx - c_3(b-a) = J(\tilde{y}_0).$$

3.2.3 The Lagrange function $F(x,y,y') = x^2(y')^2$ does not satisfy the coercivity (3.2.34)$_1$. The minimizing sequence $y_n(x) = \arctan nx / \arctan n$ is not bounded in $W^{1,2}(-1,1)$:

$$\|y_n\|_{1,2}^2 \ge \frac{1}{(\arctan n)^2} \int_{-1}^1 \left(\frac{n}{1+n^2x^2}\right)^2 dx > \frac{4}{\pi^2} n^2 \int_{-\frac{1}{n}}^{\frac{1}{n}} \frac{1}{(1+n^2x^2)^2} dx$$

$$> \frac{4}{\pi^2} n^2 \frac{2}{n} \frac{1}{4} = \frac{2n}{\pi^2}.$$

3.3.1

a) The first variation of the first constraint yields $\delta K(u_n)u_n = 2K(u_n, u_n) = 2$, and $\delta K(u_n)u_i = 2K(u_n, u_i) = 0$ for $i = 1, \ldots, n-1$. Due to linearity, the first variation of the last $n-1$ constraints is given by the functionals themselves, and in view of $K(u_i, u_k) = \delta_{ik}$ for $i, k = 1, \ldots, n-1$, the n isoperimetric constraints are not critical for u_n.

b) Scalar multiplication of the Euler-Lagrange equation by u_k in $L^2(a,b)$ gives, by virtue of the K-orthonormality of the eigenfunctions,

$$\int_a^b ((pu_n')' - qu_n)u_k dx = \frac{1}{2}\tilde{\lambda}_k.$$

On the other hand, scalar multiplication of the equation for the k-th eigenfunction by u_n gives

$$\int_a^b u_n((pu_k')' - qu_k)dx = -\int_a^b u_n \lambda_k \rho u_k dx = 0.$$

After integration by parts twice and taking into account the homogeneous boundary conditions on u_n and u_k, we see that both left sides above are equal, and hence $\tilde{\lambda}_k = 0$ for $k = 1, \ldots, n-1$.

c) Scalar multiplication of the equation $(pu_n')' - qu_n = \tilde{\lambda}_n \rho u_n$ by u_n and integration by parts yield

$$-\int_a^b p(u_n')^2 + qu_n^2 dx = \tilde{\lambda}_n \int_a^b \rho u_n^2 dx = \tilde{\lambda}_n, \quad \text{or} \quad -B(u_n) = \tilde{\lambda}_n.$$

3.3.2 Let $S = \text{span}\{u_n | n \in \mathbb{N}\}$ and let $\text{cl}_X S$ denote the closure of some space S in X. By Proposition 3.3.3, we know that $\text{cl}_{W_0^{1,2}(a,b)} S = W_0^{1,2}(a,b)$. Since convergence in $W_0^{1,2}(a,b)$ implies convergence in $L^2(a,b)$, we have

$$W_0^{1,2}(a,b) = \text{cl}_{W_0^{1,2}(a,b)} S \subset \text{cl}_{L^2(a,b)} S \subset L^2(a,b).$$

For the closure of all spaces in $L^2(a,b)$, we obtain

$$\cdot \qquad L^2(a,b) = \text{cl}_{L^2(a,b)} W_0^{1,2}(a,b) \subset \text{cl}_{L^2(a,b)} S \subset L^2(a,b).$$

Hence all spaces are equal, and the system $\{u_n\}_{n \in \mathbb{N}}$ is complete in $L^2(a,b)$.

3.3.3 We define

$$\tilde{y}'(x) = \begin{cases} y'(x) & \text{for } x \in [a,b], \\ 0 & \text{for } x \notin [a,b], \end{cases}$$

where y' is the weak derivative of y. The functions $\tilde{y}, \tilde{y}'$ are in $L^2(c,d)$, and

$$\int_c^d \tilde{y}'h + \tilde{y}h'dx = \int_a^b y'h + yh'dx \quad \text{for all} \quad h \in C_0^\infty(c,d).$$

In order to prove that $\tilde{y}'$ is the weak derivative of $\tilde{y}$ according to Definition 3.2.1, we have to show that the above integral is zero. If $\tilde{h} \in C_0^\infty(a,b)$, then $h\tilde{h} \in C_0^\infty(a,b)$, and since $y \in W^{1,2}(a,b)$, Definitions 3.2.1 and 3.2.2 imply

$$\int_a^b y'h\tilde{h} + y(h\tilde{h})'dx = 0 = \int_a^b (y'h + yh')\tilde{h} + (yh)\tilde{h}'dx, \quad \text{or}$$

$$y'h + yh' \in L^2(a,b) \quad \text{is the weak derivative of } yh \in L^2(a,b),$$

i.e., $yh \in W^{1,2}(a,b)$.

Then by Lemma 3.2.3 and since $yh \in W^{1,2}(a,b)$, there follows

$$\int_a^b y'h + yh'dx = \int_a^b \frac{d}{dx}(yh)dx = y(b)h(b) - y(a)h(a) = 0,$$

due to $y(a) = y(b) = 0$. Therefore, $\tilde{y} \in W_0^{1,2}(c,d)$, due to $\tilde{y}(c) = \tilde{y}(d) = 0$. (It is an easy exercise that, according to Definition 3.2.1, $y'h + yh'$ is the weak derivative of yh.)

3.3.4 Denote the first eigenfunctions with weight functions ρ_i by u_1^i. Then

$$B(u_1^1, u_1^2) = \lambda_1(\rho_1)K_1(u_1^1, u_1^2),$$
$$B(u_1^2, u_1^1) = \lambda_1(\rho_2)K_2(u_1^2, u_1^1).$$

By symmetry, the two left sides above are equal. Assuming that $\lambda_1(\rho_1) = \lambda_1(\rho_2)$, subtraction of the two equations yields

$$\lambda_1(\rho_1) \int_a^b (\rho_2 - \rho_1)u_1^1 u_1^2 dx = 0.$$

Under the hypotheses of Proposition 3.3.5, $\lambda_1(\rho_1) > 0$, and by Proposition 3.3.8, both first eigenfunctions are positive on (a,b) (w.l.o.g.). Thus the above integral can vanish under assumption (3.3.28) only if $\rho_2 - \rho_1 = 0$. By Proposition 3.3.5, only $\lambda_1(\rho_1) > \lambda_1(\rho_2)$ is left as a possibility, yielding $\rho_1 \neq \rho_2$.

Bibliography

Textbooks

1. Blanchard, P., Brüning, E.: Direkte Methoden der Variationsrechnung. Springer, Wien, New York (1982)
2. Bliss, G.A.: Variationsrechnung. Herausgegeben von F. Schwank. Teubner, Leipzig, Berlin (1932)
3. Bolza, J.: Vorlesungen über Variationsrechnung, 2nd edn. Chelsea Publ. Comp, New York (1962)
4. Brunt, B.v.: The Calculus of Variations. Universitext. Springer, New York (2004)
5. Buttazzo, G., Giaquinta, M., Hildebrandt, S.: One-dimensional Variational Problems. Clarendon Press, Oxford (1998)
6. Dacorogna, B.: Introduction to the Calculus of Variations. Imperial College Press, London (2004)
7. Dacorogna, B.: Direct Methods in the Calculus of Variations, 2nd edn. Springer, Berlin (2008)
8. Elsgolc, L.E.: Variationsrechnung. BI, Mannheim (1970)
9. Evans, L.C.: Partial Differential Equations. Graduate Studies in Mathematics, vol. 19. AMS, Providence, RI (1998)
10. Ewing, G.M.: Calculus of Variations with Applications. Courier Dover Publications, New York (1985)
11. Funk, P.: Variationsrechnung und ihre Anwendung in Physik und Technik, Zweite edn. Springer, Berlin (1970)
12. Gelfand, I.M., Fomin, S.V.: Calculus of Variations. Prentice-Hall, Englewood Cliffs, N.J. (1963)
13. Giaquinta, M., Hildebrandt, S.: Calculus of Variations I. Springer, Berlin (1996)
14. Giaquinta, M., Hildebrandt, S.: Calculus of Variations II. Springer, Berlin (1996)
15. Giusti, E.: Direct Methods in the Calculus of Variations. World Scientific, Singapore (2003)
16. Jost, J., Li-Jost, X.: Calculus of Variations. Cambridge University Press, Cambridge (1998)
17. Kielhöfer, H.: Bifurcation Theory. An Introduction with Applications to PDEs, 2nd edn. Springer, New York (2012)
18. Kreyszig, E.: Introductory Functional Analysis with Applications. Wiley, New York (1989)
19. Royden, H.L., Fitzpatrick, P.M.: Real Analysis, 4th edn. Prentice Hall, Boston (2010)
20. Rudin, W.: Real and Complex Analysis, 3rd edn. McGraw-Hill, New York (1987)
21. Rudin, W.: Functional Analysis, 2nd edn. McGraw-Hill, New York (1991)

© Springer International Publishing AG 2018
H. Kielhöfer, *Calculus of Variations*, Texts in Applied Mathematics 67,
https://doi.org/10.1007/978-3-319-71123-2

22. Pinch, E.R.: Optimal Control and the Calculus of Variations. Oxford University Press, New York (1993)
23. Sagan, H.: Introduction to the Calculus of Variations. McGraw-Hill, New York (1969)
24. Weinstock, R.: Calculus of Variations: With Applications to Physics and Engineering. McGraw-Hill, New York (1952)
25. Werner, D.: Funktionalanalysis. Springer, Berlin (2007)

Index

© Springer International Publishing AG 2018
H. Kielhöfer, *Calculus of Variations*, Texts in Applied Mathematics 67,
https://doi.org/10.1007/978-3-319-71123-2

Printed in the United States
By Bookmasters